# Darwin in the Archives

Papers on

ERASMUS DARWIN

&

CHARLES DARWIN

from

*JOURNAL OF THE
SOCIETY FOR THE BIBLIOGRAPHY OF NATURAL HISTORY*

&

*ARCHIVES OF NATURAL HISTORY*

2009

THE SOCIETY FOR THE
HISTORY OF
NATURAL
HISTORY

Published by Edinburgh University Press Ltd
22 George Square, Edinburgh EH8 9LF

Typeset by SR Nova Pvt Ltd, Bangalore, India, and
printed and bound in Great Britain by Page Bros, Norwich.


A CIP record for this book is available from the British Library

ISBN 978 0 7486 3888 8

# Contents

## NATURAL SELECTION AND AFTER

## METHODOLOGICAL ISSUES IN DARWIN STUDIES

# Darwin in the Archives:
# an introduction

DUNCAN M. PORTER

*Virginia Polytechnic Institute & State University, Blacksburg*

Why were the first papers on Darwin published in *Journal of the Society for the Bibliography of Natural History* about Erasmus Darwin, not his better-known (at least to those of us of the twentieth and twenty-first centuries) grandson Charles? Perhaps it was because during the time the society and journal were established in the 1930s, interest in evolution by natural selection was at a scientific and historic low point, not to be regained until the rise of the modern evolutionary synthesis in the 1940s and 1950s. Perhaps it was because Charles's biographers of the 1920s and 1930s were thought to have already mined all the gold of his life. Furthermore, their sources were mainly the six volumes of Charles's letters published in 1887, 1892 and 1903, with letters, autobiography, reminiscences and commentary by his son Francis. Little else was available to Darwin scholars.

The one-hundredth birthday of Charles and fiftieth anniversary of the publication of the *Origin* in 1909 resulted in a grand celebration at Cambridge and the resulting publications of A. C. Seward (1909) and Francis Darwin (1909). But little of historical import followed. The exception was produced by Darwin's granddaughter, Nora Barlow (1885–1989), who had access to family collections and who edited Charles's diary, letters, and manuscripts of the *Beagle* voyage (Barlow 1933, 1945). More likely, I think, is that, when Charles's library and archives were moved from the Cambridge University Botany School and family collections to the University Library beginning in 1948 (Burkhardt 1998), a new generation of scholars began to take notice. Here, they came to the attention of librarian Peter Gautrey and zoologist Sydney Smith, who recognized their importance and began to mine them. Their interests ultimately led to the Darwin Correspondence Project, founded in 1974 by the philosopher and administrator Frederick Burkhardt, with the aid of Sydney Smith (Porter 2003). The resulting *Correspondence of Charles Darwin*, which by 2009 has published 17 volumes, has revitalized Darwin scholarship and aided in the writing of numerous scholarly publications, some of which appear below.

Most of the following commentaries on the publications reprinted in the body of this *Special publication of the Society for the History of Natural History* are given in chronological order. That is, not in order of publication, but in order of where they fit in Charles Darwin's research program, from earliest to more recent projects. The first two are about grandfather Erasmus Darwin. The last three on Charles do not follow the chronology, but are given here because, unlike the others, they cover longer periods of his life. The facsimiles of the publications show the importance of *Journal of the Society for the Bibliography of Natural History* and *Archives of natural history* to Darwinian scholarship, many of them being essential to our understanding of Charles Darwin and his times. Besides these papers, over 70 books on the Darwins, Darwinism, and evolution have been reviewed in the Society's journals, and numerous notes and queries on the Darwins have appeared in its *Newsletter*.

## THE PUBLICATIONS

### STUDIES OF ERASMUS DARWIN

**1. J. W. T. Moody, 1964** Erasmus Darwin, M.D., F.R.S.: a biographical and iconographical note.

Moody's brief paper gives a short summary of the life and interests of Charles's grandfather, which have recently been given to us in magnificent detail by the Erasmus Darwin scholar Desmond

King-Hele (1999). The paper's *raison d'être* is to report Moody's acquisition of a sketch by the Derby artist James Rawlinson (1769–1848), which Moody identifies as that for Rawlinson's *circa* 1802 oil portrait of Erasmus that he states was then at the Erasmus Darwin School, Derby. Erasmus's great-great-great-great grandson, Milo Keynes (1994), takes the story further, identifying this portrait as being painted in 1802 and at High View Community School, the former Erasmus Darwin School. Keynes also identifies two earlier portraits of Erasmus by Rawlinson, one at Down House, the home of Charles Darwin in Kent now owned by English Heritage, and one at Darwin College, Cambridge. There are also several copies of Joseph Wright of Derby's (1734–1797) well-known 1792–1793 paintings of Erasmus: two copies by Rawlinson, one in a private collection, one at Derby Museums and Art Gallery; an apparent Rawlinson copy at Wolverhampton Art Gallery; and a possible Rawlinson copy at St John's College, Cambridge (Keynes 1994).

Maureen McNeil (2007) listed two Erasmus oil portraits by Rawlinson in Derby, one of *circa* 1802 owned by the Derby Corporation, and a second of 1802 at High View Community School, Breadsall, Derby (subsequently High View School and Technology Centre, and now da Vinci College). She listed a third Rawlinson portrait at the Royal College of Surgeons of England. However, the College's website www.rcseng.ac.uk lists only two engravings by James Heath (1757–1854) and Moses Haughton (1772–1846) after Rawlinson, but no Rawlinson portrait. This must be the portrait at Down House, which was owned by the College until it was purchased by English Heritage in 1996; the portrait hangs in the dining room (Overy 2004). Erasmus moved to Breadsall Priory in 1801, died in 1802, and is buried in Breadsall church. King-Hele (1999) does not mention the Rawlinson portraits.

Thanks to Lucy Salt, Keeper of Art at the Derby Museums and Art Gallery, the following can be added to our story. The painting in Derby Museums and Art Gallery may be a Wright copy of his own painting, or it may be a Rawlinson copy, opinion of experts is still divided. The Rawlinson portrait formerly at Erasmus Darwin School now resides at the Erasmus Darwin House, Lichfield, where I have seen it on display.

**2. M. McNeil, 1976** Nature, poetry and medicine in late eighteenth century England: a unified perspective of Erasmus Darwin.

This is an abstract of a paper that was not published by the Society. It was read at the Easter Meeting of the Society in London in April 1975 in the section "New trends in the history of natural history". It is the source of what Maureen McNeil later published as a chapter in Ludmilla Jordanova's (1986) *Languages of nature. Critical essays on science and literature*: "This essay has examined Erasmus Darwin's attempt to bring together science and poetry" (McNeil 1986: 202). Her essay later served as a source for the first two chapters of McNeil (1987). McNeil's book is listed in King-Hele's (1999) appendix of selected books and papers, but it was not cited by him in any footnotes.

GEOLOGICAL JOURNEYS

**3. P. Lucas, 2002** Jigsaw with pieces missing: Charles Darwin with John Price at Bodnant, the walking tour of 1826 and the expeditions of 1827.

Bodnant (or Bodnod) was the Price family home in North Wales. John Price (1803–1887) was a Shrewsbury School friend of Erasmus Alvey Darwin (1804–1881), Charles's older brother. There are very few extant sources of information relating to what Darwin was up to at this time in his young life, even his *Autobiography* (Barlow 1958) is short on details. However, from the few lines in Barlow, and a few letters written in 1826 and 1827 (Burkhardt and Smith 1985) and 1863 (Burkhardt *et al.* 1999), Peter Lucas has used inference and conjecture to fill in a few blanks on Charles's visits to North Wales in 1826 and 1827. The ultimate paragraphs briefly discuss the better-known 1831 geological tour of North Wales with the Woodwardian Professor of Geology at Cambridge, Adam Sedgwick (1785–1873), which is discussed in more detail in the next four papers in our selection.

Professor Richard Bambach, now a Research Associate at the Smithsonian Institution, has pointed out to me that Charles's 1831 field studies with Sedgwick were at the start of one of the great advances in understanding geological history. Sedgwick was studying the oldest known sedimentary rocks in order to work out their relationships to rocks that were already understood. So Charles was intimately involved in a project intended to work out what the oldest known sedimentary rocks at that time meant for earth history. Richard feels that this was an important aspect of winning Charles over to being interested in the historical aspects of science, and which ultimately led to an interest in evolution.

**4. M. B. Roberts, 2000** I coloured a map: Darwin's attempts at geological mapping in 1831.

> As I had at first come up to Cambridge at Christmas, I was forced to keep two terms after passing my final examination, at the commencement of 1831; and Henslow [John Stevens Henslow (1796–1861), Professor of Botany and Darwin's mentor at Cambridge] then persuaded me to begin the study of geology [Henslow earlier (1822–1827) had been Professor of Mineralogy]. Therefore on my return to Shropshire I examined sections and coloured a map of parts round Shrewsbury. (Barlow 1958: 68–69.)

Michael Roberts illustrates, describes and discusses four maps that Darwin copied from earlier maps and to which he added his own observations. The maps are of Shrewsbury, Kinnerley, Llanymynech, and Anglesey. "These four maps of Shropshire and North Wales show Darwin's earliest attempts at field geology and plans for future work" (p. 69). The first three maps obviously were aided by fieldwork, but there is no solid evidence that Darwin was in Anglesey at this time (see nos **6** and **7** below). Roberts concludes that Darwin's "topographic map of Anglesey probably indicates his future plans" (p. 77).

**5. S. Herbert and M. B. Roberts, 2002** Charles Darwin's notes on his 1831 geological map of Shrewsbury.

In the previous paper (no. **4** above), Michael Roberts stated that "There appear to be no extant geological notes which refer to these maps . . ." (p. 70). However, this short article reports Darwin's notes on the geology of four localities marked on his map of Shrewsbury, found in the Darwin Archive at Cambridge University Library. Those on one of the localities are extensive, indicating Darwin's growing experience in meticulous descriptions of geological features through experience in the field, which he continued to good advantage on the 1831–1836 global circumnavigation of HMS *Beagle*.

**6. M. B. Roberts, 1998** Darwin's dog-leg: the last stage of Darwin's Welsh field trip of 1831.

"Professor Sedgwick intended to visit N. Wales in the beginning of August to pursue his famous geological investigation amongst the older rocks, and Henslow asked me to accompany him" (Barlow 1958: 69). Although Michael Roberts's paper emphasizes only the end of Darwin and Sedgwick's fieldtrip to North Wales in August 1831, it is placed before Peter Lucas's treatment of the entire excursion, both because it predates Lucas and because it is discussed in Lucas's paper (see no. **7** below).

Roberts begins, as I have, by quoting from Darwin's autobiography:

> At Capel Curig I left Sedgwick and went in a straight line by compass and map across the mountains to Barmouth, never following any track unless it coincided with my course. I thus came on some strange wild places and enjoyed much this manner of travelling. I visited Barmouth to see some Cambridge friends who were reading there, and thence returned to Shrewsbury and to Maer for shooting; for at that time I should have thought myself mad to give up the first days of partridge-shooting for geology or any other science. (Barlow 1958: 70–71.)

Maer Hall was the Staffordshire home of Charles's maternal uncle and, later, father-in-law, the potter Josiah Wedgwood (1769–1843). Roberts's quote is from De Beer (1983), which is identical to Barlow (1958). However, Roberts has mistranscribed "on" as "to" and "some" as "more", and omitted the last phrase. The transcriptions make no difference to the sense of the quote; some may say that the most interesting part of it has been removed, but that which Roberts retains is integral to his story.

Roberts, whose maps are clear and quite easily understood, traces Darwin's probable route from Capel Curig to Barmouth. He has walked, biked and driven it himself and argues convincingly that Darwin could not have travelled overland in a straight line. Roberts is less convincing in arguing that Darwin and Sedgwick spent two days on Anglesey (also see Roberts 2001). Until Darwin's missing notes for 12 and 13 August are found, this remains only an intriguing hypothesis. However, Roberts rightly concludes that this fieldwork with Sedgwick, and later on his own, well prepared Darwin for his future as a geologist.

**7. P. Lucas, 2002** "A most glorious country": Charles Darwin and North Wales, especially his 1831 geological tour.

Peter Lucas gives us a day-by-day account of the time Darwin and Sedgwick spent together geologizing in North Wales during August 1831. This gives a vivid portrait of where Darwin and Sedgwick were and what they did. Sections of "Walker's map" from the Sedgwick Museum at Cambridge University are used to illustrate what is happening in the text. Unfortunately, the sites and features discussed are not highlighted or underlined, and it is thus difficult to follow their routes on them, unlike with Michael Roberts's outline maps (see no. **6** above).

Both Lucas and Roberts make the point that Darwin must have been in Anglesey because of his comparing rocks observed on Quail Island in the Cape Verde Islands during the first stop on the *Beagle* voyage in January 1832 with those on Anglesey. They presume that Darwin was there with Sedgwick, but if so why did not Sedgwick mention him in his notes on the island? Lucas does note that Darwin had been on Anglesey in 1827, while travelling from Dublin to Shrewsbury. He provides a useful list of "Darwin's known and probable visits to Wales" in an appendix, which number 16, including two after 1831. They indicate that Darwin had plenty of opportunities to observe the Welsh landscape. His last visit, in June 1869, is discussed by Lucas in a paper recently published in *Archives of natural history* (Lucas 2007).

In an interesting aside, Lucas relates the profound affect that Darwin had on Oxford student Robert Lowe (1811–1892), later an influential politician who became Viscount Sherbrooke, as they walked from Barmouth; "... I saw a something in him [Darwin] which marked him out as superior to anyone I ever met .... I walked twenty-two miles with him when he went away, a thing which I never did for anyone else before or since" (p. 19).

<br>

## THE *BEAGLE* SPECIMENS

**8. D. M. Porter, 1980** Charles Darwin's plant collections from the voyage of the *Beagle*.

This is the first of five papers to be here discussed that is devoted to Darwin's specimens collected on the 1831–1836 circumnavigation of the earth by HMS *Beagle*. This paper was presented at the International conference on the history of museums and collections in natural history, in the section "Travellers and explorers", held in London in April 1979. It relates how Darwin gathered plants for Henslow, how Henslow was ultimately unable to identify many of them, and how botanist Joseph Dalton Hooker (1817–1911) stepped in to save the day. Some of the plants from southern South America were cited in Hooker's (1844–1847) *Flora antarctica*. Most importantly, the Galapagos Islands specimens were the basis of the first flora of the archipelago (Hooker 1847a) and its first phytogeographic study (Hooker 1847b). Soon after the paper's publication, Darwin's mentioned missing plant notes were discovered by Mrs Rita I'Ons at Cambridge University Herbarium (Porter 1981).

Following the reading of the paper, the author went to lunch with three Darwin experts (Dr Sydney Smith, Mr David Stanbury, Dr David Kohn), who, knowing that he was completing a paper identifying Darwin's Galapagos plants (Porter 1980), prevailed upon him to identify the rest of the *Beagle* plants. This resulted in a series of papers, culminating in Porter (1986, 1999) and Porter *et al.* (2009).

**9. G. Chancellor, A. DiMauro, R. Ingle and G. King, 1988** Charles Darwin's *Beagle* collections in the Oxford University Museum.

Gordon Chancellor *et al.* provide an interesting history of how some of Darwin's invertebrate, mostly crustacean, collections, and a few vertebrate and mineralogical specimens, ended up in the Oxford University Museum. They provide annotated excerpts from Darwin's "Catalogue for Animals in Sprits of Wine" for the 110 specimens found. Also provided is a useful index of identifications. When I examined the dried crustaceans in their Figure 1 at Oxford in the late 1970s or early 1980s, I pointed out to Gordon that the attached numbered metal tags indicated that they had originally been preserved in brandy (spirits of wine). The authors also note that there are a number of Darwin's Australian insects in the Hope Entomological Collections at the Museum, which were discussed in Smith (1987).

**10. K. G. V. Smith, 1996** Supplementary notes on Darwin's insects.

Ken Smith reports on the insects collected by Darwin on the *Beagle* voyage that have been discovered since the publication of his *magnum opus* on them (Smith 1987). The majority are in the Natural History Museum, London; others are in various collections in Australia, Canada, Fiji and Ireland. The most interesting to me is "Insects sweeping near Sydney, S. Covington." Syms Covington (*c.* 1816–1861), "fiddler & boy to Poop-cabin" (Keynes 1988: 84), was Darwin's servant. Very few Darwin specimens of any kind have been found that he indicated were collected by others (Porter 1985). Also noted are two new Darwin insect eponyms, one a genus and one a species, plus two new publications that discuss some Darwin insect specimens.

**11. D. M. Porter, 1983** More Darwin *Beagle* notes resurface.

During the latter part of the *Beagle* voyage, Syms Covington compiled lists of the various groups of organisms collected by Darwin for the taxonomists who were to identify and study them. These were copied from Darwin's Specimen Notebooks (Porter 1985: 987–988). Those that have been subsequently published are for birds (Barlow 1963), plants (Porter 1987) and insects (Smith 1987). This short paper relates the story of my finding and identifying the insect notes in the Entomology Library of the Natural History Museum, London. It also includes Darwin's notes on encounters with a biting fly and the "Great Black Bug of the Pampas", which some believe to be the source of his later illness (Colp 2008).

**12. W. R. P. Bourne, 1992** FitzRoy's foxes and Darwin's finches.

When the three volumes of *Narrative of the surveying voyages of His Majesty's Ships Adventure and Beagle, between the years 1826 and 1836* were published in 1839, FitzRoy's volume contained 31 pages devoted to describing the Falkland Islands, while Darwin's contained only 13 pages on these islands. Bourne's thesis is that, although Darwin later used the Falkland Island fox as an early possible example of evolution on islands (Barlow 1935), FitzRoy (1839) was more perceptive in the information he recorded on the fox while in the islands. Darwin's (1839) little over one page on the fox is dwarfed by FitzRoy's three-and-a-half pages. FitzRoy speculates on its origin and discusses its variation, Darwin does not. The fox becomes more important to Darwin after his visit to the Galapagos Islands, where he had another chance to examine inter-island distribution and variation, especially in the mockingbirds (Barlow 1935). Bourne also discusses the now well-known myth that it was seeing the Galapagos finches that gave Darwin the clue that evolution takes place. The mythical aspect of this idea was brilliantly shown by Sulloway (1982).

### Darwin's data gathering: *Questions about the breeding of animals*

**13. G. De Beer, 1968** *Charles Darwin. Questions about the breeding of animals.* [*1840*].

This is the first of three publications here discussed that is devoted to Darwin's questionnaire on animal breeding. The 18-page pamphlet provided an eight page facsimile of Darwin's questionnaire, with a seven page introduction by Sir Gavin De Beer, who dates it as being published in 1840. It consists of 21 numbered paragraphs containing 44 questions. No answers to them are included. At the time of the facsimile's publication, this was the only known copy. Its contents were subsequently published in Vorzimmer (1969) and Burkhardt and Smith (1986).

De Beer provides a short essay on how Darwin used his correspondents to good effect in obtaining data for his research and publications. He hypothesizes that there are two more printed questionnaires, one on expression of the emotions, which exists in numerous forms, and one on sexual selection in humans, which has not been found. He points out that some of the questions on breeding can be related to sections in Darwin's manuscript essay on evolution of 1844 (F. Darwin 1909; Darwin and Wallace 1958). De Beer claims that the recipients of the questionnaire and their responses will never be known, because "all letters received by him before 1862 were destroyed by him" (p. ix). De Beer's claim is based on a statement by Francis Darwin in his edition of Charles's correspondence (F. Darwin 1887: **1**: v):

> Of letters addressed to my father, I have not made much use. It was his custom to file all letters received, and when his slender stock of files ("spits" as he called them) was exhausted, he would burn the letters of several years, in order that he might make use of the liberated "spits." This process, carried on for years, destroyed nearly all letters received before 1862. After that date he was persuaded to keep the more interesting letters, and these are preserved in an accessible form.

This statement is shown to be incorrect with the publication of the first ten volumes of *The Correspondence of Charles Darwin*, which cover the years 1821 to 1862 and include numerous letters to Darwin.

**14. R. B. Freeman and P. J. Gautrey, 1969** Darwin's *Questions about the breeding of animals*. With a note on *Queries about expression*.

Another copy of Darwin's breeding of animals questionnaire was discovered in the Darwin Archive at Cambridge University Library, soon after the previous publication (no. **13** above) appeared. Answers were supplied for the questions, and Darwin dated them 10 May 1839. In addition, answers for five of the questions, dated 6 May 1839, were also found. So the pamphlet was published in 1839, not 1840 as hypothesized by De Beer. The answers were supplied by friends of the family living in Staffordshire, the first by the agricultural reformer George Tollet (1767–1855), the second by the farmer Richard Sutton Ford (1785– c. 1850). Both sets of answers are given in the text of the paper and presented as letters in Burkhardt and Smith (1986).

Richard Freeman and Peter Gautrey also comment on De Beer's two other hypothesized printed questionnaires. That on sexual selection in humans is probably the same as that on the expression of the emotions. They report the finding of two copies of the latter in the Darwin Archive and state that they intend to publish a facsimile edition with commentary (see no. **23** below).

**15. R. B. Freeman and P. J. Gautrey, 1977** Charles Darwin's *Questions about the breeding of animals*, [1839].

Darwin's personal account book indicates that this pamphlet was printed by the London firm of Stewart & Murray. He probably received copies shortly before leaving London for Staffordshire on 26 April 1839.

**16. J. Browne, 1986** Darwin again.

This query quotes from a letter to Darwin from Bartholomew James Sulivan (1810–1890), one of the officers on the *Beagle*. The letter was printed in Burkhardt and Smith (1986). Sulivan's description of the Falkland Island pig was included in Darwin (1868: **1**: 77n).

**17. D. M. Porter, 1986** Charles Darwin and 'ancient seeds'.

Darwin's interest in seed vitality dates at least to 1837 (see letter to J. S. Henslow, 28 March 1837, in Burkhardt and Smith 1986). This interest led to a series of experiments during the 1850s, which culminated in Darwin (1856). The earliest correspondence between Darwin and the Scottish amateur geologist William Kemp has not been found. It is not known why Kemp initially wrote to Darwin; it may have been in answer to an undiscovered letter of Darwin's seeking information, printed in a journal. Darwin often used the *Gardeners' chronicle* in this way, beginning in 1841.

The mentioned 1843 correspondence with the botanist and horticulturist John Lindley (1799–1865), J. S. Henslow, the geologist Sir Charles Lyell (1797–1875) and the botanist Charles Cardale Babington (1808–1895) on these seeds is printed in Burkhardt and Smith (1986). This volume also has six letters from Kemp and an appendix ("Darwin and William Kemp on the vitality of seeds"), which shows that the manuscript for Kemp (1844) was drafted by Darwin. In July 1855 Darwin was deeply immersed in botanical studies. His mentioned 14 July 1855 letter to J. D. Hooker is printed in Burkhardt and Smith (1989).

Jane Shen-Miller *et al.* (1995) reported an even older germination of a sacred lotus seed than the one cited in my paper, radiocarbon dated to about 1,300 years BP. However, the new record is of a date palm (*Phoenix dactylifera*) seed, radiocarbon dated to about 2,000 years BP (Sallon *et al.* 2008).

## Natural selection and after

**18. J. W. T. Moody, 1971** The reading of the Darwin and Wallace papers: an historical "non-event".

The 1 July 1858 meeting of the Linnean Society of London was the result of the final scheduled meeting of the session of 17 June having been postponed because of the death of botanist Robert Brown (1773–1858), a past-president and member of the council, on 10 June. The 1 July meeting was held in order to elect a new member of council, which had to be done within three months of a councilman's death, and to read the papers that had been postponed from the 17 June meeting. Papers that would have been read at the next meeting of 4 November were read, including the Darwin and Wallace "joint paper", after a long eulogy of Brown. The six papers presented were read by the Under-secretary, George Busk (1807–1886), who was not included by Moody in his list of attendees. The Darwin-Wallace paper was read first and was subsequently published on 20 August (Darwin and Wallace 1858). Darwin was not in the audience, because he was in Downe grieving over the death of his youngest child, Charles Waring Darwin, aged 18 months, who had died on 28 June. Wallace also was not in the audience, because he was still in the East Indies, not to return to England until 1862.

Moody refers to the meeting as a "non-event" because it does not seem to have resulted in an immediate rush of biologists to accept natural selection as the explanation as to how evolution is driven. However, some attention was paid to the resulting paper over the next 15 months before *Origin* was published in November 1859. Positive reactions were published by the comparative anatomist Richard Owen (1804–1892), ornithologist Henry Baker Tristram (1822–1906), botanist Hewett Cottrell Watson (1804–1881), American botanist Asa Gray (1810–1888), botanist Joseph Hooker and geologist Sir Charles Lyell; negative reactions came from Irish geologist Samuel Haughton (1821–1897), entomologist Thomas Boyd (1829–1913), naturalist Arthur Hussey (1794–1862) and Scottish botanist Andrew Murray (1812–1878). Hooker and Lyell, of course, had been at the Linnean Society meeting. They, as well as Owen, Watson and Gray, had prior knowledge of Darwin's interest in variation or transmutation (Porter 1993).

I was interested to see that at "The driving forces of evolution. From Darwin to the Modern Age", a meeting of the Linnean Society of London on 3–4 July 2008, which in part celebrated the 200th anniversary of the 1 July 1858 meeting, most of the speakers were critical of both Darwin and natural selection, while Jean-Baptiste de Lamarck fared better than Darwin.

**19. T. Lie, 1985** The reception of Darwinism in Norway: the early years 1861–1900.

Thore Lie discusses several Norwegian botanists, geologists and zoologists who introduced Darwinism into Norway beginning in 1861. He also discusses translation of *Origin* into Norwegian. Glick (1974) is not cited in Lie's paper, probably because it does not mention Norway. However,

Glick (1988), a reprint with a new preface, does so and adds: "In Norway, for example, although the marine biologist Asbornsen [Asbjørnsen] was an early and enthusiastic Darwinian, academic philosophy was uniformly Hegelian and antiempiricist and constituted, along with religious opposition, a formidable barrier to reception" (Glick 1988: xx).

Lie also does not cite Freeman (1977), who listed one Norwegian translation of Francis Darwin (1887), which appeared in 1889. This translation is mentioned by Lie, but Freeman states that the first volume was translated by Peder Ulleland and the second and third volumes by M. Sørass. Ingebret Suleng's 1889–1890 translation of the sixth edition of *Origin*, cited by Lie, is not listed by Freeman (1977).

**20. G. Kritsky, 1992** Darwin's *Archaeopteryx* prophecy.

Gene Kritsky hypothesizes that Darwin became aware of *Archaeopteryx lithographica*, the fossil "missing link" between reptiles (actually, dinosaurs) and birds in 1862. However, now we know from his correspondence that this happened in 1863. Indeed, there is a flurry of letters about the fossil following that to Darwin from the palaeontologist Hugh Falconer (1808–1865) of 3 January 1863, from which Kritsky quotes, as he does from Darwin's answer of 5 and 6 January. Besides a continuing dialogue with Falconer in early 1863, Darwin also mentioned *Archaeopteryx* in letters to the American geologist James Dwight Dana (1813–1895), the Swiss botanist Alphonse de Candolle (1806–1893), the New Zealand geologist Julius von Haast (1824–1887), and Sir Charles Lyell in January, February, and March 1863 (Burkhardt *et al.* 1999). Kritsky is correct in stating that Darwin had prophesied the morphology of the wing of *Archaeopteryx* in his letter to Lyell of 11 October 1859 (Burkhardt and Smith 1991).

After the foregoing was written, on 7 December 2008 the following addendum to his paper was received from Gene:

> Additional letters written by Darwin, published since this paper first appeared, have confirmed Darwin's private views of the importance of the *Archaeopteryx* fossil. A week after Darwin wrote to J. D. Dana, he wrote to Alphonse de Candolle on 14 January 1863, "Considering that Birds are the most isolated group in the animal kingdom. What a splendid case is this Solenhofen bird-creature with its long tail & fingers to its wings!" (Burkhardt *et al.* 1999: 39–40).
>
> On 22 January 1863, Darwin wrote to Julius von Haast asking that he inform him when he discovered the identity of the creature that produced a fossil trackway in New Zealand. Darwin speculated, "Perhaps they may turn out something like the Solen-hoven bird creature with its long tail & fingers with claws to its wings!" (Burkhardt *et al.* 1999: 67–68).
>
> Darwin's enthusiasm for this fossil is further documented in his 20 February 1863 follow-up letter to Dana. Darwin apparently enjoyed giving Richard Owen another jab for his sloppy work, and he confirmed his pleasure with this fossil. Darwin wrote (Burkhardt *et al.* 1999: 155–156):
>
>> I daresay Owen will work out everything carefully before he publishes in detail; but there is little doubt he was at *first* very careless. He overlooked a jaw with teeth which however may possibly not have belonged to this marvelous Bird, with its long tail & fingers to its wings. As Birds are so isolated this case, as you may suppose, has pleased me.

On 29 May 1866 J. D. Hooker wrote Darwin about William Robert Grove's request for evidence that supported Darwin's views for Grove to use in his presidential address to the British Association for the Advancement of Science. In his reply to Hooker, written on 31 May 1866, Darwin stated, "The Eozoon is one of the most important facts, & in much lesser degree the Archeopteryx [*sic*] 'Fritz Muller Fur Darwin' [Müller 1864] is perhaps the most important contribution" (Burkhardt *et al.* 2004: 192). Though Darwin publicly clung to his view that *Archaeopteryx* demonstrated how little is known about the fossil record, he clearly did not discourage others from promoting this important transitional fossil.

**21. R. B. Freeman, 1973** Offprints of Darwin's "Climbing plants", 1865.

Richard Freeman's note reports the two distinct issues of offprints for this paper. A more expanded version, plus a history of the second edition (Darwin 1875), reprints and translations, is given in Freeman (1977).

**22. A.-K. Mayer, 1999** Note on the Fritz Müller–Charles Darwin correspondence.

Anna Mayer discusses the circumstances under which their correspondence began and the nature of their correspondence in this short paper. Müller's life, his correspondence with Darwin, and the close relationship between them is given in detail in West (2003). West is currently revising his book and has published some further information about Darwin and Müller in West (2007) recently published in *Archives of natural history*.

**23. R. B. Freeman and P. J. Gautrey, 1975** Charles Darwin's Queries about expression.

Richard Freeman and Peter Gautrey (1972) previously had published a facsimile of Darwin's questionnaire on expression and discussed its several known versions. This is a more up-to-date discussion of the earlier known versions and several more that subsequently had been discovered. They are also discussed in Freeman (1977). The authors conclude that the earlier versions of the questionnaire were handwritten. The first printed version was circulated by Asa Gray beginning in March 1867. They conclude (p. 261) that Darwin did not have a manuscript version sent to the printer,

> until very late in 1867 or even early in January 1868; and that there was only one English printed version which reached Darwin from the printers not long before 31 January 1868. The evidence is circumstantial, but we feel that, taken together, it carries considerable weight. It can be disproved if a printed English version, in leaflet form and with the earlier text is found, or if a letter conclusively dated 1867 contains the word printed.

There are three printed copies of the questionnaire in the Darwin Archive at Cambridge University Library. The final version is printed as an appendix in Burkhardt *et al.* (2005). It is dated 1867.

**24. J. W. Valentine, 1997** The early American printings of Darwin's Descent of man . . . .

While preparing a handlist of US issues of Darwin's works, Jim Valentine discovered some differences between early John Murray and D. Appleton & Co. English and American issues of *The descent of man, and selection in relation to sex* (Darwin 1871). The Appleton printings are not precise reprints of the English issues, but have been "Americanized", with changes in spelling and punctuation. Valentine discusses these changes and also the English sources of the different Appleton printings.

**25. K. G. V. Smith and R. E. Dimick, 1976** Darwin's American neighbour.

The English-born lawyer Wallis Nash (1837–1926) and his wife, Louisa, lived near Downe from 1873 to 1877. They became acquainted with the Darwins, and both later recorded their impressions of the village and the Darwins (L. Nash 1890; W. Nash 1919). They named one of their sons Louis Darwin Nash. This paper quotes a few lines from these references about the Darwins, but most of it is devoted to Nash's importance in the developing state of Oregon, where he moved in 1879. Two letters from Darwin to Wallis Nash are published here for the first time. The Darwin Correspondence Project's website, www.darwinproject.ac.uk, lists ten letters exchanged between them from September 1871 to March 1881. Nash apparently began their correspondence by sending Darwin information on the inheritance of behavioural and physical traits in hunting dogs, probably after reading *Variation under domestication* (Darwin 1868).

**26. B. Coleman, 1974** Samuel Butler, Darwin and Darwinism.

To celebrate Charles Darwin's seventieth birthday, in February 1879 a Festschrift was published in the German journal *Kosmos* that contained an essay on Erasmus Darwin by botanist and science writer Ernst Krause (1839–1903). Upon reading it, Charles wrote to Krause, asking him if he would object to the essay's translation and publication in English. Krause agreed, and Charles suggested that he would write a short introduction to it. Both essays were published in November as *Erasmus Darwin* (Krause 1879), with Darwin's introduction one-and-a-half times as long as Krause's essay. In the meantime, author and artist Samuel Butler (1835–1902) in May had published *Evolution old and new*

(Butler 1879), unaware of Krause's February essay. Unfortunately, Krause made changes in the *Erasmus Darwin* essay based on Butler's book that Butler found derogatory towards himself. Butler blamed Darwin, because Darwin had not indicated in his essay that Krause had done so. Brian Coleman gives the details of the public (Butler) and private (Darwin) controversy that erupted. Many of the letters pertaining to the controversy are printed in an appendix to Barlow (1958). Coleman's conclusion is that Butler misunderstood evolution because he was not trained as a scientist. The talk referred to in Coleman's note 7 was published as a chapter in Young (1985).

Coleman writes (pers. comm. 2 October 2008) that,

> The longer quotation from [Loren] Eisley, 'William Wells delivered ... ', page 95, line four, should read: 'it was not published ... '" Thus, the sentence should read, "There is no record that the paper ["An account of a white female, part of whose skin resembles that of a Negro"] aroused any particular attention and it was not published again until 1818 after the author's death.

This fuss between Butler and Darwin might have been avoided if Darwin's daughter, Henrietta Litchfield, had not edited his manuscript. The original, unabridged manuscript, which was found in the Darwin Archive, Cambridge University Library, has recently been published, edited by Erasmus Darwin scholar Desmond King-Hele (2003). Two quotes will suffice:

> [Charles's] *Life of Erasmus Darwin* was shortened by 16% before publication in 1879, and several of the cuts were directed at its most provocative parts. The cutter, with Charles's permission, was his daughter Henrietta – an example of the strong hidden hand of meek-seeming Victorian women. (King-Hele 2003: ix.)

and, a paragraph deleted from the published book,

> Dr. Krause has taken great pains, and has added largely to his essay as it appeared in 'Kosmos'; and my preliminary notice, having been written before I had seen the additions, unfortunately contains much repetition of what Dr. Krause has said. In fact the present volume contains two distinct biographies, of which I have no doubt that by Dr. Krause is much the best, I have left it almost wholly to him to treat of what Dr. Darwin has done in science, more especially in regard to evolution. (King-Hele 2003: 6.)

The steadily worsening relationship between Butler and Darwin is discussed in detail by Raby (1990).

**27. H. P. Moon, 1982** Charles Darwin at Glenridding House, Ullswater, Cumbria.

The second Darwin family visit to the English Lake District took place in 1881. Moon quotes daughter Henrietta Litchfield as stating, "On June 2nd, we all went to a house at Patterdale taken for a month." Emma Darwin's diary in the Darwin Archive at Cambridge University Library (unavailable at the time to Moon), indicates that this is the day that the family left Downe, arriving at Glenridding House the next day. Darwin is later quoted as writing that, "We have just returned home after spending five weeks on Ullswater." Emma's diary gives this date as 5 July, having left Ullswater on the fourth. The diary entries, while brief, give a fascinating look at life with the Darwins. Emma indicates where they went, who went with her (not always Charles), his state of health, and terse comments on the weather. Moon quotes Henrietta on Charles and Emma's strolls along the lake, mentioned in the diary, which also mentions drives further afield and boating on the lake.

Besides Henrietta (1843–1927), Moon correctly places eldest son William Darwin (1839–1914), and only grandchild at the time Bernard Darwin (1876–1961), at Ullswater. He is mistaken in stating that Henrietta is the only surviving daughter, as her younger sister Elizabeth (1847–1926) still lived. Indeed, Emma's diary shows that Bessy, as she was known by the family as an adult (Keynes 2001), was at Ullswater, too. In addition, second son George (1845–1912) and fourth son Leonard (1850–1943) also were there. The illness that caused a doctor to be summoned, discussed by Moon, does not appear in Emma's diary. He cites Colp (1977) as the source of this information, which is repeated in Colp (2008). However, on 30 May, before the trip to Ullswater began, Emma refers to Charles' heart problem.

The Darwin's first visit to the Lake District took place in August 1879. In a footnote, Moon states that, "There is no known record of where Darwin stayed in Coniston." Again, the reference works that Moon needed to date the visit and determine where the Darwin's stayed were unavailable to him. Emma's diary shows that they left Downe on 2 August 1879, returning on 27 August. It also indicates that George and Leonard accompanied their parents for part of this time. Of the eleven letters known to have been posted by Darwin on this trip, most are from Coniston, several are on stationary with his Downe address, and one is from the "Waterhad Hotel" (Burkhardt and Smith 1994). This letter, dated 25 August 1879, is to Victor Alexander Ernest Garth Marshall (1841–1928), and thanks him for courtesies extended to the Darwin family during their stay at Coniston. The website of the Waterhead Hotel, Hawkshead Road, Coniston Water (www.waterhead-hotel.co.uk) states that Darwin had stayed there as a guest.

**28. R. B. Freeman, 1986** Darwin in Chinese.
This is an addendum to Freeman (1977), which listed eight entries for Chinese language translations of *On the origin of species* (Darwin 1859) dated 1902 to 1972. Half include a few chapters, half the entire book. Through the good offices of Mr Chou Pang-li of Shanghi, who had translated five of Darwin's books from Russian into Chinese, and his widow Madame Ku Yuan, this paper lists 13 publications of Darwin in 41 different editions or printings. In addition, there are ten translations of Darwin's autobiography (Darwin 1887 or Barlow 1958), two printings of his *Life and letters* (Darwin 1887), one transcript of manuscripts (Barlow 1933), one collection of letters (Barlow 1945), five original biographies, and one exhibition catalogue. Entries for *Origin* are brought up to date. The 61 listed publications are dated 1902 to 1983. Freeman's paper is cited by Glick (1988: xxi, n. 34).

**29. P. J. P. Whitehead, 1988** Darwin in Chinese: some additions.
During a hurried trip to China, Whitehead was able to add three new Chinese translations of Darwin's publications to Freeman's list (no. **28** above). He also added information about four other works in Freeman's list.

Methodological issues in Darwin studies

**30. J. Browne, 1978** The Charles Darwin–Joseph Hooker correspondence: an analysis of manuscript resources and their use in biography.
This and the following two papers are placed at the end of our list because, unlike the foregoing, which are discussed in chronological order, they cover a number of years in Darwin's life. The Darwin Correspondence Project, on which Janet Browne was later an Assistant Editor, has gathered about 14,500 letters that Darwin exchanged with almost 2,000 people during his lifetime. According to the Darwin Correspondence Project, about 1,400 of these (almost 10%) are between Darwin and Hooker. We now know that their correspondence began with a letter from Hooker on 13 or 20 November 1843 (Burkhardt and Smith 1986) and, so far as we know, ended with a letter to Hooker on 20 January 1882 (Burkhardt and Smith 1994). There may be later ones in the so far undated letters in the Darwin Archive.
Long before the Darwin Correspondence Project began in 1974, about 20% of the letters to and from Darwin had been published, mostly edited by his son Francis (Darwin 1887, 1892; Darwin and Seward 1902). At the beginning, Francis stated (1887: **1**: iv), "In printing the letters I have followed (except in a few cases) the usual plan of indicating the existence of omissions or insertions." Anyone familiar with the volumes of *The correspondence of Charles Darwin* will know that this is incorrect. Many of the letters edited by Francis have been heavily redacted, especially those mentioning personal relationships or religion. Francis was not the only one to censor his father's letters, others in the family also did so. Browne gives examples of Darwin's close friends J. D. Hooker and the zoologist Thomas Henry Huxley (1825–1895) being given opportunities to do so by Francis. One of these is the famous encounter between Huxley and Samuel Wilberforce (1805–1873), Bishop of Oxford, at the annual meeting of the British Association for the Advancement of Science in Oxford in 1860. Browne points

out that the much-cited accounts by Huxley and Hooker of the encounter were penned after Darwin's death, some 25 years after the incident took place. To me, the most interesting comments on the encounter were by Hooker, in a letter to Darwin of 2 July 1860 apparently unknown to Browne at the time, that is too long to give here *in toto* (see Burkhardt *et al.* 1993: 270–271):

> ... Well Sam Oxon [Wilberforce] got up & spouted for half an hour with inimitable spirit ugliness & emptyness & unfairness, I saw he was coached up by Owen [Richard Owen] & knew nothing & he said not a syllable but what was in the Reviews [of *Origin*] – he ridiculed you badly & Huxley savagely– Huxley answered admirably & turned the tables, but he could not throw his voice over so large an assembly, nor command the audience; & did not allude to *Sam's* weak points nor put the matter in a form or way that carried the audience. ... my blood boiled, I felt myself a dastard; now I saw my advantage – I swore to myself I would smite that Amalekite Sam hip & thigh if my heart jumped out of my mouth & I handed my name up to the President (Henslow) as ready to throw down the gauntlet ... it moreover became necessary for each speaker to mount the platform & so there I was cocked up with Sam at my right elbow, & there & then I smashed him amid rounds of applause – I hit him in the wind at the first shot in 10 words taken from his own ugly mouth – & then proceeded to demonstrate in as few more 1 that he could have never read your book & 2 that he was absolutely ignorant of the rudiments of Bot. Science – I said a few more on the subject of my own experience, & conversion & wound up with a very few observations on the relative position of the old & new hypotheses, & with some words of caution to the audience – Sam was shut up – had not one word to say in reply & the meeting was dissolved forthwith leaving you master of the field after 4 hours battle. Huxley who had borne all the previous brunt of the battle & who never before (thank God) praised me to my face, told me it was splendid, & that he did not know before what stuff I was made of ...

This volume (Burkhardt *et al.* 1993) contains other letters on the meeting and an appendix with reports on the meeting from *The Athenaeum*. In the report on the encounter in *The Times*, several paragraphs are devoted to Hooker; Huxley is not mentioned.

By using examples from the Darwin-Hooker correspondence and from other collections of lives and letters, Browne well defends her statement (p. 361) that "a re-evaluation of the *Life and Letters* type of biography is essential." This paper was read at the 1977 Easter Meeting of the Society.

### 31. T. Veak, 2003 Exploring Darwin's correspondence: some important but lesser known correspondents and projects.

In this paper, Tyler Veak has published a very important guide for Darwin scholars. He has examined Burkhardt and Smith (1994) and found that there are more than 50 correspondents that exchanged 30 or more letters with Darwin, not including well-known ones such as Gray, Hooker, Huxley and Lyell. Two correspondents very important to Darwin's research, the naturalist William B. Tegetmeier (1816–1912) and the gardener/botanist John Scott (1836–1880), are discussed in some detail. In an appendix, Veak lists 231 correspondents who exchanged six or more letters with Darwin and indicates which of 13 Darwin projects, chronologically from entomology to worms, these letters discuss. "A quantitative analysis of his correspondence reveals that many of Darwin's most important sources and projects have not been researched" (p. 118). As examples, Veak discusses Charles Nelson's (2000) book on seed dispersal and viability and earlier sources, which show how significant such projects were to Darwin throughout his working life. A table (Table 2, not Table 3 as given in the text) of Darwin's "seed salting" experiments is provided, as are two appendixes on seeds and dispersal listing letters by topics discussed and by correspondents involved.

This paper guides the researcher to prospective sources and projects to pursue: "The primary intent of this paper is to offer fresh research possibilities for Darwin's correspondence" (p. 125). I have found only three papers, published after Veak's paper appeared, that utilize the *Correspondence of Charles Darwin* as he suggests: Davis (2004) and Sayer (2007) who do not cite Veak in their papers, and DeArce (2008) who does.

### 32. M. A. Di Gregorio, 1986 Unveiling Darwin's roots.

Mario Di Gregorio points out that there are three kinds of materials available to researchers in the Darwin Archive: notebooks, letters, and library (books and pamphlets). Use of the letters has been explored in a number of the preceding papers. The letters, "represent Darwin talking to his friends and colleagues". The notebooks, some of which have been published subsequent to the paper (for example Barrett *et al*. 1987), "represent Darwin talking potentially to a public". The marginalia, to which we are introduced by this paper, "represents Darwin talking to himself, and thus probably constitute the least guarded and filtered commentary available on the development of his thought – often no doubt giving us the benefit of his immediate reactions to individual statements made by his contemporaries and predecessors" (p. 323). Di Gregorio gives us a number of examples of Darwin's marginal notations from books and pamphlets published in English, French and German, illustrating his many research interests. Two writers whose books were heavily annotated by Darwin are specially recognized for their influence: the Swiss botanist Alphonse de Candolle and the Prussian naturalist and traveller Alexander von Humboldt (1769–1859). All the marginalia in their books in Darwin's library, and many more, were later published in Di Gregorio and Gill (1990). As one who worked on the Darwin Correspondence Project for 15 years, I can't tell the reader enough about how helpful this book is to the researcher on Darwin. A work on the marginalia in Darwin's pamphlets is forthcoming.

## ACKNOWLEDGMENTS

Lucy Salt (Keeper of Art, Regeneration and Community Department, Derby Museums and Art Gallery, Derby) kindly supplied information on James Rawlinson's and Joseph Wright's portraits of Erasmus Darwin, as did Simon Chaplin (Director of Museums and Special Collections, The Royal College of Surgeons of England), on the College and Down House. My colleagues Richard Bambach, David A. West, and Peter W. Graham, and my wife Sarah, have made many helpful comments on the manuscript, as have reviewers Juliet Clutton-Brock and Arthur Lucas.

## REFERENCES

BARLOW, N. (editor), 1933 *Charles Darwin's diary of the voyage of H. M. S. "Beagle"*. Cambridge.

BARLOW, N., 1935 Charles Darwin and the Galapagos Islands. *Nature* **136**: 391.

BARLOW, N. (editor), 1945 *Charles Darwin and the voyage of the Beagle*. London.

BARLOW, N. (editor), 1958 *The autobiography of Charles Darwin 1809–1882. With original omissions restored*. London.

BARLOW, N. (editor), 1963 Darwin's ornithological notes. *Bulletin of the British Museum (Natural History), historical series*, **2**: 203–278.

BARRETT, P. H., GAUTREY, P. J., HERBERT, S., KOHN, D. and SMITH, S. (editors), 1987 *Charles Darwin's notebooks, 1836–1844. Geology, transmutation of species, metaphysical enquires*. London & Ithaca.

BURKHARDT, F., 1998 The Darwin papers, pp 118–135 in FOX, P. (editor), *Cambridge University Library. The great collections*. Cambridge.

BURKHARDT, F., PORTER, D. M., BROWNE, J. and RICHMOND, M. (editors), 1993 *The correspondence of Charles Darwin*. Volume **8**: *1860*. Cambridge.

BURKHARDT, F., PORTER, D. M., DEAN, S. A., TOPHAM, J. R. and WILMOT, S. (editors), 1999 *The correspondence of Charles Darwin*. Volume **11**: *1863*. Cambridge.

BURKHARDT, F., PORTER, D. M., DEAN, S. A., EVANS, S., INNES, S., SCLATER, A., PEARN, A. and WHITE, P. (editors), 2004 *The correspondence of Charles Darwin*. Volume **14**: *1866*. Cambridge.

BURKHARDT, F., PORTER, D. M., DEAN, S. A., EVANS, S., INNES, S., PEARN, A., SCLATER, A. and WHITE, P. (editors), 2005 *The correspondence of Charles Darwin*. Volume **15**: *1867*. Cambridge.

BURKHARDT, F. and SMITH, S. (editors), 1985 *The correspondence of Charles Darwin*. Volume **1**: *1821–1836*. Cambridge.

BURKHARDT, F. and SMITH, S. (editors), 1986 *The correspondence of Charles Darwin*. Volume **2**: *1837–1843*. Cambridge.

BURKHARDT, F. and SMITH, S. (editors), 1989 *The correspondence of Charles Darwin*. Volume **5**: *1851–1855*. Cambridge.

BURKHARDT, F. and SMITH, S. (editors), 1991 *The correspondence of Charles Darwin*. Volume **7**: *1858–1859*. Cambridge.

BURKHARDT, F. and SMITH, S. (editors), 1994 *A calendar of the correspondence of Charles Darwin, 1821–1882*. Second edition. New York & London.

BUTLER, S., 1879 *Evolution old and new. Or the theories of Buffon, Dr. Erasmus Darwin and Lamarck compared with that of Mr. C. Darwin*. London.

COLP, R., 1977 *To be an invalid. The illness of Charles Darwin*. Chicago.

COLP, R., 2008 *Darwin's illness*. Gainesville.

DARWIN, C., 1839 *Narrative of the surveying voyages of His Majesty's Ships Adventure and Beagle, during the years 1826 to 1836. Describing their examination of the southern shores of South America and the Beagle's circumnavigation of the globe*. Volume **3**. *Journal and remarks, 1832–1836*. London.

DARWIN, C., 1856 On the action of sea-water on the germination of seeds. *Journal of the proceedings of the Linnean Society (botany)* **1**: 130–140.

DARWIN, C., 1859 *On the origin of species by means of natural selection, or the preservation of favoured races in the struggle for life*. London.

DARWIN, C., 1868 *The variation of animals and plants under domestication*. London.

DARWIN, C., 1871 *The descent of man, and selection in relation to sex*. London.

DARWIN, C., 1875 *The movements and habits of climbing plants*. Second edition. London.

DARWIN, C. and WALLACE, A. R., 1858 On the tendency of species to form varieties; and on the perpetuation of varieties and species by natural means of selection. *Journal of the proceedings of the Linnean Society of London (zoology)* **3**: 45–62.

DARWIN, C. and WALLACE, A. R., 1958 *Evolution by natural selection*. Cambridge.

DARWIN, F. (editor), 1887 *The life and letters of Charles Darwin, including an autobiographical chapter*. London.

DARWIN, F. (editor), 1892 *Charles Darwin: His life told in an autobiographical chapter, and in a selected series of his published letters*. London.

DARWIN, F. (editor), 1909 *The foundations of The origin of species. Two essays written in 1842 and 1844*. Cambridge.

DARWIN, F. and SEWARD, A. C. (editors), 1903 *More letters of Charles Darwin. A record of his work in a series of unpublished letters*. London.

DAVIS, S., 2004 Darwin, Tegetmeier and the bees. *Studies in history and philosophy of biological and biomedical sciences* **35**: 65–92.

DeARCE, M. 2008 Correspondence of Charles Darwin on James Torbitt's project to breed blight-resistant potatoes. *Archives of natural history* **35** (2): 208–222.

DE BEER, G. (editor), 1983 *Charles Darwin, Thomas Henry Huxley. Autobiographies*. Oxford.

DI GREGORIO, M. A. and GILL, N. W. (editors), 1990 *Charles Darwin's marginalia*. Volume **1**. New York & London.

FITZROY, R., 1839 *Narrative of the surveying voyages of His Majesty's Ships Adventure and Beagle, between the years 1826 and 1836. Describing their examination of the southern shores of South America, and the Beagle's circumnavigation of the globe*. Volume **2**. *Proceedings of the second expedition, 1831–1836, under the command of Captain Robert Fitz-Roy, R. N*. London.

FREEMAN, R. B., 1977 *The works of Charles Darwin. An annotated bibliographical handlist*. Second edition. Folkestone & Hamden, Connecticut.

FREEMAN, R. B. and GAUTREY, P. J., 1972 Charles Darwin's Queries about expression. *Bulletin of the British Museum (Natural History), historical series*, **4**: 205–219.

GLICK, T. F. (editor), 1974 *The comparative reception of Darwinism*. Austin.

GLICK, T. F. (editor), 1988 *The comparative reception of Darwinism. With a new preface*. Chicago.

HOOKER, J. D., 1844–1847 *The botany of the Antarctic voyage of H. M. discovery ships Erebus and Terror in the years 1839–1843, under the command of Captain Sir James Clark Ross*. Part 1. *Flora antarctica*. London.

HOOKER, J. D., 1847a An enumeration of the plants of the Galapagos Archipelago; with descriptions of those which are new. *Transactions of the Linnean Society* **20**: 163–233.

HOOKER, J. D., 1847b On the vegetation of the Galapagos Archipelago, as compared with that of some other tropical islands and the continent of America. *Transactions of the Linnean Society* **20**: 235–262.

JORDANOVA, L. J. (editor), 1986 *Languages of nature. Critical essays on science and literature*. London.

KEMP, W., 1844 An account of some seeds, buried at a great depth in a sand-pit, which germinated; by Mr. William Kemp of Galasheils in a letter to Charles Darwin Esqr. *Annals and magazine of natural history* **13**: 89–91.

KEYNES, M., 1994 Portraits of Dr Erasmus Darwin, F.R.S., by Joseph Wright, James Rawlinson and William Coffee. *Notes and records of the Royal Society of London* **48** (1): 69–84.

KEYNES, R., 2001 *Annie's box. Charles Darwin, his daughter and human evolution*. London.

KEYNES, R. D. (editor), 1988 *Charles Darwin's Beagle diary*. Cambridge.

KING-HELE, D., 1999 *Erasmus Darwin. A life of unequalled achievement*. London.

KING-HELE, D. (editor), 2003 *Charles Darwin's The life of Erasmus Darwin*. Cambridge.

KRAUSE, E., 1879 *Erasmus Darwin. With a preliminary notice by Charles Darwin*. London.

LUCAS, P., 2007 Charles Darwin, "little Dawkins" and the platycnemic Yale men: introducing a bioarchaeological tale of the descent of man. *Archives of natural history* **34** (2): 318–345.

McNEIL, M., 1986 The scientific muse: the poetry of Erasmus Darwin, pp 159–203 in JORDANOVA, L. J. (editor), *Languages of nature. Critical essays on science and literature*. London.

McNEIL, M., 1987 *Under the banner of science: Erasmus Darwin and his age*. Manchester.

McNEIL, M., 2007 Darwin, Erasmus (1731–1802), in GOLDMAN, L., (editor), *Oxford dictionary of national biography*. Online edition [http://www.oxforddnb.com/view/article/7177 accessed 21 February 2009]. Oxford.

MÜLLER, F., 1864 *Für Darwin*. Leipzig.

NASH, L. A., 1890 Some memories of Charles Darwin. *Overland monthly* (San Francisco) October: 404–408.

NASH, W., 1919 *A lawyer's life on two continents*. Boston.

NELSON, E. C., 2000 *Sea beans and nicker nuts: a handbook of exotic seeds and fruits stranded on beaches in north-western Europe*. London.

OVERY, C., 2004 *Down House. Home of Charles Darwin. Making a visit. Information for teachers*. Colchester.

PORTER, D. M., 1980 The vascular plants of Joseph Dalton Hooker's *An enumeration of the plants of the Galapagos Archipelago; with descriptions of those which are new*. *Botanical journal of the Linnean Society* **81**: 79–134.

PORTER, D. M., 1981 Darwin's missing notebooks come to light. *Nature* **201**: 13.

PORTER, D. M., 1985 The *Beagle* collector and his collections, pp 973–1019 in KOHN, D. (editor), *The Darwinian heritage*. Princeton.

PORTER, D. M., 1986 Charles Darwin's vascular plant collections from the voyage of *HMS Beagle*. *Botanical journal of the Linnean Society* **93**: 1–172.

PORTER, D. M. (editor), 1987 Darwin's notes on *Beagle* plants. *Bulletin of the British Museum (Natural History), historical series*, **14**: 146–233.

PORTER, D. M., 1993 On the road to the *Origin* with Darwin, Hooker, and Gray. *Journal of the history of biology* **26**: 1–38.

PORTER, D. M., 1999 Charles Darwin's Chilean plant collections. *Revista Chilena de historia natural* **72**: 181–200.

PORTER, D. M., 2003 The Darwin Correspondence Project. *The Virginia Tech scholarly review* **1**: 3–7.

PORTER, D. M., MURRELL, G. and PARKER, J., 2009 Some new Darwin vascular plant specimens from the *Beagle* voyage. *Botanical journal of the Linnean Society* **159**: 12–18.

RABY, P., 1990 *Samuel Butler: a biography*. London.

ROBERTS, M., 2001 Just before the *Beagle*: Charles Darwin's geological fieldwork in Wales, summer 1831. *Endeavour* **25** (1): 33–37.

SALLON, S., SOLOWEY, E., COHEN, Y., KORCHINSKY, R., EGLI, M., WOODHATCH, I., SIMCHONI, O. and KISLEV, M., 2008 Germination, genetics, and growth of an ancient date seed. *Science* **320**: 1464.

SAYER, K., 2007 "Let nature be your teacher": Tegetmeier's distinctive ornithological studies. *Victorian literature and culture* **35**: 589–605.

SEWARD, A. C. (editor), 1909 *Darwin and modern science. Essays in commemoration of the centenary of the birth of Charles Darwin and of the fiftieth anniversary of the publication of "The origin of species"*. Cambridge.

SHEN-MILLER, J., MUDGETT, M. B., SCHOPF, J. W., CLARKE, S. and BERGER, R., 1995 Exceptional seed longevity and robust growth: ancient sacred lotus from China. *American journal of botany* **82**: 1367–1380.

SMITH, K. G. V. (editor), 1987 Darwin's insects. Charles Darwin's entomological notes. *Bulletin of the British Museum (Natural History), historical series*, **14**: 1–143.

SULLOWAY, F. J., 1982 Darwin and his finches: the evolution of a legend. *Journal of the history of biology* **15**: 1–53.

VORZIMMER, P. J., 1969 Darwin's Questions about the breeding of animals. *Journal of the history of biology* **2**: 269–281.

WEST, D. A., 2003 *Fritz Müller, a naturalist in Brazil. Based on Fritz Müllers Werke, Briefe, und Leben by Alfred Möller*. Blacksburg.

WEST, D. A., 2007 Fritz Müller's first copy of Darwin's *Origin* discovered. *Archives of natural history* **34** (2): 352–354.

YOUNG, R. M., 1985 *Darwin's metaphor. Nature's place in Victorian culture*. Cambridge.

# ERASMUS DARWIN, M.D., F.R.S.:
# A BIOGRAPHICAL AND ICONOGRAPHICAL NOTE

By J. W. T. Moody, F.L.S.*

(*With 2 Plates*)

The name of Erasmus Darwin [1731–1802] has become synonymous with the historical image of the great eighteenth-century philosophers—a man of brilliance in speculative science and of equal talent in practical science and applied technology. Dr. Erasmus Darwin is certainly the epitome of such qualities. He is, perhaps, most renowned for his ideas on evolution as expressed in his *Zoonomia*, *The Botanic Garden*, and *The Temple of Nature*. It is often forgotten that Dr. Darwin was considered one of the leading medical minds as well as one of the great poets of his day.

Dr. Darwin's medical and scientific reputation made him one of the best known philosophers in the latter half of the eighteenth century. A portion of Darwin's scientific reputation came from his association with the Lunar Society of Birmingham. As one of the founding members, Darwin became a close friend and scientific colleague of some of the leading men in the field of science and applied technology. These included the great chemist Joseph Priestley, the inventor James Watt and his partner Matthew Boulton, the botanist Dr. William Withering, the renowned potter of Etruria, Josiah Wedgwood, and numerous others.

Darwin's interest in promoting scientific ideas and spreading useful scientific knowledge is evidenced in his founding two other scientific societies. The first of these was the Lichfield Botanic Society. This society was established in Lichfield at a period when there was a great deal of intellectual activity, centred for the most part around Anna Seward, " the Swan of Lichfield ", Dr. Butler, Bishop of Lichfield, Dr. Erasmus Darwin, and the sometime resident, Dr. Samuel Johnson. From this circle, however, the Lichfield Botanic Society acquired only three members: Dr. Darwin, Sir Brooke Boothby, and Mr. Jackson. In spite of the obviously small response of Lichfield, Dr. Darwin developed his plans to spread Linnaean Botanical Philosophy. With the assistance of a botanic garden near Lichfield, the added influence of the publications of his Lichfield Botanic Society, and through his own poetic works, Dr. Darwin promoted the popular acceptance and understanding of the Linnaean System.

When Darwin moved from Lichfield to Derby he quickly established himself as one of Derby's leading citizens. In 1783 he founded another scientific society, The Philosophical Society of Derby, which proved to be a popular centre for the discussion of science and applied technology. The Society arranged a yearly series of lectures and amassed a library for the use of its members. This collection is now part of the Derby Museum and Library.

Darwin's fertile mind produced a staggering number of ideas, designs, and inventions for the practical application of science to the technological needs of his day. Among these are several " improved designs for carriages ", the

* Greenville College, Illinois.

*In the possession of the Erasmus Darwin School, Derby*

ERASMUS DARWIN.    Portrait by James Rawlinson, *c.* 1802

*In the possession of J. W. T. Moody, Esq.*

ERASMUS DARWIN.   Sketch for a portrait by James Rawlinson

designs for a polygraph, a number of designs for windmills—one of which Josiah Wedgwood used successfully, an improved grain drill which he describes in *Phytologia*, an improved oil lamp, and, in 1765 (!), a design for a steam carriage. His interests ranged over those of the astronomer, agriculturalist, botanist, geologist, horticulturist, meteorologist, physician, plant physiologist, and zoologist, to name just a few. All these and more are reflected in his great works. Darwin appeared to have a theory for everything, from medicine to the lowly task of corn planting. This endless supply of theories gives some grounds for Coleridge's coining the term " Darwinising " to express the excesses of speculation.

If we look to the portraits of Erasmus Darwin to reveal the active mind of a great philosopher, we are somewhat disappointed. The best known portrait of Darwin is that done by James Wright of Derby. In this we see a rather taciturn, ponderous Dr. Darwin, a sedentary philosopher, but not the vigorous man of medicine and science. It is, however, the work of another Derby artist, James Rawlinson, that is of immediate importance to this paper. One of Rawlinson's portraits of Erasmus Darwin was popularised through an engraving which was published in Dr. Robert John Thornton's *New Illustrations of The Sexual System of Carolus von Linnaeus* (1799–1807). There is also a relatively unknown Rawlinson portrait of Dr. Darwin done *circa* 1802. This portrait was, until 1962, in the Guildhall, Derby, but is now at the Erasmus Darwin School at Derby. It is the cartoon of this portrait which I have discovered.

The artist, James Rawlinson, was a student of the great Romney, but remained a local painter, and thus quite unknown outside Derbyshire. For biographical details of this Derby artist I am indebted to Mr. A. L. Thorpe, Curator of the Derby Museum and Art Gallery, who informs me that James Rawlinson was probably born in Derby in 1769. Rawlinson's artistic career was restricted to Derbyshire, with the exception of his " Portrait of An Old Woman Knitting ", exhibited in 1798 at the Royal Academy, London. For a time Rawlinson enjoyed the patronage of the famous Strutt family, for whom he painted portraits " Washington," " Mr. Douglas ", " Mrs. Strutt ", " William Strutt, F.R.S.", " Jedidiah Strutt ", and also " A View of Willersley ". In 1807 he donated his services to paint an altarpiece for All Saints Church in Derby. It has since been destroyed, but a drawing of " The Three Marys Visiting The Tomb " was published in Mr. George Bailey's *Chronicles of All Saints Church*. Among Rawlinson's other artistic productions is *An Album of Derbyshire Views* (1822), being lithographs after drawings by his daughter Elizabeth Rawlinson. This father and daughter artistic team continued, and in 1829 they toured Italy.

In the period from 1800 to 1805, James Rawlinson experimented with the manufacture of artists' colours. In 1805 he contributed a paper to *Nicholson's Journal* (Vol. IX) entitled " Description of an improved mill for levigating painters' colours ". It was on the back of a page of a notebook on the making of artists' colours that Rawlinson sketched the cartoon for the Erasmus Darwin portrait which now hangs in the Erasmus Darwin School at Derby. Whether Rawlinson was consulting Dr. Darwin about his experiments in the manufacture of artists' colours is not known, but from what we know of Dr. Darwin it is not improbable.

This sketch was obviously not a planned sitting, but rather an " on the spur of the moment " suggestion of the artist, thus the use of his notebook. It

13*

certainly was a fortuitous sitting, as Rawlinson himself wrote on the cartoon: " This sketch of Dr. Darwin I took from him in his study with the table desk and back ground exactly as they were at the time, J. Rawlinson, and this was the last picture he ever sat for." Dr. Erasmus Darwin died at Derby on 18 April 1802.

On 2 August 1848 James Rawlinson of Derby died, and the obituary notice in the *Derby Mercury* for that day is an excellent summation of the artist: " Rawlinson well understood his profession, but being of unassuming and retired habits, and living at a distance from the Metropolis, his name has not been enrolled among those of his fellow labourers in art, to the extent to which his talents might have claimed. His likenesses, many of which are to be found in this county, are distinguished for their fidelity to their originals."

From 1848 the history of this Rawlinson cartoon is unknown. In October 1963 I discovered it on sale at Maggs Brothers of Berkeley Square, London, W.1. They did not have a record of its previous history, but they assumed it came to them as part of a collection of letters and manuscripts. With the assistance of Mr. Kerslake of the National Portrait Gallery, and Mr. Thorpe, Curator of the Derby Museum, I was able to determine the authenticity of the cartoon and to gather the historical information about it and the artist.

The following list includes original and secondary sources relating to Dr. Erasmus Darwin.

### A. *Darwin Manuscripts:*

Downe House, Down, Kent.
British Museum Manuscript Collections.
The Royal Society.
The Wellcome Historical Medical Library.

### B. *References:*

BARLOW, NORA, 1959. " Erasmus Darwin." *Notes and Records of the Royal Society of London.* Vol. 14, no. 1, pp. 85–98.

BROWN, THOMAS, 1798. *Observations on the Zoonomia of Erasmus Darwin, M.D.* Edinburgh, Murdell & Son.

DARWIN, ERASMUS, 1782–83. *A System of Vegetables. . . . Translated from the Thirteenth Edition of Systema Vegetabilium. . . . By a Botanical Society at Lichfield* (Vols. 1–2). London, Leigh & Sotheby.

DARWIN, ERASMUS, 1784. *Dr. Darwin's Address to the Philosophical Society of Derby on their First Meeting, July 18, 1784.* Derby.

DARWIN, ERASMUS, 1787. *The Families of Plants. . . . Translated from the Last Edition of the Genera Plantarum and of The Mantissae Plantarum. . . .* Lichfield, *etc.* (2 vols.)

DARWIN, ERASMUS, 1789. *The Botanic Garden, Part II. Containing the Loves of the Plants, a Poem.* London, J. Johnson.

DARWIN, ERASMUS, 1791. *The Botanic Garden, Part I. Containing the Economy of Vegetation, a Poem.* London, J. Johnson.

DARWIN, ERASMUS, 1794–96. *Zoonomia; or the Laws of Organic Life.* (2 vols.) London, J. Johnson.

DARWIN, ERASMUS, 1800. *Phytologia; or the Philosophy of Agriculture and Gardening.* London, J. Johnson.

DARWIN, ERASMUS, 1803.  *The Temple of Nature; or the Origin of Society.* London, J. Johnson.

DOWSON, JOHN, 1861.  *Erasmus Darwin: Philosopher, Poet, and Physician.* London, H. K. Lewis.

EMERY, CLARK, 1941.  " Scientific Theory in Erasmus Darwin's *The Botanic Garden* ", *Isis*, Vol. 33, pp. 315–325.

GARFINKLE, NORTON, 1955.  " Science and Religion in England, 1790–1800: the Critical Response to the Work of Erasmus Darwin." *Journal of the History of Ideas*, Vol. 16, pp. 376–388.

KING-HELE, DESMOND, 1963.  *Erasmus Darwin.*  London, Macmillan & Co.

KRAUSE, ERNST, 1879.  *Erasmus Darwin.*  London, John Murray.

PEARSON, HESKETH, 1930.  *Doctor Darwin.*  London, J. M. Dent.

ROBINSON, ERIC, 1953.  " The Derby Philosophical Society ", *Annals of Science*, Vol. 9, pp. 359–367.

SCHOFIELD, R. E., 1963.  *The Lunar Society of Birmingham.*  O.U.P.

SEWARD, ANNA, 1804.  *Memoirs of the Life of Dr. Darwin.*  London, J. Johnson.

THORPE, T. E., 1906.  *Joseph Priestley.*  London, Dent.

WEDGWOOD, JULIA, 1915.  *The Personal Life of Josiah Wedgwood.*  London, Macmillan & Co.

*J. Soc. Biblphy nat. Hist.* (1976) 7 (4): 537

# Nature, poetry and medicine in late 18th century England: a unified perspective of Erasmus Darwin

By M. McNEIL

Darwin College,
Cambridge.

ABSTRACT

Erasmus Darwin was the focus and embodiment of provincial England in his day. Renowned as a physician, he spent much of his life at Lichfield. He instigated the founding of the Lichfield Botanic Society, which provided the first English translation of the works of Linnaeus, and established a botanic garden; the Lunar Society of Birmingham; the Derby Philosophical Society; and two provincial libraries. A list of Darwin's correspondents and associates reads like a "who's who" of eighteenth century science, industry, medicine and philosophy. His poetry was also well received by his contemporaries and he expounded the evolutionary principles of life. Darwin can be seen as an English equivalent of Lamarck, being a philosopher of nature and human society. His ideas have been linked to a multitude of movements, including the nosological movement in Western medicine, nineteenth century utilitarianism, Romanticism in both Britain and Germany, and associationist psychology.

The relationships between various aspects of Darwin's interests and the organizational principles of his writings were examined. His poetical form and medical theory were not peripheral to his study of nature but intrinsically linked in providing his contemporaries with a panorama of nature. A richer, more integrated comprehension of Erasmus Darwin as one of the most significant and representative personalities of his era was presented.

*Archives of natural history* **29** (3): 359–370. 2002 © The Society for the History of Natural History

# Jigsaw with pieces missing: Charles Darwin with John Price at Bodnant, the walking tour of 1826 and the expeditions of 1827

PETER LUCAS

Pwllymarch, Llanbedr, Gwynedd LL45 2PL, Wales.

ABSTRACT: Darwin seems to have paid two visits to John Price, Welsh speaking school friend of both Darwin brothers, at his family's home, Bodnant in North Wales. In June 1826 with Nathan Hubbersty he stayed at Bodnant on the walking tour of North Wales which took them to Ffestiniog, Snowdon and other places. In July 1827, the fragmentary evidence strongly suggests, both Darwin brothers, Erasmus and Charles, stayed at Bodnant with expeditions to the Great Ormes Head, near Llandudno, and west across the Conwy river to the Carneddau range. The visits serve as a reminder of the wealth of topographical experience which Darwin brought to the geological tour with Adam Sedgwick in 1831.

KEY WORDS: North Wales – Ffestiniog – Great Ormes Head – Adam Sedgwick – Erasmus Alvey Darwin

## INTRODUCTION

"I am astonished at your vivid recollection of old times", Charles Darwin told John Price in the autumn of 1863, writing in tactful response to the early instalments of *Old Price's remains* (Price, 1863–1864)[1], "an extremely odd monthly periodical (April 1863–March 1864) of great interest to dwellers in the Llandrillo-yn-Rhos district" (Lloyd and Jenkins, 1959: 787). "I wonder", Darwin continued, "you do not give us more about the wilder parts of North Wales, which I admire with the fervour of a Welchman" (Burkhardt *et alii*, 1999: 627). But Price's youth had been spent in houses near the coast of North Wales, at a distance from the "wilder parts" which remained so dear to Darwin.

Some experiences they had shared in "old times" can be pieced together from a number of mostly cryptic references scattered over more than half a century. These make a small addition to our understanding of Darwin's movements in 1826 and 1827 and of his geological tour with Adam Sedgwick in August 1831, providing also an introduction to two quite unusual members of a Welsh clerical dynasty.

## JAMES AND JOHN PRICE

The father, James Price (or Pryce), enjoyed half a century, from 1800 to 1850, as "non-resident rector of Llanfechain, Mont[gomeryshire]" (Lloyd and Jenkins, 1959: 787), a valuable and much coveted living in the gift of the Bishop of St Asaph, blessed in 1831 with a net annual income of £530 and a glebe house "fit for residence" (Anonymous, 1836: 124). "Allowed on account of ill-health to be non-resident and recuperate at the seaside", for the next 50 years he "proceeded to occupy substantial residences on the North Wales coast with an adequate supply of servants" (Porter, 1938: 39–40). At length, after he had "complained of his health all his life" (Lloyd and Jenkins, 1959: 787), "a surgical accident when he was still full of

24

life and vigour ... cut his life short" in his ninety-fourth year (Porter, 1938: 41). This may be too good a story to be taken quite at face value and one short sentence in Price's letter to Darwin of July 1826 – "We expect my Father from his living tonight" (Burkhardt and Smith, 1985: 1, 43) – suggests the father was not entirely inattentive.

"Cleric, naturalist, and eccentric", according to his entry in *The dictionary of Welsh biography* (Lloyd and Jenkins, 1959: 787), John Price was born in 1803 "at Pwll-y-crochan, Colwyn Bay, of a long and long-lived line of clerics." "Though he became a master at Shrewsbury and other schools", the same source tells us, "he developed an instability and an oddity in dress and behaviour which unfitted him for a normal career" (Lloyd and Jenkins, 1959: 787). Eventually he "settled down as coach and tutor in Chester, a poor climax for a man of such promise" (Porter, 1938: 42). "Old Price", in his own description, was "transplanted from Pwll-y-Crochon to Bodnant ... to Shrewsbury School ... detained in North Wales twice, for twelve months at a time, and subsequently for three whole years at his Father and Mother's residence, Plas yn Llysfaen[2], near Abergele" (Price, 1863–1864: 2). Born in 1803, the year before Erasmus, the elder of the Darwin brothers, he died in 1887, five years after Charles.

## THREE "SUBSTANTIAL RESIDENCES"

A school now stands at the top of Pwllycrochan Avenue (grid reference SH 842787) below the Pwllycrochan woods, facing north down to the sweep of Colwyn Bay and the Iron Age fort of Bryn Euryn with the limestone headland of the Little Ormes Head beyond. Price's childhood was spent in the "smallish brick mansion" demolished when Jane Silence Williams, the heiress of Pwllycrochan, and other properties[3], married Sir John Erskine in 1821: the family's sale of the estate in 1865 would signal the development of the resort of Colwyn Bay (Porter, 1938: 14, 22). Bryn Euryn, a regular sunset walk for "a very silent party, from Pwllycrochon, in the olden time" (Price, 1875: 66), dominates Llandrillo-yn-Rhos (Rhos-on-Sea), "Gros near Abegele", the scene of some of Darwin's earliest memories, where in 1813 (aged 4½), with his younger sister Catherine and their Wedgwood cousins, Fanny and Emma, he spent the first of the two less than wholly successful seaside holidays of his childhood (Lucas, 2002b). Rhos was perhaps a recommendation of James Price who was among those "financially obliged to Dr Darwin" (Browne, 1995: 25).

The marriage of the Pwllycrochan heiress brought the move of "the temporary tenants, the Reverend James Price and family ... a few miles into Bodnant" (Porter, 1938: 22), or Bodnod[4], including perhaps the 116 acres of demesne shown in the tithe commutation records.[5] The house (Figure 1), high above the tidal Conwy, had been "built in 1792 ... of typical Georgian design with sash windows and white stucco covering" (Anonymous, 1997: 37)[6] in a magnificent setting, now familiar to visitors from all over the world, facing west to the great, grassy ridges of the Carneddau. "I walk out every day" – Price wrote to Darwin in July 1826, convalescing after apparently near fatal illness – "& the view of these glorious mountains & smell of their air is life to me" (Burkhardt and Smith, 1985: 1, 43).[7]

After seven years at Bodnant James Price moved for the last time, to Plas yn Llysfaen (SH 901772), with its 358 acres of which the tithe commutation records show him in occupation.[8] The house lies to the east (as Pwllycrochan to the west) of Hen Golwyn, Old Colwyn, beyond Mynydd Marian then with its semaphore station, one of the chain for shipping between Liverpool and Holyhead. Almost as close to the coast as Pwllycrochan, the house

Figure 1. Watercolour of Bodnant by "William Hanmer Esq". (Reproduced by courtesy of Lord and Lady Aberconway.) (Photograph Carrie Hitchcock.)

faces in the opposite direction, inland to the Denbighshire hills. Some two miles to the south at Bettws-yn-Rhos – on Darwin's route on 8 August 1831 – another James Pryce, John's grandfather, had been vicar from 1746 to 1758 (Lloyd and Jenkins, 1959: 787).

## JOHN PRICE AND THE DARWINS

Price was a friend of Erasmus Darwin, his slightly younger contemporary, at Shrewsbury school and Cambridge. With the younger brother, Charles, "full six years my Junior"[9], the relationship was necessarily more remote (Browne, 1995: 25–26). Price met other members of the family while an assistant master at Shrewsbury school in 1826 and 1827. "Erasmus' Friend, Mr Pryce" – Catherine Darwin wrote to her brother on 11 April 1826, towards the end of his first year at Edinburgh – "and 2 other unfortunate masters ... dined here yesterday – We all liked Pryce; he seems very gentlemanly and agreeable" (Burkhardt and Smith, 1985: 1, 40). At around the same time Price was consulting Darwin's father, Dr R. W. Darwin, about his health (Burkhardt and Smith, 1985: 42), while entries in his commonplace book in October 1827, around the time of his resignation from Shrewsbury school (Fisher, 1899: 474), show him, on 14 October, apparently receiving advice from the Doctor on the propagation of mulberries on a south wall and, "on a dull foggy evening", musing on the refraction of light "at y distance of several paces" from the Doctor's gate, "the ghost of a gate".[10]

It is to within Price's time as a master at the school that the evidence for Darwin's visits to Bodnant quite strongly points.

## BODNANT 1826: THE WALKING TOUR WITH HUBBERSTY

It was "with two friends", Darwin remembered late in life, that "during the summer of 1826 I took a long walking tour ... with knapsacks on our backs through North Wales. We walked thirty miles most days, including one day the ascent of Snowdon" (Barlow, 1958: 53–54). Much earlier, in a journal entry made retrospectively in 1838, he had noted only "Walking tour into North Wales with Hubbersty" – Nathan Hubbersty (1803–1881) – one of the "unfortunate masters" of the letter of 11 April 1826 (Burkhardt and Smith, 1985: 1, 538, 635). In the same letter Caroline Darwin described the masters as "3 of Eras's friends" (Burkhardt and Smith, 1985: 1, 41) but there is nothing to suggest that Charles had made Hubbersty's acquaintance at this time, nor would the tour cement a friendship: Hubbersty has no place among the serried ranks of Darwin's correspondents.[11] As sole companion he is on the face of it a puzzling choice.

The tour began on 15 June (Burkhardt and Smith, 1985: 1, 538) and must have ended before 28 July when "Hubberstie" was reported as "very well after his fatigues in Wales" (Burkhardt and Smith, 1985: 1, 43). Some light is thrown on its early days in two letters from Price to Darwin more than half a century apart. It seems that Darwin and Hubbersty called on Price at Bodnant perhaps at the beginning of their tour; possibly Price was the *intended* third member of the party, and the link between the other two. "I am glad you derived some pleasure from your tour", Price told Darwin in the first of the letters, undated but attributed to July 1826[12], "wh. I much wish I could have joined & should have been glad to assist you by speaking the British language" (Burkhardt and Smith, 1985: 1, 42). Confirmation that the letter must have been written from Bodnant comes 55 years later. "Poor Nathan Hubbersty", Price wrote on 17 September 1881, "is, I fear, in a very poor way, confined to bed, & his Wife all but blind". "He and you", Price continued, "had a troublesome journey, after you left me to die at my old home, as seemed likely; Ffestiniog, I think, puzzled you".[13] The inference that from Bodnant the destination was Ffestiniog has an added interest from Darwin's stopover there in August 1831. Probably the two travellers, and Darwin again in 1831, stayed at The Pengwern Arms, enjoying the hospitality of Martha Owen (Lucas, 2002a: 14), reputedly one of only two in the parish able to converse in the foreign English tongue.[14] Here at least Price's fluency in "the British language" would not be missed.

Darwin's pocket diary for 1826 shows glimpses of his dissecting activities – on 26 June he "shot a cormorant", finding "the capacity of the stomach was very great", and on 3 July it was the turn of a natterjack to have the contents of its stomach examined[15] – but, the ascent of Snowdon apart, nothing further is known of the geography of the tour nor whether it was limited to Snowdonia or took in also the Ardudwy mountains and Cadair Idris, later so familiar to Darwin from his Barmouth visits (Lucas, 2002a: 26). That he gave Price the impression of deriving "some pleasure" from the tour suggests this was not undiluted.

## BODNANT 1827: THE GREAT ORMES HEAD AND THE CARNEDDAU

Some scattered references bear on episodes with apparently disparate dates, "about 1824" and "July 1827", for which Bodnant is the obvious starting point.

"When we were at the Sea ... at Rhyl", Catherine Darwin told her brother in a letter she began on 25 July 1832, "we made the usual little Tour to Bangor and Conway, and also to your old acquaintance, the Orme's Head" (Burkhardt and Smith, 1985: 254). Her letter of the following month to her cousin Fanny Wedgwood, written in the last days of Fanny's short life[16], provides some further glimpses. The visit to "the Rhyl" included "a charming 2 days' expedition to Bangor and Conway", from where they went "an expedition of about 4 miles to the two Elizabethan Houses of Sir Thomas Mostyn's", Bodysgallen and Gloddaeth.[17] "It has very long been the object of my ambition to see them ... we went a round from there to the Orme's Head, which I had seen before."[18]

The North Wales coast was evidently a familiar hinterland for more than one member of the Shrewsbury family. Though brother and sister may have been to the "Orme's Head" in the course of the holiday at Rhos in 1813 (when Catherine was only three), this does not fully account for the brother's "old acquaintance". Price's "handy guide" to Llandudno refers to another episode: "asperugo procumbens grew here when we went down with Charles Darwin about 1824, but seems smothered by long nettles since" (Price, 1875: 58). "We" are unspecified; "here", another passage makes clear, was "above the Llech" – a fissure in the cliffs of the Great Ormes Head some 30 feet above sea level – where "the scarce plant Asperugo Procumbens, [was] found ... with Charles Darwin, and never afterwards" (Price, 1875: 102). *Asperugo procumbens* – noted in Price's list of "the more uncommon plants" in Williams (1835: 157)[19] as "German madwort. On Llandudno rocks; at Llech, turning to the left" – is "an established alien known since 1848" (Clement and Foster, 1994: 251). This may explain the excitement recalled by Price in his letter to Darwin of 5 March 1868: "I never was <u>so mad</u> on any subject (nor were <u>you</u>, even at the Ormes Head ...! <u>New plant</u>'!!)".[20]

Price's list illustrates the botanical attractions, already then well recognised (Jones, 1996: 106–107), of the limestone areas around the Ormes Heads. These attractions were no doubt one reason for the expedition which must also have been the occasion of a second episode, between the Great Ormes Head and Deganwy on the western shore of the tombolo on which Llandudno would be developed by the Mostyn family in the 1850s –

> The writer, in company with Charles Darwin, caught a Viper in the Warren, about 1824, favoured by Wellington boots and very strong gardening gloves. Holding him short by the neck, we let him bite at the glove, and emit a drop of clear fluid along the fang, which sank instantly into the leather. When this had been done about five times, no more poison was left, and we killed, but did not eat him, a fact never satisfactorily explained (Price, 1875: 133).

That the "we" of the Great Ormes Head included both Darwin brothers can be inferred from a letter Darwin wrote to Price on 8 September in around "1875–80" (Burkhardt and Smith, 1994: 578, letter 13836). "Your Guide Book", he must be referring to the "handy guide", "seems to me a very interesting one to any visitor. But", Darwin continued, "I shall never visit the place again; & as for Erasmus he has not left his house in Queen Anne Street for 3 years & I fear never will again."[21]

Erasmus was certainly present on another occasion of which Price evidently reminded Darwin in a letter of condolence after the death of Erasmus on 26 August 1881. "I am sorry that I cannot answer your question about the Dotterels", Darwin responded on 3 September with as close to asperity as he allowed himself, "for my memory is a very poor one for small past events." "I remember", he continued, "going with you & Erasmus on ponies a long excursion to some extraordinary wild mountainous scenery but I can remember no more."[22]

The occasion can be dated with relative confidence from a marginal note, "Again on y same hill with Darwin July 1827", to an entry headed "<u>Dotterel</u>" in Price's commonplace book: "I saw a pair of these birds on Galedffordd in Caernarfonsh on July 17th 1824. They were extremely tame, having probably a nest ... on a steep bank covered with the rushy grass called *crawcellt*."[23] Galedfordd, "hard way", is probably "the strangely bare and stony little plateau called Gledrfordd" (Condry, 1972: 36), SH 649705.

The 1824 sighting had been made "returning with Warter[24] from Carnedd Llewelyn".[25] Price describes the route from Bodnant in the "handy guide" (Price, 1875: 86). To reach "the top of Carnedd Llewelyn", he explained, "is easily done, either by skirting Moel Fras from Llanfairfechan, or taking Llanbedr-cennin" – on Darwin's route with Sedgwick on 10 August 1831 – "from Talycafn Station and Ferry", the Conwy crossing only two miles from Bodnant (Figure 2). "The top of Galedfordd", near which "were once seen a brood of adult Dotterel", was common to both routes (Price, 1875: 86–87). As in 1824, the goal in 1827 may have been Carnedd Llewelyn, at 3485 feet the highest summit of the Carneddau, or further south towards Carnedd Dafydd, 3426 feet, where the summit ridge better fits Darwin's "extraordinary wild mountainous scenery".

One reference only explicitly links Darwin to Bodnant. "After trying most things, including Hydropathy" – Price wrote to him on 5 March 1868, after a period of intense depression – "I set myself, one night, to reflect upon God's mercies ... you remember one of these, about 1828, at Bodnant."[26] The allusion, perhaps to his own dangerous illness, is irretrievable, the date unlikely to be 1828, James Price's last year at Bodnant, when Darwin's summer vacation at Barmouth is quite well documented. A sentence in Price's letter of July 1826 implies that Darwin's recent visit, with Hubbersty, was his first: "My Mother begs to be kindly remembered & is sorry you had such a comfortless visit, but you knew the cause & I hope will not be deterred from a second trial under better auspices" (Burkhardt and Smith, 1985: 43). If the expeditions to the Great Orme and the Carneddau were undertaken from Bodnant, this must have been on a "second trial" for which the sighting of the dotterels at "Galedffordd" in "July 1827" provides an unusually firm date.

The present fragmentary evidence thus points to a single visit to Bodnant by both Darwin brothers[27] in the summer before Price's resignation from Shrewsbury school in October. This is consistent with what is known of Darwin's movements that year.

## 1827: DARWIN'S GAP YEAR

1827 brings a gap in Darwin's correspondence – with just one letter, incomplete and ascribed to the very last day of the year, as against 17 for 1826 and 18 for 1828 (Burkhardt and Smith, 1985: ix, 46–48) – and also the one gap in six otherwise uninterrupted years of university education. Instead of proceeding to Cambridge in October, Darwin remained at Shrewsbury with a private tutor, recovering the classics mislaid in the two years since he had left school. The task was not an arduous one, "I soon recovered my school standard of knowledge" (Barlow, 1958: 58), and it is not obvious why it could not have been undertaken and completed before October. Possibly this contributed to his father's famous outburst against his "turning an idle sporting man ... a disgrace to yourself and all your family" (Barlow, 1958: 56, 28).

After leaving Edinburgh in April, Darwin was back in Shrewsbury from his "tour" in Scotland and Ireland by 14 May[28], before making brief visits to London and Paris by the end of the month (Burkhardt and Smith, 1985: 538–539). The autumn, and the shooting season, brought "delightful" visits to the Wedgwoods at Maer (Barlow, 1958: 55–56) and "many visits to Woodhouse" (Burkhardt and Smith, 1985: 539) in "the bright old Woodhouse times" evoked by Sarah Owen many years later: "I can recall the Beetle, & the Fungus hunting,

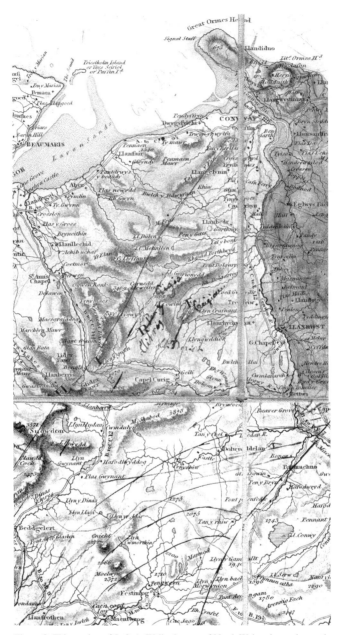

Figure 2. This section of J. & A. Walker's map of North Wales shows the north coast from Llandrillo-yn-Rhos to Beaumaris, the lower Conwy valley (including "Bodnod" and "Tal y Cafn Ferry" and, west of the Conwy, Llanbedr and Carnedd Llewelyn) and, through Dolwyddelan or Penmachno, the approaches to "Festiniog" which converge with those from Capel Curig (Lucas, 2002a: 10–13, and Figure 3). (Reproduced, from Sedgwick's own copy, by permission of Sedgwick Museum, Cambridge.)

& above all the glee with which 'Charles Darwin' used to be descried, cantering up to the house ... you being always the most influential favourite".[29]

A visit to Bodnant in July, before the start of the shooting season, fits well within this context, and also with Dr Darwin's letter to Josiah Wedgwood of 3 August, "Charles is from home"[30], and so not at Maer, nor at Woodhouse either (or the Doctor would have said so, as on a previous occasion[31]).

## 1831: THE TOUR WITH SEDGWICK

### The road to Conwy

On her expedition to "the Rhyl" in 1832, Catherine Darwin "had a good deal of enjoyment in the journey there & back".[32] So also, it may be assumed, Darwin with Hubbersty in 1826 and Erasmus in 1827. Did they, like Darwin with Sedgwick in 1831, take the Holyhead road to Llangollen, then north down the Vale of Clwyd and west in the direction of Conwy? If so, this was the first of a number of anticipations of the tour of 1831.

On 8 August 1831, after parting from Sedgwick at St Asaph (or shortly afterwards), Darwin continued on his own, west to Betws-yn-Rhos, north-east to Abergele and west again to Conwy. There, on the evidence of their geological notes, he rejoined Sedgwick only the next day. Between Betws and Abergele, and again between Abergele and Conwy, Darwin must have passed quite close to Plas yn Llysfaen. Rather than continuing on to spend the night at Abergele, he may have found a reunion with the Price family a welcome alternative after a strenuous day from Denbigh.

### The road to Aber

On 10 August 1831 Darwin set out from Conwy with Sedgwick initially preoccupied with the "'d—d' waiter" (Lucas, 2002a: 5). By the time they had "ascend[ed] the valley of Llanrwst to the chapel of Llanbedr"[33], the thunder must have gone from Sedgwick's brow. Amid the distractions did Darwin remember his earlier passage through Llanbedr y Cennin? Then he had been heading with Price and Erasmus for the heights of the Carneddau. In 1831 he would turn in the opposite direction, over the summit of the great headland of Penmaenmawr, where the Carneddau reach the sea, and down to Aber.

### The road to Ffestiniog

Darwin's walk from Capel Curig to Ffestiniog in August 1831 appears to combine puzzling elements of a direct route by Bwlch y Gorddinan and a more easterly one by Carreg y Fran, Rock of the Crow (Lucas, 2002a: 12). The same routes were used from Bodnant. In March 1821 Samuel Holland travelled from Liverpool to Ffestiniog, mostly on foot, to take charge of his father's slate quarry at what is now Blaenau Ffestiniog. At Llanrwst, five miles beyond Bodnant, he was warned not to go "over the Mountain past Dolwyddelan" by Bwlch y Gorddinan – as "there was no regular road only to Dolwyddelan and merely a *footpath* afterwards over the Mountain which my informant was sure I could not find" – but to make instead for Penmachno "and then to Festiniog ... by a mountain cart road" (Davies, 1952; 5), the pre-turnpike trackway which goes past Carreg y Fran and down Cwm Teigl (Lucas, 2002a: 13).

It is intriguing that Darwin's "journey" from Bodnant with Hubbersty in June 1826 was "troublesome", that Ffestiniog "puzzled" him and that he and Hubbersty may then have headed for Capel Curig, the reverse of his journey in 1831, for their ascent of Snowdon. It would be idle to speculate whether his experiences in 1826 affected his choice of route five summers later but the earlier visit to Ffestiniog does underline that the walk from Capel Curig to Barmouth in August 1831 was indeed "less a venture into an unknown interior than a valedictory passage through familiar landscapes" (Lucas, 2002a: 7).

## CONCLUSION

Darwin visited Bodnant with Hubbersty in June 1826, probably at the start of their walking tour. With his brother, Erasmus, he made a single visit to Bodnant in July 1827: this at least is the most plausible conclusion from the fragmentary evidence which survives. The visits underline the active part Darwin must have taken in 1831 in any preliminary discussions of the tour with Sedgwick during the Easter term at Cambridge and also as energetic guide during the tour itself.

## ACKNOWLEDGEMENTS

Quotations are by permission of the Syndics of Cambridge University Library for the Darwin papers; of the Sedgwick Museum, Cambridge for the transcript of Sedgwick's journal; and of the National Library of Wales, Aberystwyth for John Price's commonplace book. Quotations from the Wedgwood archives in Keele University Library are by courtesy of the Trustees of the Wedgwood Museum, Barleston, Stoke-on-Trent, Staffordshire, England. Permission of the Darwin Correspondence Project, Cambridge University Library, for access to unpublished electronic transcripts is gratefully acknowledged.

## NOTES

[1] Darwin's copy, unbound, some pages still uncut, is in the Darwin Library, Cambridge University Library (CUL).

[2] It was "the Rev. J. Price, of Llysfaen" who "first directed" the attention of "J. E. Bowmall to the silurian rocks to the west of Abergele ... brought under the notice of the Geological Society ... in an able paper" (Hicklin, 1848: 10). According to *Proceedings of the Geological Society of London* 1833–1838, **II**: 666–667, the paper was read on 25 April 1838 by "J. C. Bowman". Both versions of his name are incorrect: the author was evidently John Eddowes Bowman (1785–1841), fellow of the Linnean and Geological Societies and a frequent visitor to North Wales (Jones, 1996: 114).

[3] Including the Porthaethwy Ferry across the Menai Strait for which, after the opening of Telford's suspension bridge, her trustees were awarded £26,394 7s 6d in 1826 (Davies, 1966: 274–275).

[4] The tithe records, note 5 below, have Bodnod (demesne) and the associated Bodnod Hen, Bodnod uchaf and Bodnod bach. Walker's map, Figure 2, has Bodnod. Successive OS editions have various permutations of Bodnod and Bodnant. Price himself refers always to Bodnant.

[5] National Library of Wales, Aberystwyth (NLWA), Parish of Eglwys Bach, Township of Bodnod, A/C 1081 (1844). The landowner is recorded as William Hanmer (born 1792, married 1820 Euphemia Mary, only daughter and heir of John Forbes of Bodnod Hall, died 1872 (Townend, 1963: 1127)). The watercolour of Bodnant (Figure 1) is most likely his work or that of his eldest son, William of Bodnod (1824–1894).

[6] The house was enlarged and altered by H.D. Pochin, father-in-law of the first Lord Aberconway, who bought the estate in 1874.

[7] At the end of the same year he noted "a bat ... near Bodnant on Xmas Day, and ... the last day of 1826": J. Price, Commonplace book, c. 1817–1883, NLWA 22129B, f. 45. The manuscript, henceforth cited as commonplace book, much of it apparently written up in 1836 from earlier notes, was purchased from a Swansea bookstore in 1985.

[8] NLWA, Parish of Llysfaen, A/C 45 (1843).

[9] Recollections sent to Francis Darwin after Darwin's death (1882), CUL, Darwin Papers (DAR) 112: B102.

[10] NLWA, commonplace book, f. 47v, 50v.

[11] Hubbersty has been identified as the "Habberley" to whom Darwin, on the evidence of the "4th notebook on transmutation", "suggested ... that he should do some plantbreeding experiments" (Freeman, 1978: 167). "Habberley" was actually Abberley, Dr Darwin's gardener (Kohn, 1987: 455).

[12] The general accuracy of the attribution, made on the basis of references to the Welsh tour and to Price's expected return to Shrewsbury (Burkhardt and Smith, 1985: 43n1), is confirmed by the second letter, note 13 below.

[13] CUL, DAR 174:76.

[14] This, with other useful information, is given in the historical note in the lobby of The Pengwern Arms.

[15] CUL, DAR 129.

[16] Fanny died on 20 August, "she had no idea of her own danger" (Burkhardt and Smith, 1985: 269).

[17] A third Mostyn house, Corsygedol in Arduddwy, became familiar to Darwin in his Barmouth days (Lucas, 2002a: 15).

[18] Keele University Library, Wedgwood archive, W/M 215, henceforth cited as KUL, WA.

[19] The whole of the section on natural history (Williams, 1835: 144–162) was contributed by Price (Lloyd and Jenkins, 1959: 787).

[20] CUL, DAR 174: 74.

[21] CUL, DAR 147: 278.

[22] CUL, DAR 174: 282.

[23] NLWA, commonplace book, f. 20.

[24] Probably Darwin's schoolfellow John Wood Warter, who wrote to him on 23 December 1824 (Burkhardt and Smith, 1985: 9–10).

[25] NLWA, commonplace book, f. 20.

[26] CUL, DAR 174: 74.

[27] As does the absence in Price's recollections to Francis Darwin (note 9 above) of any reference to visits to Bodnant, suggesting these were not recurring events.

[28] Dr R.W. Darwin to Josiah Wedgwood, 14 May 1827, KUL, WA, E35–26674.

[29] Sarah Haliburton, née Owen and twice widowed, to Darwin, 3 November 1872, CUL, DAR 166: 85.

[30] KUL, WA, E35–26675.

[31] "Caroline Charles & Catherine are at Woodhouse", letter of 13 July 1821 to Josiah Wedgwood, KUL, WA, E35–26596.

[32] Letter of August 1832 to Fanny Wedgwood, KUL, WA, W/M 215.

[33] Transcript by Professor Owen T. Jones of Sedgwick's Journal XXI, 1831 no. I (2–29 August), "Wednesday 10th", Sedgwick Museum, Cambridge.


## REFERENCES

ANONYMOUS, 1836 *The clerical guide and ecclesiastical directory*. London: J. G. & F. Rivington. Pp 287.

ANONYMOUS, 1997 *The garden at Bodnant*. National Trust guide. Norwich: Jarrold Publishing. Pp 43.

BARLOW, N. (editor), 1958 *The autobiography of Charles Darwin 1809–1882*. London: Collins. Pp 253.

BROWNE, J., 1995 *Charles Darwin, voyaging*. London: Jonathan Cape. Pp 605.

BURKHARDT, F. H. and SMITH, S. (editors), 1985 *The correspondence of Charles Darwin, volume* **1** *(1821–1836)*. Cambridge: Cambridge University Press. Pp 702.

BURKHARDT, F. H. and SMITH, S. (editors), 1986 *The correspondence of Charles Darwin, volume* **2** *(1837–1843)*. Cambridge: Cambridge University Press. Pp 603.

BURKHARDT, F. H. and SMITH, S. (editors), 1994 *A calendar of the correspondence of Charles Darwin with supplement (1821–1882)*. Cambridge: Cambridge University Press. Pp 690+46.

BURKHARDT, F. H. *et alii* (editors), 1999 *The correspondence of Charles Darwin, volume* **11** *(1863)*. Cambridge: Cambridge University Press. Pp 1080.

CLEMENT, E.J. and FOSTER, M. C., 1994 *Alien plants of the British Isles. A provisional catalogue of vascular plants (excluding grasses) with guidance on nomenclature by D. H. Kent*. London: Botanical Society of the British Isles. Pp 590.

CONDRY, W. M., 1972 *Exploring Wales*. London: Faber & Faber. Pp 304.

DAVIES, H. R., 1966 *The Conway and the Menai ferries*. Cardiff: University of Wales Press. Pp 342.

DAVIES, W. L. (editor), 1952 *The memoirs of Samuel Holland one of the pioneers of the North Wales slate industry*. The Merioneth Historical and Record Society Extra Publications **1**, 1. Bala: R. Evans & Son. Pp 32.

FREEMAN, R. B., 1978 *Charles Darwin: a companion*. Folkestone: Dawson. Pp 309.

HICKLIN, J., 1848 *Excursions in North Wales*. London, Whittaker & Co. Pp 208.

JONES, D., 1996 *The botanists and guides of Snowdonia*. Llanrwst: Gwasg Carreg Gwalch. Pp 170.

KOHN, D., 1987 Notebook E, pp 395–455 in BARRETT, P. H. *et alii* (editors) *Charles Darwin's notebooks, 1836–1844*. Cambridge: Cambridge University Press. Pp 747.

LLOYD, J. E. and JENKINS, R. T. (editors), 1959 *The dictionary of Welsh biography down to 1940*. London: Cymmrodorion Society. Pp 1157.

LUCAS, P., 2002a "A most glorious country": Charles Darwin and North Wales, especially his 1831 geological tour. *Archives of natural history* **29**: 1–26.

LUCAS, P., 2002b (in press) "Three weeks which now appears like three months": Charles Darwin at Plas Edwards, 1819. *National Library of Wales journal* **33**.

PORTER, G., 1938 *Colwyn Bay before the houses came*. Manchester: Sherratt & Hughes. Pp 109.

PRICE, J., 1863–1864 *Old Price's remains; praehumous, or during life*. 12 parts. London: Virtue, Brothers & Co. Pp 600.

PRICE, J., 1875 *Llandudno and how to enjoy it. Being a handy guide to the town and neighbourhood*. London: Simkin, Marshall. Pp 152.

TOWNEND, P. (editor), 1963 *Burke's genealogical and heraldic history of the peerage, baronetage and knightage*. 103rd edition. London: Burke's Peerage Limited. Pp 2986.

WILLIAMS, R., 1835 *The history and antiquities of the town of Aberconwy and its neighbourhood*. Denbigh, T. Gee. Pp 200.

Received: 12 June 2001. Accepted: 9 August 2001.

EDITORIAL NOTE.

In the original printing, page 370 was blank.

*Archives of natural history* **27** (1): 69–79. 2000 © The Society for the History of Natural History

# I coloured a map: Darwin's attempts at geological mapping in 1831

MICHAEL B. ROBERTS

Chirk Vicarage, Trevor Road, Chirk, Wrexham, LL14 5HD.

ABSTRACT: In his autobiography describing his geology of 1831 Darwin wrote, "on my return to Shropshire I coloured a map of parts around Shrewsbury." There are four extant maps in the Cambridge University Library, which fit this description. Two, at a scale of ⅞ inch to 1 mile, are of Anglesey and Llanymynech and are hand-drawn copies of Evan's map of North Wales, and are without geological annotation. The other two of Shrewsbury and Kinnerley have a scale of 1 inch to 1 mile and are copied from Baugh's *Map of Shropshire* (1808). These contain orange shading to the west of Shrewsbury indicating New Red Sandstone, but make no allowance for drift. The Shrewsbury map includes some attempted stratigraphic boundaries and marks four sites; A, B, C and D. These maps demonstrate Darwin's grasp of geology before his Welsh tour with Sedgwick in August 1831. They show his realisation of the need of a topographic base map, an acquaintance of the conventions of geological mapping in shading and the marking of boundaries. These maps form an early part of Darwin's considerable geological activity in the summer of 1831.

KEY WORDS: history of geology – Shropshire – Anglesey – Adam Sedgwick

## INTRODUCTION

One of the first things a student geologist has long had to do as part of his training is to make a geological map. For most geologists, the present author included, the first attempt is best forgotten and probably destroyed! For many the learning process is more by error than anything else. In the course of learning to make a geological map, the aspirant geologist has to grasp geologists' conventions about shading or colouring a map to indicate rock types, ways of indicating boundaries and faults and also dip and strike. Recently some early maps of Charles Darwin have come to light at Cambridge University Library, having been transferred recently from the archive at Downe House. These four maps of Shropshire and North Wales show Darwin's earliest attempts at field geology and plans for future work. They indicate his awareness for the need of a basic topographic map as he proceeded to make them himself, copying off available maps. As they were made in the early summer of 1831 before his field trip with Adam Sedgwick they shed light both on Darwin's origin as a geologist and on the state of geology in 1831.

As he recorded in his autobiography, Darwin's first attempt at field geology in late June and early July 1831 was to attempt to make a geological map, "Therefore on my return to Shropshire I examined sections and COLOURED A MAP of parts round Shrewsbury" (de Beer, 1983: 39). This seems to be his only attempt to produce a geological map with a largish scale (about 1 mile to 1 inch). Charles Darwin's later published works on geology are marked by the absence of geological maps as discussed by Stoddart (1995: 7). Instead Darwin produced a large number of geological sections in both his notes and published works. As Stoddart points out this is because of the absence of contoured maps. Maps of any kind were virtually non-existent for South America and those of Britain were devoid of indications of altitude, including hachuring.

Darwin's geological mapping took place sometime after his return to Shrewsbury in mid-June and before the arrival of Sedgwick on 2 August. From a consideration of his visit to Llanymynech and absence of dip and strike on his maps, it is most likely that the maps were made before Darwin received his clinometer on 10 July. Shortly afterwards, he visited first Llanymynech on his own and then in August went round North Wales accompanying Sedgwick as far as Bethesda, if not Holyhead. He then continued to Barmouth on foot on his own (Barrett, 1974; Roberts, 1996, 1998; Secord, 1991). Though Sedgwick made a geological map and Darwin claimed to have helped him when he returned from a parallel traverse when he was asked "to mark the stratification on a map" (de Beer, 1983: 39), none of Darwin's maps from these field trips appear to be extant, at least in the Darwin archives in Cambridge University Library.

At Cambridge University Library there are four maps of Shropshire and North Wales, which are hand-drawn copies at a scale mostly of about 1 inch to 1 mile.[1] Three are overlapping maps of the areas around Shrewsbury, Kinnerley and Knockin, and Llanymynech, and the fourth is a larger one of Anglesey. The first two are partly coloured in brownish orange signifying New Red Sandstone. There appear to be no extant geological notes which refer to these maps, and thus a description of these maps is largely inferential from reminiscences in his autobiography and the two separate sets of notes he made at Llanymynech[2] and on the Welsh trip with Sedgwick[3] in July and August 1831 respectively.

Darwin gave no indication on the maps from where he made the copies. In 1831 the Ordnance Survey was well under way with its preparation for the publication of its first edition, which appeared in the 1830s. Sheet 32 for Llangollen was published in 1837, Sheet 78 for Bangor in 1840 and Sheet 77 for Holyhead in 1841. The leading surveyor was Robert Dawson who Sedgwick had arranged to meet on 5 August 1831 at the beginning of the Sedgwick-Darwin tour. The best and only maps available in 1831 were Baugh's *Map of Shropshire* (1808) at a scale of 1 inch to 1 mile (1:63,360), and Evans's *Map of the six counties of North Wales* (1795) at a scale of about ⅞ inch to 1 mile (approx. 1:80,000). As these are the respective scales of Darwin's maps, which have every appearance of being traced by hand complete with the milestones on the Shropshire maps, it would be perverse to doubt that Darwin used these maps.

Both Baugh and Evans (North *et al.*, 1949: 154) lived at Llanymynech near Oswestry, which was then an important industrial centre. Baugh began his cartography as Evans's assistant (Dodd, 1990) and engraved the small-scale *Map of North Wales* for Evans in 1775. Evans lived at Llwynygroes to the east of Llanymynech. Baugh was born in 1748 and baptised at Llanymynech in 1749, and lived at Llwyntidman, a tiny hamlet 2km east of Llanymynech, with his wife Catherine and family. As Leighton (1917: v) wrote, "Baugh was a copperplate engraver ... He worked in conjunction with the great engineers, Telford and Stephenson, and assisted in the surveys for the Holyhead Road and the aqueduct at Pontcysylltau ... He did not neglect local responsibilities ... He filled the office of Parish Clerk, and for some years was Church Warden." By the standards of the first Ordnance Survey map, Baugh's *Map of Shropshire* was slightly inaccurate in places. However it is an admirable piece of work in clarity, detail and accuracy. There was also available a smaller-scale *Map of North Wales*, at 1 inch to 3.6 miles (1:17,600), produced by J. and A. Walker (1824). Sedgwick used both maps in 1831 and recorded that below Elidir Fawr one "small lake represented in Walker but omitted by Evans" demonstrating the unreliability of the maps. Darwin used Walker's map on his 1842 glacial trip to Snowdonia.[4] From a comparison of scales one may safely conclude that the two 1 inch to 1 mile maps of the area around Shrewsbury were traced from Baugh's map, while the maps of Anglesey and Llanymynech were traced from Evans' map.

## PARTLY COLOURED MAPS OF SHREWSBURY

These two maps[5] are of a 1 inch to 1 mile scale (1:63,000), one of the area west of Shrawardine, and the other centred on Shrewsbury. The maps slightly overlap. From a comparison with Baugh's map of 1808 they include all milestones and also minor inaccuracies in the courses of the River Severn and the Rea Brook found in Baugh's map, which differ from recent maps. A large area on Darwin's two copied maps, from Meole Brace (written as Brace Meole) 1½ miles (2.4km) south of Shrewsbury town centre to Kinnerley in the west, is coloured orange indicating New Red Sandstone. On the Shrewsbury map[6] are four locations marked A, B, C, and D. There seem to be no extant notes for these sites.

The first three sites occur between 2km and 4 km south of Shrewsbury town centre in the Meole Brace area, marked by both Baugh and Darwin as Brace Meole. These sites lie on the old Bishop's Castle road through Nobold and Longden. (The present road to Bishop's Castle (A488) through Minsterley was authorised only in 1834.) The area is of drift underlain by the uppermost Carboniferous. There are a large number of gravel pits in the vicinity and in 1837 and 1838 Darwin visited several of these.[7]

Site A (GR 482108) is now the site of a school. No outcrops are visible and it appears to be landscaped land on the northeast side of the crossroads. Sometime in the 1840s and probably 1841 according to the editors of Darwin's correspondence (Burkhardt and Smith, 1986: 294–295), Darwin visited the gravel pit at GR 474478 and on a sketch map he sent to Leighton[8] marked an old coal pit to the south west of the turnpike. Thus A may refer to Coal Measures found in this pit, which were somewhat red and thus ascribed to the New Red Sandstone. It is also the site of a turnpike gate marked on Baugh's map.

Site B (GR 469096) is aptly by Red Hill Lane and is a long 15ft (4.6m) high outcrop of red flaggy sandstone, dipping at a shallow angle to the north. The beds were identified by Darwin as New Red Sandstone, and are, in fact, the Keele Beds of the Upper Carboniferous. These are variably purple and reddish-brown mudstones, marls, and sandstones (Toghill, 1990:131). These late Carboniferous "red beds" are the earliest development of the New Red Sandstone facies, and are, thus, rightfully marked as New Red Sandstone.

Site C (GR 465093) lies just to the south of Rea Brook at Hook-a-Gate. Today there are no obvious exposures visible. From the "coal pitts" marked on Baugh's map I conclude the Darwin was probably noting the strata of these and marked in the boundary of the New Red Sandstone (which included the Keele Beds) and the Coal Measures, but never got round to colouring in the Coal Measures. These are the Coed-yr-Allt group of the Shrewsbury Coalfield which lies to the south of Shrewsbury.

The fourth site, D, is marked on the map close to Great Ness. There are no outcrops in the vicinity and it may be that Darwin incorrectly marked the location, as 2 miles (3.2km) away at Nesscliffe there are excellent exposures in cliffs of New Red Sandstone. These are Wilmslow Sandstone (Upper Mottled Sandstone) and are deep red in colour and have been extensively quarried for building stone.

Most of the area coloured in to indicate New Red Sandstone is clearly just that, as any exposures jutting through the drift are Permo-Triassic or Uppermost Carboniferous.

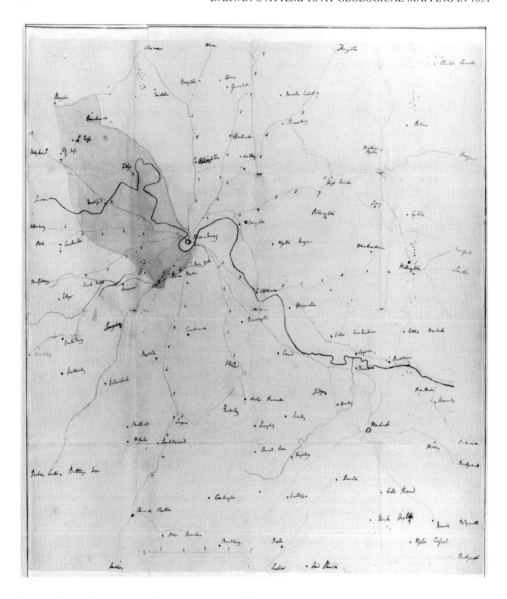

Figure 1. Darwin's map of the Shrewsbury area traced from Baugh's map of Shropshire (1808) (CUL DAR 265, DH/GPD 10; iv). On the original the shaded area was coloured orange to signify New Red Sandstone. A, B, C, and D are sites marked by Darwin. The scale of the original map was 1 inch to 1 mile and is here reduced. Milestones are recorded. (Reproduced by permission of the Syndics of Cambridge University Library.)

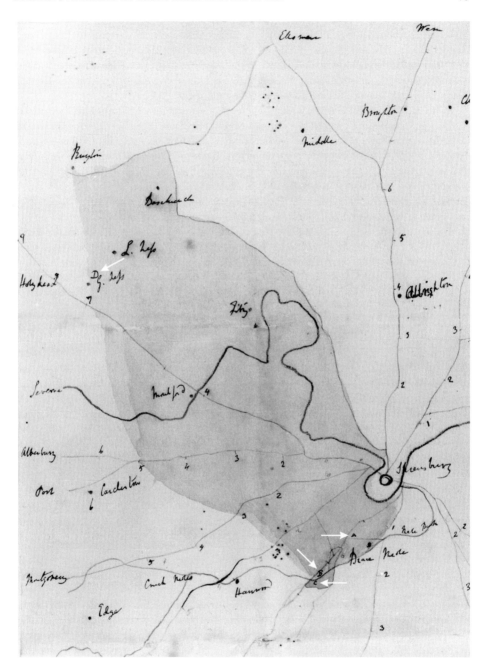

Figure 2. Portion of Darwin's map of Shrewsbury (CUL DAR 265, DH/GPD 10; iv). The shaded area on the original was coloured orange to signify New Red Sandstone. A, B, C, and D are sites marked by Darwin. The quarry where the volute shell was found was 500m west of site A (lower right). The scale of the original map was 1 inch to 1mile, and the milestones are recorded on most roads. (Reproduced by permission of the Syndics of Cambridge University Library.)

However, as recent geological maps show, the area is almost entirely covered by Pleistocene boulder clay, sands and gravels, which are of a distinctively red hue. Actual exposures of New Red Sandstone tend to be on higher ground as at Nesscliffe. One suspects that Darwin was not being very discriminating in the use of his orange crayon. However a *Solid* rather than a *Drift* map would indicate only Permo-Triassic rocks.

In his autobiography Darwin wrote: "Therefore on my return to Shropshire I examined sections and coloured a map of parts round Shrewsbury" (de Beer, 1983: 39). These two maps are clearly those maps and the sections probably include his later visit to Llanymynech (Roberts, 1996). Later in the same paragraph Darwin tells of Sedgwick dismissing his finding of a volute shell. There is simply insufficient evidence to locate the gravel pit precisely and it could be any one of many which lie on the south side of the River Severn, many of which Darwin subsequently visited in 1837 and 1838, where he found further shells. A possible, and tentatively the most likely, location is the large gravel quarry near the turnpike on the old Bishop's Castle road (GR 474478). This was probably the largest pit in the area and was visited by Darwin in the 1840s when he found weld (*Reseda luteola*) and probably the rarer form of houndstongue (*Cynoglossum sylvaticum*).[9] Darwin visited this quarry while working on the gravels around Shrewsbury in 1837 and 1838. On the second occasion he recognised the existence of glaciers in an estuarine environment, meaning icebergs. As a result I conclude that this is the most probable site of the famous volute shell.

The work on these maps represents several days in the field and one can safely assume that he travelled on horseback, or possibly on foot to the area south of Shrewsbury. Darwin also visited Cardeston, as indicated in his notes for 6 August for Penstryt Quarry, near Denbigh; he compares the lime-rich fault breccia on the west side of the fault with the strata at Cardeston.[10] A visit to Cardeston indicates that Darwin was probably referring to a quarry, now partially filled in, ½ mile (0.8km) west of Cardeston itself, and marked on Baugh's map as lime rock. This part of the Alberbury Breccia or Cardeston Stone, ascribed by the Geological Survey map of 1932 as uppermost Carboniferous and by Toghill and recent British Geological Survey maps as Permian, consists of angular purple and brown breccias with fragments of Carboniferous limestone (Toghill, 1990: 141–143). This has a general visual similarity to the fault breccia at Penstryt quarry made up of Carboniferous Limestone and New Red Sandstone, despite the one being tectonic in origin and the other sedimentary. This has to be inference as Darwin did not record anything on his map at Cardeston and the only reference is what he wrote at Ruthin on 6 August. However it is more probable that Darwin visited Cardeston on 3 August with Sedgwick.[11]

## MAP OF THE LLANYMYNECH AREA

This map[12] covers the area from Shelve and Chirbury in the south to Oswestry in the north. and is at a scale of ⅞ inch to 1 mile (1:80,000). Just from the scale alone it is clear that Darwin did not use Baugh's map as a base map, but Evans's map. It is clearly incomplete as the roads around Llanymynech are not connected up. Incidentally Woodhouse is marked, despite the fact that that is a country house rather than a village. Its interest to Darwin was that it was the home of Fanny Mostyn Owen whom Darwin described as "the prettiest, plumpest charming personage that Shropshire possesses".[13]

However, not one geological detail is transcribed onto the map, despite Darwin visiting Llanymynech Hill in the latter half of July (Roberts, 1996). This lack of information on

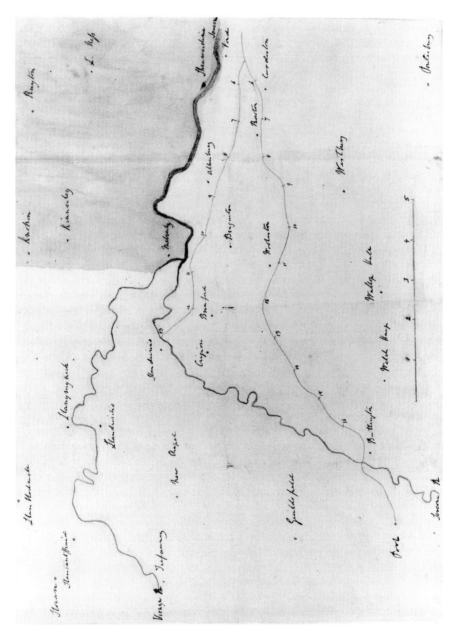

Figure 3. Darwin's map of the Kinnerley area traced from Baugh's map of Shropshire (1808) (CUL DAR 265, DH/GPD 10; ii). On the original the shaded area was coloured orange to signify New Red Sandstone. The scale of the original map was 1 inch to 1 mile and is here reduced. Milestones are recorded. (Reproduced by permission of the Syndics of Cambridge University Library.)

the map is probably an indication that Darwin went to Llanymynech after attempting to produce the maps around Shrewsbury and did not have time to colour in the map before Sedgwick arrived on 2 August.

## MAP OF ANGLESEY

This map[14] is at a scale of ⅞ inch to 1 mile (1:80,000) and is probably based on Evans' map. The locations of most of the larger villages are indicated; typically Welsh names as Llandegfan were always written as "L-degfan".

The most likely explanation of why for North Wales there are maps of Anglesey but none for the intervening area between Anglesey and Llangollen is probably to be found in Darwin's friendship with Henslow. Henslow had written an early paper on the geology of Anglesey (Henslow, 1822) and presumably Darwin wished to visit Anglesey armed with his mentor's article as part of his geological preparation for his visit to Tenerife. With easy access to Anglesey from Shrewsbury by the London-Holyhead road, this would have made a straightforward expedition for Darwin to improve his geology with Henslow in absentia as his guide. This is probably what Darwin meant when he wrote to Whitley on 12 July: "I am at present mad about Geology & daresay I shall put a plan which I am now hatching, into execution sometime into August, viz of riding through Wales & staying a few days at Barmouth on my road".[15] However at Henslow's suggestion Sedgwick took Darwin round part of North Wales and thus in August 1831 Darwin passed within 7 miles (11km) at the very most of the suspension bridge from the mainland to Anglesey. A study of Darwin's notes taken on that field trip, compared with Sedgwick's notes, gives the very strong impression that Darwin made no diversion to Anglesey as the notes for Cwm Idwal follow immediately after those made at Bethesda slate quarry; he simply continued directly to Barmouth after leaving Sedgwick near Bethesda but no more travelled in a straight line than did Belloc's drunkard travelling to Birmingham by way of Beachy Head (Roberts, 1998). In Barmouth in mid-August he met up with Whitley and others.[16] However the notes Darwin made on Quail Island on his first day's fieldwork on the *Beagle* voyage liken the conglomerates found there to those on Angelsey: "I could have scarcely credited that rocks nearly as hard as the conglomerates of older formation (viz of red-sandstone formation Anglesey) could daily be increasing under my own eyes".[17] There are two main possibilities to explain Darwin's reference to Old Red Sandstone in his Quail Island notes. The first, favoured by Secord,[18] is that Darwin was simply referring to Henslow's paper on Anglesey much is the same way as he referred to the Temple of Serapis on Sao Tiago shortly afterwards (cited Herbert, 1991: 170). The alternative is that Darwin crossed Anglesey with Sedgwick. On 12 August Sedgwick travelled to Holyhead, recording limestone at Llangefni and "O.R." (Old Red Sandstone), which in fact is of Ordovician rather than Devonian age, further on, probably to the northeast of Llanfihangel yn Nhowyn. As far as time constraints are concerned Darwin could easily have gone to Holyhead and then returned to Cwm Idwal before travelling to Barmouth.

There is a possibility that Darwin took notes on Anglesey as his notes indicate that a page has been cut out and removed immediately after his notes on Bethesda.[19] Further the notes Darwin made on Quail Island cited above give the distinct impression that Darwin had actually seen and felt the strata on Anglesey, which are extremely hard. There is no evidence how far Darwin actually accompanied Sedgwick, but from the absence of any reference to his visiting Ireland one may conclude that he returned to Bethesda as Sedgwick boarded the ferry. It would have been a straightforward journey by stagecoach. Darwin may well have stayed at the coaching inn at Tyn y Maes, between Bethesda and Llyn Ogwen. From there it was a short walk to Cwm Idwal.

## CONCLUSION

These four maps show Darwin's earliest attempts at field geology and plans for future work. They indicate his awareness for the need of a basic topographic map as he proceeded to make them himself, copying from available maps.

His work around Shrewsbury is an uncompleted attempt at fieldwork and his topographic map of Anglesey probably indicates his future plans, which were swallowed up, first by his accompanying Sedgwick in August, and then the invitation to join the *Beagle*. In themselves they tell us nothing about the development of geology, but show us the first attempts of a keen young naturalist to teach himself geology in 1831. Their interest is twofold. Firstly, they illustrate the geological development of Darwin in those months before he was invited to join the *Beagle*. Secondly, they give an insight into how a moderately informed beginner in geology set about the task of practising field geology in the early 1830s. On one level these maps can be seen as simply the indifferent attempts of a beginner, and were it not for this beginner's later importance, would attract little attention. As well as the personal interest in helping to understand the early development of one of the leading nineteenth century scientists, Darwin's attempts show that in 1831 a total beginner would be aware of basic principles of geological mapping. He was aware of the need for a base map, as well as knowing the conventions of using particular colours for each rock-type. Darwin did not find it easy as he admitted to his cousin, W. D. Fox, on 9 August: "I am trying to make a map of Shrops: but dont find it so easy as I expected".[20] However he found it interesting, unlike Spanish which he thought stupid. His visit to Llanymynech is another attempt which reflects his limited grasp of geology and he needed the guidance of a competent field geologist to set him on the right road. This he found in Adam Sedgwick.

## ACKNOWLEDGMENTS

This paper would not have been possible without Adam Perkins of Cambridge University Library informing me of these maps during a visit to the library. The Darwin manuscripts (DAR) are quoted with permission of the Syndics of Cambridge University Library. Thanks to David Pannet and Peter Toghill who advised me on old maps of Shrewsbury and the local geology respectively, and whose local knowledge prevented several errors. Thanks also to Neville Herdsman of Chirk Bank for the loan of several books on local history. Also to Dr J. Secord whose critical comments, whether in conversation or as referee, have been most constructive. This research has been supported by the Isla Johnston Trust (administered by the Church in Wales).

## NOTES

[1] CUL DAR (Cambridge University Library; Darwin Manuscripts.) 265, DH/GPD 10; Maps drawn by Charles Darwin:

     i. Angelsey, 1 large map.

     ii. North Wales (in fact of area from Cardeston in the east to Llansantffraid (-ym-Mechain) in the west).

     iii. North Oswestry (actually of area from Oswestry in the north and Shelve in the south).

    iv. Shrewsbury district.
Map ii will be referred as the Kinnerley Map and Map iii as the Llanymynech Map.

[2] CUL DAR 5 (series 2), 1–2.

[3] CUL DAR 5 fols 5–14. There are 20 sheets, the first being numbered 5 and those are referred to here as 12i, 12ii, 13i etc., fols 10ii and 11i.

[4] Adam Sedgwick, entries for 23 August, Sedgwick's journal, No. XXI (1831), Sedgwick Museum, Cambridge. Darwin Glaciation Notes, CUL DAR 27 fol. 21.

[5] CUL DAR 265, DH/GPD 10; iv & ii.

[6] CUL DAR 265, DH/GPD 10; iv 7 ii.

[7] CUL, DAR 5 fols 19–29.

[8] Darwin to Leighton, (1–23 July 1841) (Burkhardt and Smith, 1986: 294–295)

[9] See note 11.

[10] CUL, DAR 5 fol. 6 i.

[11] Adam Sedgwick, entries for 3 August, Sedgwick's journal, No. XXI (1831), Sedgwick Museum, Cambridge.

[12] CUL DAR 265, DH/GPD 10; iii.

[13] Darwin to Fox, 24 December 1828 (Burkhardt and Smith, 1985: 72).

[14] CUL DAR 265, DH/GPD 10; i.

[15] Darwin to Whitley, 12 July 1831 (Burkhardt and Smith, 1991: 466).

[16] Darwin to Whitley, 9 September 1831 (Burkhardt and Smith, 1985: 151).

[17] CUL DAR 34 (i) fol. 19.

[18] J. Secord, personal communications 1998 and 1999.

[19] CUL DAR 5 fol. 10; ii. A page appears to have been excised.

[20] Darwin to Fox, 9 July 1831 (see Burkhardt and Smith, 1985: 124).


## REFERENCES

BARRETT, P. H., 1974 The Sedgwick-Darwin geologic tour of North Wales. *Proceedings of the American Philosophical Society* **118**: 146–164.

BAUGH, R., 1808 *Map of Shropshire*. Reprinted 1983 by the Shropshire Archaeological Society, edited by B. Trinder. Shropshire: Alan Sutton.

BURKHARDT, F. and SMITH, S. (editors), 1985 *The correspondence of Charles Darwin*, **I** (1821–1836). Cambridge: Cambridge University Press. Pp 702.

BURKHARDT, F. and SMITH, S. (editors), 1986 *The correspondence of Charles Darwin*, **II** (1837–18430). Cambridge: Cambridge University Press. Pp 603.

BURKHARDT, F. and SMITH, S. (editors), 1991 *The correspondence of Charles Darwin*, **VII** (1858–1859, supplement 1821–1857). Cambridge: Cambridge University Press. Pp 671.

DE BEER, G. (editor), 1983 *Charles Darwin, Thomas Henry Huxley: autobiographies*. London: Oxford University Press. Pp 123.

DODD, A. H., 1990 *The Industrial Revolution in North Wales*. Wrexham: Bridge Books. Pp 439.

EVANS, J., 1795 *Map of the six counties of North Wales*. (Inscribed to Sir Watkin Williams-Wynn of Wynnstay Hall, Ruabon, 1 June 1795.) Liverpool and London.

HENSLOW, J. S., 1822 Geological description of Anglesey. *Transactions of the Cambridge Philosophical Society* **1**: 359–452.

HERBERT, S., 1991 Charles Darwin as a prospective geological author. *British journal for the history of science* **24**: 159–192.

LEIGHTON, R., 1917 *Shropshire Parish Register Society – Llanymynech*. Shrewsbury.

NORTH, F. J., CAMPBELL, B. and SCOTT. R., 1949 *Snowdonia*. London: Collins. Pp 469.

ROBERTS, M. B., 1996 Darwin at Llanymynech: the evolution of a geologist. *British journal for the history of science* **29**: 469–478.

ROBERTS, M. B., 1998 Darwin's dog-leg. *Archives of natural history* **25**: 59–73.

SECORD, J. A., 1991 The discovery of a vocation: Darwin's early geology. *British journal for the history of science* **24**: 133–157.

STODDART, D. R., 1995 Darwin and the Seeing Eye. *Earth sciences history* **14**: 3–22.

TOGHILL, P, 1990 *Geology in Shropshire*. Shrewsbury: Swan Hill. Pp 188.

WALKER, J. and WALKER, A., 1824 *Map of North Wales*. London and Liverpool.

[Received 2 January 1998: Accepted: 28 October 1998]

EDITORIAL NOTE.

For more on Darwin's development as a geologist in the fortnight to three weeks spent with Sedgwick in the first half of August 1831, see

ROBERTS, M. B., 2001 Just before the Beagle: Charles Darwin's geological fieldwork in North Wales. *Endeavour* **25**: 33–37.

*Archives of natural history* **29** (1): 27–30. 2002 © The Society for the History of Natural History

# Short notes

### Charles Darwin's notes on his 1831 geological map of Shrewsbury

As a young geologist Charles Darwin made use of hand-drawn maps and sections (Herbert, 1991; Rhodes, 1991; Secord, 1991; Stoddart, 1995; Roberts, 2000). One of his early maps (reproduced in Roberts, 2000), made in 1831 during the interval between the completion of his university education and his departure on the voyage of HMS *Beagle*, is of the environs around his home in Shrewsbury. On this map Darwin noted four sites, which he labelled A, B, C and D. We offer here a transcription of notes kept by Darwin relating to these four sites. Roberts (2000: 71) pointed out that a school presently occupies site A, and no outcrop is visible. Site B is a long outcrop of red flaggy sandstone (5–6 metres high by 80 metres long), which has probably been quarried. In present-day terms, these late Carboniferous "red beds" are the earliest development of the New Red Sandstone facies, and are, thus, appropriately marked as New Red Sandstone. Site C has no obvious exposures visible today. Site D was mismarked on Darwin's map (see Roberts, 2000: 71). Darwin's notes confirm that he was indeed at Nesscliffe, one mile to the northwest of Great Ness, and, importantly, where, in an area covered by pine and deciduous woodland, there are excellent exposures of cliffs of New Red Sandstone. In his notes Darwin also remarked on the appearance of the sandstone. For example, he described its carious nature at Hopton.

Darwin's notes to his Shrewsbury map show him using a clinometer for angular measurements of dip of strata (as at site B). This helps to date Darwin's map and his notes, for on 11 July 1831 Darwin wrote to his mentor J. S. Henslow that he had recently purchased a clinometer (Burkhardt and Smith, 1985: 125). Another sign of the date of the notes is that in them Darwin employed the term "direction" rather than the term "strike" that he began to use after his excursion with Adam Sedgwick in August 1831 (Barrett, 1974: 155). Thus these notes were likely produced sometime between mid-July and the first week of August 1831 when Darwin departed with Sedgwick on a geological excursion through North Wales (Roberts, 1998, 2001; Lucas, 2002). These notes are thus roughly contemporary with Darwin's notes on Llanymynech in Wales (Roberts, 1996). Finally, the notes in their inclusion of comments and queries oriented towards theory (as in the last sentence for site D) suggest a continuity with Darwin's later note-taking style (Herbert, 1987).

The notes transcribed here are contained in a bound, vellum-covered notebook now catalogued as DAR 210.11.37 (formerly DAR 210.17) in the collection of Darwin manuscripts housed at Cambridge University Library. The notebook seems to have been originally the property of Albert Way (1805–1874), a fellow student with Darwin at the university. The cover is inscribed "AV. Trin. Coll." (the "V" suggests the direction of Way's interests, the classical Latin alphabet having no W). Possibly Darwin had in mind returning Way's notebook to him, for in the letter to Henslow, previously quoted, he asked "Do you know A. Ways direction?" In any event, the notebook remained with Darwin.

ACKNOWLEDGEMENTS

The authors are grateful to the Syndics of Cambridge University Library for permission to publish the manuscript. One of us (S. Herbert) thanks Marsha Richmond for having drawn her attention to these notes some years ago, and we both thank Sheila Ann Dean for proofreading the transcription.

## REFERENCES

BARRETT, P. H. *et alii* (editors), 1987 *Charles Darwin's notebooks, 1836–1844: geology, transmutation of species, metaphysical enquiries*. London: British Museum (Natural History). Pp 747.

BARRETT, P. H., 1974 The Sedgwick–Darwin geologic tour of North Wales. *Proceedings of the American Philosophical Society* **118**: 146–164.

BURKHARDT, F. and SMITH, S. (editors), 1985 *The correspondence of Charles Darwin*. Cambridge: Cambridge University Press. Volume **1**. Pp 702.

HERBERT, S. (editor), 1987 Notebook A (geology), pp 83–139 in BARRETT, P. *et alii* (editors). *Charles Darwin's notebooks, 1836–1844: geology, transmutation of species, metaphysical enquiries*. London: British Museum (Natural History).

HERBERT, S., 1991 Charles Darwin as a prospective geological author. *British journal for the history of science* **24**: 159–192.

LUCAS, P., 2002 "A most glorious country": the 1831 "Sedgwick–Darwin geologic tour" of North Wales revisited. *Archives of natural history* **29**: 1–26.

RHODES, F. H. T., 1991 Darwin's search for a theory of the Earth: symmetry, simplicity and speculation. *British journal for the history of science* **24**: 193–229.

ROBERTS, M. B., 1996 Darwin at Llanymynech: the evolution of a geologist. *British journal for the history of science* **29**: 469–478.

ROBERTS, M. B., 1998 Darwin's dog-leg: the last stage of Darwin's Welsh field trip of 1831. *Archives of natural history* **25**: 59–73.

ROBERTS, M. B., 2000 I coloured a map: Darwin's attempts at geological mapping in 1831. *Archives of natural history* **27**: 69–79.

ROBERTS, M. B., 2001 Just before the *Beagle*: Charles Darwin's geological fieldwork in Wales, summer 1831. *Endeavour* **25**: 33–37.

SECORD, J., 1991 The discovery of a vocation: Darwin's early geology. *British journal for the history of science* **24**: 133–157.

STODDART, D. R., 1995 Darwin and the seeing eye: iconography and meaning in the Beagle years. *Earth sciences history* **14**: 3–22.

## TRANSCRIPTION OF DARWIN'S TEXT

Following the editorial conventions used in Barrett *et alii* (1987), in this transcription Darwin's deletions are indicated by ‹ › and his insertions by « ». Square brackets enclose the present authors' insertions.

(A)

A Large gravel pit. worked for mending the roads: the lower beds chiefly sand: under which is a <u>red marl</u>. — The gravel like that of the rest of Shropshire. consists of various sorts of rocks. — Limestone. Trap. Clay Slate. Grey Wacke[.] Quartz. Granite. In the gravel there are numerous balls of sandstone, dark red and almost spherical. — I have not observed them in other pits. —

(B)

An escarpements of sandstone, which follows the course of Meole Brook. — Within a few yards on the other side «of the stream» are old coal pits. — Its direction is W.N.W. & E S.E. — dipping at an Đ of 20° to N.N E. The sandstone is much stratified & of a very hard consistence: — is covered on parts of its face by a stalactite: —

(C)

On the Southern bank of brook, strata of Slate clay appear; they are irregularly stratified & full of bits of shining coal. —

(D)

The Nessclif hills «consisting of red Sandstone» begin«ning» about 1/4 of mile south of Great Ness & run«ning» in a N E by N direction towards Boreatton Park. In this range I include the hills called Clive & Hopton. — Their escarpement generally bears W.S.W. facing the plain. which lies between the Breddin & LLanemynech [Llanymynech]. The escarpement in different portions of the range appears to vary. sometimes it is even N E. — The direction of the Strata very difficult to be ascertained. The best observation I made was on the top of hill

(D)

called Hopton. where the sandstone is more distinctly stratified. of a darker colour, much harder but carious. — here the D was N by W & S by E dipping at 8° to E by N. — Nearer to Nessclif. the D is more Western being W.N.W — Again in other places W S W & E N E     &        S W & N E [.] Generally speaking the Strata are nearly horizontal; but both the inclination & direction is difficult to be ascertained owing to the number of fissures or seams which run through the whole bed of rock. These are generally horizontal but occasionally

(D)

vertical. & divide the rock into beds of various thickness generally about 6 feet. — The stone is soft, but used for inferior buildings, it is generally red. dappled with white; but occasionally altogether white. — The surface is in many places ‹is› honeycombed. ‹&› or rather corroded into holes of various sizes; when a piece is broken off. there may be observed patches of a darker hue: it is apparently from the easier decomposition of the blacker spots. that the surface takes the

(D)

described appearance. — The stone is very little covered with Lichens. The general red colour, abrupt escarpement & wooded top of these hills gives much picturesque beauty to that part of Shropshire. — The most curious thing I observed in this rock were numerous veins of a harder sandstone running in straight lines often quite through the main bed of rock. From their resisting decomposition longer than the surrounding stone they project outwards & are very visible. — These veins are about a line in thickness. are generally vertical. but sometimes oblique &

(D)

occasionally curved. — They are more numerous in the upper beds. & these frequently cut each other: But what is most important the[y] ‹cut› pass through the seams or stratifications in the rock. This indeed first led me to suppose that most of the regular lines of apparent stratifications were not really such. but merely fissures caused by some force after‹wards› their deposition. These veins of harder sandstone I do not think could be formed by cracks afterwards filled up by infiltration[,] their remarkable uniformity in thickness

(D)

& ‹consti› appearance. together with the manner in which they cut each appear to me to preclude this idea. — It is on this supposition that ‹I think› the continuity of the veins through the seams in the rock, prove that [they] were caused by some force & not by a succession of depositions. —

Received: 2 March 2001. Accepted: 13 May 2001.

SANDRA HERBERT

Department of History, University of Maryland Baltimore County, 1000 Hilltop Circle, Baltimore, MD 21250 USA.


MICHAEL B. ROBERTS

The Vicarage, 5 Lancaster Road, Cockerham, Lancaster, LA2 0EB, UK.

EDITORIAL NOTE.

In the original printing, the text on page 30 was not about Erasmus or Charles Darwin.

*Archives of Natural History* (1998) **25** (1): 59–73

# Darwin's dog-leg: the last stage of Darwin's Welsh field trip of 1831

By MICHAEL B. ROBERTS

Chirk Vicarage,
Trevor Road,
Chirk,
Wrecsam LL14 5HD.

"At Capel Curig I left Sedgwick and went in a straight line by compass and map across the mountains to Barmouth, never following any track unless it coincided with my course. I thus came to some strange wild places and enjoyed much this manner of travelling. I visited Barmouth to see more Cambridge friends who were reading there, and thence returned to Shrewsbury and to Maer for shooting;" (Darwin and Huxley, 1983: 40).

Darwin's reminiscences about the last part of his field trip of 1831 have passed into Darwinian folklore. It sounds romantic and heroic that the young geologist should, in a style worthy of Orde Wingate, simply take a compass bearing from Capel Curig and then go up hill and down dale until he reached Barmouth. On 5 August 1831 Darwin had left Shrewsbury with Adam Sedgwick, visiting first Llangollen, then the Vale of Clwyd and finally the country south and west of Conwy. After a week Sedgwick went to Holyhead and then to Ireland leaving Darwin to find his own way home, by following a compass bearing to his old haunts at Barmouth (see Figure 1).

In fact, Darwin's *Autobiography* is faulty on several points. From a consideration of letters sent by Sedgwick to Darwin in September 1831 and the first notes Darwin made on Quail Island in January 1832, Darwin and Sedgwick left Penrhyn Quarry on 11 August and journeyed across Angelsey together and separated sometime before Sedgwick set sail from Holyhead on 13 August and not at Capel Curig as claimed in his *Autobiography* (Darwin and Huxley, 1983: 40). Darwin first visited Cwm Idwal on his own, and spent two nights at Capel Curig, before heading south to Barmouth (see Figures 1 and 2).

The conventional picture that Darwin followed a compass bearing appears in most biographies, but is highly unlikely when one considers the nature of the terrain, even without considering his geological notes or letters from Sedgwick. A route following a bearing from Capel Curig to Barmouth is 27 miles over difficult terrain, involving 2,500 metres (8,000 feet) of ascent. Spread over two days this would be a good undertaking, but Barrett's (1974: 152) suggestion that "Darwin probably walked it in one day" is simply absurd. A fairly direct route along the Moelwynion skyline for the first stage to Maentwrog over Moel Siabod and then the Moelwynion would give a fairly energetic hillwalk of 14 miles over grassy mountains. However, if Darwin was "never following any track unless it coincided with my course", he would have had an exhausting day ploughing his way through bogs and coarse grass as well as ascending some 1,400 metres. A fit young person would need 6 hours for the ridge-

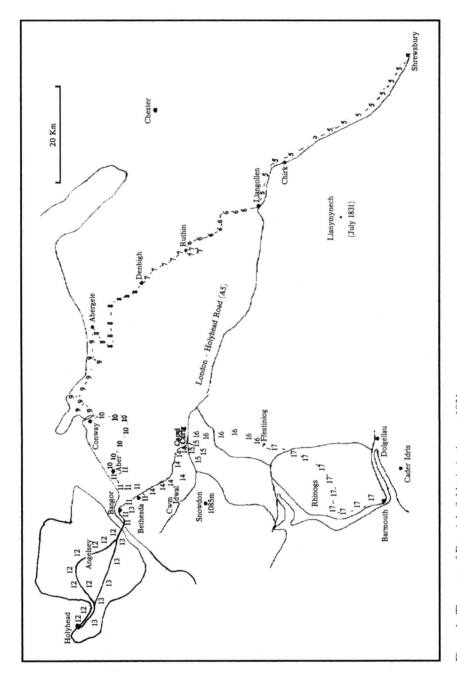

Figure 1. The route of Darwin's field trip in August 1831.
[The numbers (5–17) refer to dates in August as in the chronology of Appendix 1.]

walk and well over 8 for the compass route excluding the several hours needed for making notes.

The second stage to Barmouth over the Rhinog mountains would give a 20-mile trek over the roughest terrain in Wales. Although the highest point is Y Llethr at 754 metres, their lack of height is deceptive. The northern Rhinogs are even today almost devoid of north-south paths and consist largely of boulders covered in 2-foot high heather. As John and Ann Nuttall (1989: 151, 156) write in their walking guide to Wales, "The northern Rhinogs present some of the roughest and toughest walking to be found anywhere in Wales", in which "the resulting succession of alternating ascents and descents reduces progress to a crawl". This is no exaggeration as anyone who has walked in the Rhinogs can testify. Today a "skyline" walk from Maentwrog to Barmouth would take at least 10 hours following poor footpaths but a yomp on a bearing would give an exceedingly long and exhausting day. Darwin was well used to long walks in North Wales as his 1826 walking tour testifies, but direct routes in this terrain would require a perversity only exceeded by Orde Wingate in Ethiopia during the Second World War where he insisted on marching a compass bearing through bamboo jungle and lava fields, thus gaining scathing condemnation from Wilfred Thesiger (1987: 327). Claims of walking 30 miles a day in Snowdonia in 1826 notwithstanding (Darwin and Huxley, 1983: 29), Darwin's statements in his *Autobiography* must be seen as a faulty memory some 50 years on. The reasons for a less direct route are even clearer when one considers Darwin's 1831 notes and identifies the precise location of the outcrops described. From these and a consideration of the topography and with reference to both ancient and modern maps the route taken by Darwin in 1831 can be worked out with a high degree of probability and much of it with certainty. Since moving to Chirk, in the crypt of whose church lies Fanny Mostyn Owen, who sent Darwin several love letters in 1828 and "a *leetle* purse . . . In remembrance of the *housemaid* of the *Black Forest*" in October 1831 (Burkhardt and Smith, 1985: 166 & passim), I have walked, cycled and driven extensively in North Wales, and am thus familiar with the topography having "climbed every mountain" from Conwy to Barmouth several times, and have followed the entire length of Darwin's route, mostly on foot, but also by car and bicycle.

Darwin recorded his notes for the 1831 field trip in a quarto notebook, taking 20 *surviving* pages of notes, beginning at Valle Crucis (Vale of Crucis) near Llangollen. Dates are given only up to 8 August, but dates can be worked out by comparison with Sedgwick's *dated* notes up to the time they visited Penrhyn Quarry on 11 August. After Penrhyn Quarry Sedgwick's and Darwin's notes cannot be correlated at all, thus it would be reasonable to conclude that they separated at that point. However, when describing the conglomerates at Quail Island in the Cape Verde Islands on 17 and 18 January 1832—the first fieldwork of the *Beagle* voyage—Darwin compared them to "red-sandstone formation Anglesey". Sedgwick recorded some Old Red Sandstone northwest of Llangefni in his notes for 12 August and thus Darwin travelled at least as far as that with Sedgwick. Most probably Darwin left Sedgwick the following morning at Holyhead and returned to Bethesda, presumably by stage coach which would have carried Darwin to Bethesda or Ogwen, a mile below Cwm Idwal, in a few hours. Further, Darwin's notes are not continuous as immediately after his notes on Penrhyn Quarry there is clear evidence that one page has been removed.[1]

Presumably the excised page contained Darwin's notes of Angelsey. After Darwin left Sedgwick no dates are given in Darwin's notes and outcrop localities are not always given precisely. The dates and route Darwin took can be worked out with fair accuracy, by comparing his notes with the local geology and the topography. After Capel Curig Darwin made notes at some ten localities. In his notes some are underlined, and several have different spellings from today's conventions. Simply by plotting these localities on the map (Figure 2) it is very clear that Darwin did not take a compass direction of 190 deg., but took a fairly direct route in the form of a dog-leg avoiding high ground. Several localities can be obviously identified, interpreting Darwin's odd Welsh spelling, e.g. Moel Siabod (Shiabred), Dolwyddelan, Carreg y Fran, Festineog, and Drus Ardidy. The localities of Darwin's *Moelwyn, Northern Moelwyn* and Pont Cwmnantcol, i.e. *W end of valley* below *Drus Ardidy* are identified on several lines of evidence put forward below.[2]

## DARWIN'S ROUTE FROM CAPEL CURIG TO BARMOUTH

Darwin gave little indication in his notes of the route he followed. Though he did not mention the inn he stayed at in Capel Curig, the most likely place is what is now the mountain centre at Plas Y Brenin (lit. *King's Mansion*), which in the early nineteenth century was known as the Royal Hotel. This was a large coaching inn and during the last century had extensive stabling facilities. It is almost certainly the inn where Sedgwick stayed in September 1831, finding a letter waiting for him from Darwin. In Sedgwick's reply he referred to "the road turning off (the turnpike) to the Inn".[3] Darwin probably stayed here in 1826 and returned here in 1842 on his glaciation trip. The inn is given a fine description in George Borrow's (1905: 178) *Wild Wales*.

There are two possibilities for Darwin's journey on leaving Capel Curig. The first is that he went up Moel Siabod en route to Dolwyddelan. If so, then having left the Inn, Darwin followed a line direct to the top of Moel Siabod, which was probably close to the present path from Plas y Brenin. As the next identifiable outcrop described was the Dolwyddelan quarries (GR 743521), Darwin simply descended the broken ground from the summit of Moel Siabod to the village. As paths are not clearly visible even today Darwin proceeded in a south-south-easterly direction. He went through Dolwyddelan and visited the quarries to the south-east of the village, and then carried on to the south to Ffestiniog via Carreg y Fran.

The second alternative is the more probable. Darwin stayed over a second night at the inn and spent a whole day on Moel Siabod. From his notes he ascended the mountain by the north-west flank making copious notes, far more than he did on any other occasion on the way to Barmouth. The last notes describe "The beds comprising NE of mountain" (site 3) and this would be on the obvious route to descend to Capel Curig avoiding his ascent route. This route leads to Pont Cyfyng, close to the main road (A5) a mile east of the centre of Capel Curig. From his original notes, the notes on Moel Siabod appear to have been written with a blunt pencil and the next notes on *"South of Shiabed"* are in a sharp pencil.[4] It seems more than possible that Darwin began the second day with either a new pencil or a sharp one. When Sedgwick wrote to Darwin from Tremadoc on 4 September, he complained that "the stupid red nosed waiter did not shew me your letter till a few hours before I

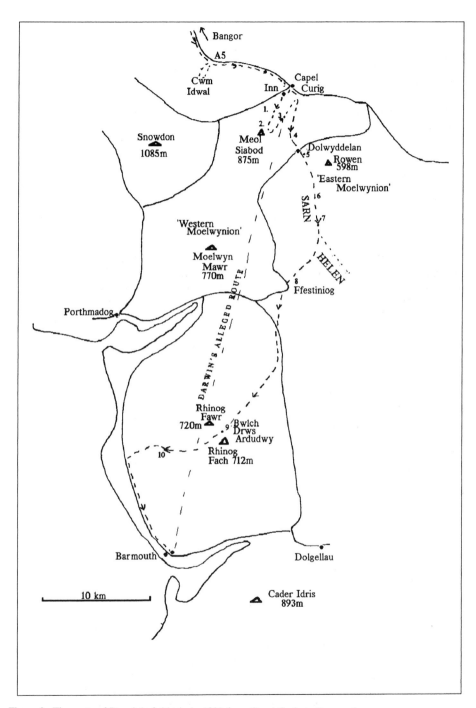

Figure 2. The route of Darwin's field trip in 1831 from Capel Curig to Barmouth.
Numbered locations refer to site numbers described in the text.

started."[5] This means that Darwin had left this letter with the waiter to give to Sedgwick on arrival. According to Sedgwick's letter to Darwin of 4 September, Sedgwick had stayed at this inn twice, once for lunch on 25 August and then for three nights from 30 August to 2 September.[5] The red-nosed waiter obviously failed to give Sedgwick Darwin's letter until early on the 2nd. In this letter, which is lost, Darwin had described both Cwm Idwal and Moel Siabod, proving beyond a doubt that Darwin returned to the inn after climbing Moel Siabod, and also that Darwin had not visited Cwm Idwal with Sedgwick as claimed (Darwin and Huxley, 1983: 38). This gives a short day for Darwin as no more than four hours are needed for the climb and another three for taking notes, but Darwin had had a good day the previous day when he walked 15 miles from Bethesda taking in Cwm Idwal, where he made several pages of notes. Despite Darwin's claims to have covered 30 miles a day in 1826, which are probably exaggerated, it is more reasonable to assume less heroic days. Thus, Darwin spent 11 full days journeying from Shrewsbury to Barmouth. Assuming he travelled by gig when with Sedgwick, Darwin still walked a good 100 miles, and 60 in the 4 days from Bethesda to Barmouth.

The obvious route for Darwin to walk from Capel Curig to Dolwyddelan would be down the London–Holyhead Road (A5) to Pont Cyfyng and then over a track (marked on the First Edition O.S. map) skirting Moel Siabod to the east leading to Dolwyddelan village. The notes made "South of Shiabed" were most likely made on the craggy hummocks half a mile north of Dolwyddelan. The quarries visited are on the south side of the River Lledr less than half a mile from the village centre.

After Dolwyddelan, Darwin described three sites before visiting Ffestiniog (Festineog). A brief glance at the notes with its references to the Moelwyn supports the contention that Darwin followed the compass and went close to, if not over, Moelwyn Mawr and Moelwyn Bach. Barrett (1974: 512) wrote "Near Blanau Ffestiniog, he described the geology of the Moelwyn peaks, and of Carreg y fran, the two sites which, in an east-west direction, are about five miles apart" and "obviously strayed from his compass line". Barrett also suggests that Darwin did this at Sedgwick's request. However, Darwin probably used *Moelwyn* in a loose, but acceptable, geographical sense, to include all the hills above Blaenau Ffestiniog from Moelwyn Bach in the south-west to Rowen in the north-east. Thus Darwin's comments on *Moelwyn* and *Northern Moelwyn* refer to the moorland immediately north of Carreg y Fran and the Manod quarries above Blanau Ffestiniog. Darwin's few notes do describe this area and from there it is a mile or so to the next outcrop Carreg y Fran. This is better than Barrett's unfounded suggestion that Darwin and Sedgwick met again on about 20 August and that Darwin made a 5-mile detour at Sedgwick's instigation. The whole of this route lies on the *Sarn Helen*, an ancient road or track dating back to Roman times (North *et al.*, 1949: 320) going from Kanovium (halfway between Conwy and Llanrwst) through Dolwyddelan and Ffestiniog to Maentwrog. *Sarn Helen* is marked on most O.S. maps including the first edition of the late 1830s. Today the Sarn Helen can be followed easily from Dolwyddelan to Bwlch Garreg y Fran and then descends as a tarmac lane south-west to Ffestiniog. This is the most probable route for Darwin's first day from Capel Curig and one may reasonably surmise that he spent the night in Ffestiniog after a fairly easy 16-mile walk. It is equally possible that Darwin walked another 4 or 5 miles to Trawsfynydd and found a tavern there.

Darwin gave no clues as to his route south of Ffestiniog. To join the Drovers' Road over Bwlch Drws Ardudwy he would have passed through Trawsfynydd. The recent O.S. map indicates a fairly direct track between the two villages, which is a bridleway for much of its distance. The first edition O.S. map also marks the track and thus it is a possible route, especially if he took the most direct country route, which his memories in the *Autobiography* imply. En route Darwin would have passed a Roman Amphitheatre and Fort (GR 708388). However, it is surprising no notes were made. From just south of Trawsfynydd a Drovers' Road led over the Rhinogs to join the Barmouth Road south of Llanbedr. The next site or area where notes were made was at *Drus Ardidy*, which is probably Bwlch Drws Ardudwy, a pass of 350 metres, as Darwin referred to 'the hill through which the road passes," and his descriptions tally with the geology of the bwlch. Darwin then described "At W end of valley the rock is jointed . . . as to give the appearance of beds dipping".[6] This is a good description of the beds at the west end of Cwm Nantcol by Pont Cwmnantcol (GR 617268). As Darwin's outcrops on the Rhinogs can be identified with high certainty, one can conclude that this was Darwin's route and that he joined the Barmouth road at Llanbedr.

The distance from Ffestiniog to Barmouth is 24 miles. One can only hazard a guess at whether Darwin did this in one day, or spent the night at Llanbedr. The paucity of geological notes indicates that he pressed on making few stops and thus most possibly did the whole journey from Capel Curig in two days. If, as is most probable Darwin left Sedgwick at Holyhead on the 13th August, he would have visited Cwm Idwal the next day, spent two nights at Capel Curig, left for Ffestiniog on Tuesday 16 and thus arrived in Barmouth in the late evening of 17 August or 18th at the latest. Darwin's claims in his *Autobiography* to have left Sedgwick at Capel Curig are contradicted by Sedgwick's letter to Darwin of 4 September and his dated Fieldnotes. These make it clear that Sedgwick first visited Capel Curig on 25 August when he stopped for lunch. Despite Darwin's statement that "We spent many hours in Cwm Idwall" Sedgwick did not visit Cwm Idwal until Wednesday 31 August, when Darwin was at Maer writing to his father about the value of the *Beagle* voyage.

From this revised chronology Darwin could easily have spent several days at Barmouth, and then returned to Shrewsbury. A few days shooting at Maer would have taken him to Monday 29 August when he found the famous letter on returning home. This is implied by a quick reading of Darwin's *Autobiography* (Darwin and Huxley, 1983: 40), "I visited Barmouth . . . and thence returned to Shrewsbury and to Maer for shooting". In fact, it is far more likely that Darwin returned to Shrewsbury on Monday 29 August with the *intention* of going shooting at Maer. Darwin wrote (Darwin and Huxley, 1983: 40) "at that time I should have thought myself mad to give up the first days of partridge-shooting for geology". The accepted first day of partridge shooting was 1 September and this became law with the Game Act, 1831, which was passed on 5 October 1831 (Mead, 1912: 285). Before 1832 it was good practice, rather than Law, which encouraged the delaying of partridge shooting until September. Thus Darwin would not have gone partridge shooting in August as no wise "countryman" would countenance it. Ironically Darwin was at Maer on 31 August, not for partridges, but to enlist the support of Uncle Jos (Josiah Wedgwood). With his support Darwin obtained his father's permission to sail with the *Beagle* voyage and left for Cambridge on 2 September. Immediately after his reference to partridges in his *Autobiography* he wrote, "On returning home from my short geological tour in N.Wales, I found a letter from Henslow . . .", which implies he did not go to Maer before receiving the letters. Several letters sent or received by Darwin in early September 1831 support this. His letter of acceptance to Francis Beaufort on 1 September said "There has been some delay owing to my being in Wales, when the letter arrived" (Burkhardt and Smith, 1985: 135); on 6 September he wrote to W.D. Fox "I returned from a geological trip with Prof Sedgwick in N.Wales on Monday 29th of August" (Burkhardt and Smith, 1985: 145); to Whitley, with whom he had stayed at Barmouth, he wrote on 9 September, "When I arrived home, after having left Barmouth, I found letters from Peacock & Henslow . . ." (Burkhardt and Smith 1985: 150). Darwin also asked Whitley to arrange for the post office to forward a letter to Sedgwick to Caernarfon, which Sedgwick replied to on 18 September. Thus, it is concluded that Darwin spent about a fortnight at Barmouth with Whitley, Lowes and others and returned to Shrewsbury on Monday 29 August arriving in the evening, in anticipation of shooting partridges at Maer on 1 September.

## THE GEOLOGICAL VALUE OF DARWIN'S NOTES

The notes made from Capel Curig to Barmouth, along with those made at Cwm Idwal are the first Darwin recorded after leaving Sedgwick at Bethesda. By 1831 Sedgwick had worked extensively in Cumbria and the north of England and decided on a North Welsh field trip to elucidate the geology there (Clark and Hughes, 1890; Speakman, 1982). His aims were to check whether Greenough was correct in recording Old Red Sandstone from Ruthin to Colwyn Bay on his 1818 Geological map (Secord, 1991)—he was not—and to map the Killas, i.e. the Lower Palaeozoic. This Sedgwick began on 21 August, presumably after church, on his return from Ireland, by which time Darwin was in Barmouth. Henslow had suggested that Sedgwick should take Darwin and thus Sedgwick stayed at the Mount before leaving with Darwin on 5 August.

Before this Darwin had shown little interest in geology, as Jameson's lectures at Edinburgh did not inspire him. However, when he returned to the Mount in June 1831 he obtained a clinometer and began to make a geological map of the Shrewsbury area. The only (extant) notes are those he made at Llanymynech Hill which he visited on his own in July. These notes reflect limited geological understanding, and when compared with those he made from 6 to 12 August provide evidence on how Darwin rapidly progressed under Sedgwick's tuition. In those few days his use of terminology became more precise, his knowledge of rock-types and mineralogy vastly improved and he was more cautious in the use of his clinometer, especially on outcrops of irregular bedding (Roberts, 1996: 474–78).

To summarise, Darwin's notes made at Llanymynech were efforts of an indifferent beginner. Those made at Cwm Idwal, the day after Sedgwick had left for Holyhead and Ireland, exhibit considerable expertise and insight in a complex area of confusing rock types, with no previous work to guide him. Though Darwin wrongly reminisced in later life, "We spent many hours in Cwm Idwall, examining all the rocks with extreme care, as Sedgwick was anxious to find fossils in them" his sentence "This tour was of decided use in teaching me a little how to make out the geology of a country" (Darwin and Huxley, 1983: 39) is typically Darwinian self-depreciation and understatement.

Apart from notes made ascending Moel Siabod and at Bwlch Drws Ardudwy, Darwin's notes are extremely brief, recording rock-type, minerals and cleavage or bedding. To claim, as Barrett (1974: 152) does, "His description of the geology along the way corresponds remarkably well with that of the British Ordnance Horizontal Sections of 1880" overlooks Darwin's total lack of stratigraphy. Whether at Idwal, Moel Siabod or further south Darwin described the lithology, without giving any stratigraphic order. The elucidation of the stratigraphy had to await Sedgwick's visits of 1831 and 1832. However the descriptions of the rock-types are competent, with detailed mineralogy along with measurements of bedding and cleavage. Darwin clearly distinguished between cleavage and bedding, following Sedgwick who introduced him to this earlier in North Wales. Even though cleavage was not widely accepted for another 20 years, Darwin was recognising it in Wales and throughout the *Beagle* voyage, both on the Falkland Islands and the South American mainland (Darwin, 1846; Barrett, 1977; Secord, 1991: 148–50). After Darwin left Sedgwick he made detailed notes at Cwm Idwal and on Moel Siabod, but thereafter his notes were brief and intermittent, indicating that geological notetaking was secondary to pressing on to Barmouth.

Darwin gave a good description of the geology of Moel Siabod (Shiabred) as one would observe it ascending from the inn (Plas y Brenin), describing it in three parts. The lowest part is an Ordovician sandstone and acid tuff, described as "a coarse quartzose Grey Wacke . . . proceeding higher a Feldspathic rock . . .". Part of the way up the north-west side Darwin describes "a line or rather a valley of separation" (site 1).[7] Barrett (1974: 152) did a Darwin here and ascribes the valley of separation to the gully on the craggy south-east side of the mountain, whereas it is on the north side. Then "the whole northern side of the mountain is composed of a blue (rock) slate" as between the tuff and the summit is much Caradocian slate. And finally "the top is of a crystalline grey coloured Trap", which "forms a precipice to the South" (site 2). Darwin thought that the slate overlaid the trap. On the whole Darwin's is a thoroughly competent description of the geology of Moel Siabod as seen on the path ascending from Plas y Brenin. On 1 September Sedgwick climbed Moel Siabod and reckoned that the trap "appears to me decidedly to be *over*" the slate, drawing a sketch to show it.[5] Darwin had mentioned the geology of Cwm Idwal and Moel Siabod in his missing letter to Sedgwick, referring to a "puzzle about M. Shabod". The outcrops at the top of Moel Siabod do *appear* to lie under the slate.

The notes Darwin made from Capel Curig to Barmouth are far less detailed than those made at Cwm Idwal and Moel Siabod, and appear to be made hastily. This was inevitable as the journey was 2 good days' walking. The first *south of Shiabed* (site 4) records the increasing amount of slate compared to "trap" as one moves away from the volcanic mountain of Moel Siabod to Dolwyddelan over 2 miles and is a good description of the terrain. At Dolwyddelan he gave a cursory description of the slate Quarry (site 5). From Dolwyddelan he followed *Sarn Helen* or a Roman road up to Cwm Penamnen, consisting mostly of slate but some beds of volcanics. Most notable among the "beds of trap" is the prominent crag of Carreg Alltrem formed of volcanic ash, lying a mile south of Dolwyddelan.

His notes on *Northern Moelwyn* and on "Eastern side of it" (site 6), simply give the cleavage and the strike of the slate around the Manod quarries. The notes on Carreg y Fran (site 7) are a fair description of the geology of the outcrop. Both the columnar nature and the breccia can be identified. However the latter is neither a breccia nor conglomerate. This Darwin realised when he recorded that it had "pebbles or a rock like itself" and is an agglomerate rather than a breccia. As Darwin wrote the rock contains much quartz, which at times occurs as a later infill giving a lattice structure.[8]

Records made for Ffestiniog (Festineog) (site 8) are brief in the extreme, simply recording Greywacke. No further notes were made for 10 miles when Darwin was crossing Merioneth siltstones. The next site or area where notes were made was at Drus Ardidy (site 9) (Bwlch Drws Ardudwy).[9] Darwin described the rocks to the east as "micaceous slate" corresponding with the Merioneth siltstones. His description of the strata at the bwlch or pass itself is better mineralogically than lithologically, identifying the Feldspar and Quartz of the Rhinog grits, noting that at times it is so coarse as to be a conglomerate. These beds are described in *British Regional Geology; North Wales* (Smith and George, 1961: 18) as "Tough and massive greywackes, sandstones and grits with shaly partings," with Greenly finding pebbles from the Mona complex in the conglomeratic layers. However Darwin's suggestion that "The pass seems to be caused by a subsidence and crack of SE part of strata" is ill-founded, although the linear nature of the upper 2 miles of Cwm Nantcol leading to the pass gives the impression of a major fault. Darwin made another compass inversion here as he recorded a "hill on NE side. The "N" is crossed out or altered. It is most likely that Darwin's intention was to write "SE", and it appears that the "N" in NE is replaced by an "S". Rhinog Fach lies slightly to the south east of the pass and is 8 metres lower than Rhinog Fawr.[10]

Near Pont Cwmnantcol (site 10) Darwin described an outcrop where "the rock is jointed with such remarkable regularity, as to give the appearance of beds dipping S by W."[11] which is a good description of unusual jointing. There is much jointing in the whole of Cwm Nantcol, but only at the lower end does the jointing become so pronounced that even from 100 yards it appears to be "bedding". The outcrops on the Barmouth road have not been identified, but as they are of "micaceous slate" they are most probably siltstones of the Comley beds, the lowest part of the Cambrian. The alternations of "porphyry and porphyritic slate" with the slate are more likely to be sandstone and grit. This may be seen to be a major error to confuse sedimentary and volcanic rocks, but it must be considered that Darwin was making these notes with no map or geological treatise to guide him. This "error" is similar to that made at Cwm Idwal where he described the volcanics of the Idwal Slabs as "altered slate" though "it resembles a basalt".[12] However Darwin made a distinction between dip and cleavage, which Sedgwick had taught him.

Most of the notes Darwin made between Capel Curig and Barmouth are too brief to give much indication of his developing geological prowess. The fullest notes are on Moel Siabod where he spent a day before heading south and give the impression that he took time to make them. The remainder are simply hurried notes in a geologically unknown area and record his brief first thoughts. Barrett's (1974: 151) comment that these could form a basis for a geological paper is based more on reverence than judgment and as he had not worked out the stratigraphic succession one cannot agree with Barrett that "His geology all along the way corresponds remarkably well with that of the British Ordnance Horizontal Sections of 1880" produced by A.C. Ramsay. Respect for Darwin is not the same as worship.

In a sense Darwin's "geology all along the way corresponds . . . with . . . Ramsay", and a comparison with the geology as described in later maps and memoirs can be used to confirm the locality of the sites as Darwin's notes are accurate and methodical, BUT they are entirely lithological and mineralogical and nowhere have any explicit concept of stratigraphic succession. How far Darwin had even begun to elucidate the stratigraphy is impossible to conclude from his notes alone. Darwin's notes for 1831 totally lack any sketch, diagram or cross-section, and are descriptive rather than theoretical. However, in his letters to Darwin on 4 and 18 September Sedgwick drew two cross-sections of Snowdonia.[13] While on the *Beagle* Darwin made extensive use of diagrams and cross-sections, often using these in preference to geological maps (Stoddart, 1995: 6–7) the first being his cross-section at Quail Island (reproduced Herbert, 1991: 166). Further his notes lack theorising, but that was the approach of many geologists in the 1820s, possibly in reaction to the over-theorising of the Diluvialists of previous years, and thus the "factual" description of rock-types and their historical order became paramount. Stoddart argues how reading Lyell's *Principles of Geology* made him far more speculative and theoretical. However, at this time Darwin had not "deserted . . . that tram-road of all solid physical truth—the truth method of induction"[14] and thus his geological notes were more descriptive than theoretical reflecting the influence of Sedgwick (Rudwick, 1988: 242–9).

Elsewhere his notes record only part of his understanding of the geology. This is seen when one compares his field notes of the glaciation of North Wales and his published paper, as the latter develops several aspects scarcely implicit in his notes. (The published paper also contains Darwin's most spectacular example of compass inversion in Nant Peris. Darwin described a cliff to the E.N.E. of Lyn Peris, where there is no cliff. The cliff lies to the W.S.W. and Darwin's localities are easily identified.) (Darwin, 1842; Barrett, 1977: 169). It would be wrong to expect an infant geologist to have worked out the stratigraphy on a complicated area in so short a time. Chronologically Sedgwick began his main work on sorting out the Lower Palaeozoic of Caernarvonshire a few days after Darwin reached Barmouth, spending some 2 months in energetic traverses of the mountains. Only in 1832 did Sedgwick visit the Harlech dome.

## CONCLUSION

Despite its inclusion in Darwin's *Autobiography* and its subsequent repetition in most studies of Darwin's early life, the heroic image of Darwin closely following a compass bearing for the 40 miles of rugged terrain from Capel Curig simply is not true. A comparison of his notes, both on topographic and geological details, with the nature of the terrain, place names and geology show that Darwin took a far less direct route, choosing one which avoided the highest and roughest ground, and followed well-worn paths.

Geologically his notes are strong on lithology, mineralogy, direction of dip and cleavage (compass inversions notwithstanding!) but are understandably weak on stratigraphy. Darwin had left Cambridge as a complete novice in geology as his notes at Llanymynech show. By the time he reached Barmouth his geological skills matched his biological ones and thus Darwin was ready to make his expedition to the Canaries. But that was not to be!

## ACKNOWLEDGEMENTS

The Darwin manuscripts (DAR) are quoted with permission of the syndics of Cambridge University Library. This work was partly funded by a grant from the Isla Johnston Trust administered by the Church in Wales. The following have helped, but are not responsible for my errors: the Most Rev. Alwyn Rice Jones, Archbishop of Wales for encouragement and historical information on Capel Curig, members of Shropshire Geological Society and North Wales Geological Society, librarians at the Sedgwick Museum and University Library at Cambridge, Jim Secord, Jim Moore and Janet Browne, Gordon Litherland for drawing the maps on his computer, and my canine companion, Topper, who always "looks on his master as a god."

*Appendix 1*

### A Darwin chronology from June to October 1831

In the absence of a diary by Darwin, details of his activities from the time he left Cambridge in June 1831 and his receiving the letters inviting him to join the *Beagle* on 29 August can be elucidated by comparison of letters, Darwin's geological notes, and Sedgwick's notes. For August there is no evidence either way that Darwin accompanied Sedgwick on 3 and 4 August, but it seems likely. His movements from 5 to 11 August are beyond doubt, mostly coming from dated notes of both Darwin and Sedgwick. From 12 to 17 August when Darwin probably arrived in Barmouth, the dates are very probable, and is the shortest time Darwin could have taken. Darwin may have arrived in Barmouth up to 20 August. It is most likely that Darwn stayed ten days in Barmouth and returned to Shrewsbury on 29 August (the route and dates of this field trip are given in Figure 1.)

| Date | DARWIN | SEDGWICK |
|---|---|---|
| mid June | Return to The Mount | — |
| June–July | Mapping nr Shrewsbury | — |
| 10th July | Receives clinometer | — |
| late July | Visits Llanymynech | — |
| August | | |
| 3rd | *?with Sedgwick?* | Alderbury |
| 4th | *?with Sedgwick?* | Pontesford |
| 5th | Llangollen, Dinas Bran | with CD |
| 6th | Velvet Hill to Ruthin | with CD |
| 7th Sun. | Ruthin to Denbigh | with CD |
| 8th | Denbigh to Abergele | Denbigh to Conwy |
| 9th | Abergele to Conwy | Penmaenmawr & Conwy |
| 10th | Conwy, Llanbedr, Penmaenmawr | with CD |
| 11th | Aber Falls, Penrhyn | with CD |
| 12th | Anglesey, ? to Holyhead | to Holyhead |
| 13th | return from Anglesey | Dublin |
| 14th Sun. | Cwm Idwal to C Curig | Dublin |
| 15th | Moel Siabod | Holyhead Mt |
| 16th | Capel Curig to Ffestiniog | Anglesey |
| 17th | Ffestiniog to prob Barmouth | Anglesey |
| 18th | Barmouth | Anglesey |
| 19th | Barmouth | Anglesey |
| 20th | Barmouth | Beaumaris to Caernarfon |
| 21st Sun. | Barmouth | Caernarfon to Dolbadarn |
| 22nd | Barmouth | Snowdon |
| 23rd | Barmouth | Elidir Fawr to Twll Du |
| 24th | Barmouth | Bad weather |
| 25th | Barmouth | Capel Curig to Tyn-y-Maes |
| 26th | Barmouth | Carnedd, Aber, Tyn-y-Maes |
| 27th | Barmouth | slates to Bangor |
| 28th Sun. | Barmouth | Rest to Aber |
| 29th | Return to Mount | Conway, Orme, Llanrwst |
| 30th | Mount | Crafnant to C. Curig. |
| 31st | Maer | Idwal, Glyders, C. Curig |
| September | | |
| 1st | Mount | C. Curig, Moel Siabod |
| 2nd | Cambridge | C. Curig to Llanrwst, Bettws |
| 3rd | Cambridge | Bettws y Coed to Tremadoc |
| 4th Sun. | Cambridge | Rest & Rain |
| 5th–11th | 17 Spring Gdns, London | |
| 11–16th | Devonport | |
| 17–20 | 17 Spring Gdns | |
| 21st | Cambridge | |
| 22–2nd Oct. | Shrewsbury | |
| 2nd | left for Devonport | |

NOTES

[1] Adam Sedgwick, entries for 10 and 11 August for their route from Conwy to Penrhyn, and entry for 12 August for Anglesey, Sedgwick's Journal, No XXI (1831), Sedgwick Museum, Cambridge. Charles Darwin, manuscript of North Wales Field trip: Darwin Papers, Cambridge University Library, DAR 5, fols 5–14. There are 20 sheets, the first being numbered '5' and those are referred to here as 12i, 12ii, 13i etc. For route from Conwy to Penrhyn see fols 9ii, 10i and 10ii. The excised page is visible on fol 10ii.; fol 11i contains notes for Cwm Idwal. For Darwin's notes on Quail Island see CUL DAR 34(i) fol 19.

[2] CUL, DAR 5 fols 12–14.

[3] Adam Sedgwick to Charles Darwin, 4 September 1831, see Burkhardt and Smith, 1985: 137.

[4] CUL, DAR 5 fols 12ii, 13i.

[5] Sedgwick to Darwin, 4 September 1831, see Burkhardt and Smith, 1985: 8.

[6] CUL, DAR 5 fols 13ii to 14ii.

[7] CUL, DAR 5 fols 12i & ii.

[8] CUL, DAR 5 fols 13i & ii.

[9] CUL, DAR 5 fols 13ii & 14i.

[10] CUL, DAR 5 fols 14i.

[11] CUL, DAR 5 fols 14i & ii.

[12] CUL, DAR 5 fols 11i & 14ii.

[13] Sedgwick to Darwin, 4 September 1831, and 18 September 1831, see Burkhardt and Smith, 1985: 137 and 157.

[14] Sedgwick to Darwin, 24 November 1859, see Burkhardt, Smith *et al*. **7**: 396.


REFERENCES

BARRETT, P.H., 1974 The Sedgwick-Darwin geologic tour of North Wales. *Proceedings of the American Philosophical society* **118**: 146–164, Transcription of Darwin's notes on 155–62.

BARRETT, P.H., 1977 *The collected papers of Charles Darwin*. Chicago/London. Pp 326.

BORROW, G., 1905 (1862) *Wild Wales*. London. Pp 733.

BURKHARDT, F. and SMITH, S. (Eds), 1985 *The Correspondence of Charles Darwin*. Vol. 1 1821–1836. Cambridge. Pp 702.

CLARK, J.W. and HUGHES, T.M., 1890 *The life and letters of the Reverend Adam Sedgwick*. Cambridge. 2 Vols.

DARWIN, C., 1842 Notes on the Effects produced by the ancient glaciers of Caernarvonshire, and on boulders transported by floating ice. *London, Edinburgh and Dublin Philosophical Magazine and Journal of Science* **21**: 180–88 (also in BARRETT, P.H., 1977: 163–71).

DARWIN, C., 1846 On the geology of the Falkland Islands. *Proceedings of the Geological Society* pt. 1. **2**: 264–74 (Reprinted: BARRETT, P.H., 1977: 203–211.)

DARWIN, C. and HUXLEY, T.H., 1983 *Autobiographies*. Oxford. Pp 123.

MEAD, L., 1912 5th Ed *Oke's Game Laws*. London. Pp 374.

NORTH, F.J., CAMPBELL, B. and SCOTT, R., 1949 *Snowdonia*. London. Pp 469.

NUTTALL, J. and A., 1989 *The mountains of England and Wales, Volume I: Wales*. Milnethorpe. Pp 256.

ROBERTS, M.B., 1996 Darwin at Llanymynech: the evolution of a geologist. *British Journal for the History of Science* **29**: 469–78.

RUDWICK, M.J.S., 1988 A year in the life of Adam Sedgwick and company, geologists. *Archives of Natural History* **15**: 243–68.

SECORD, J.A., 1991 The discovery of a vocation: Darwin's early geology. *British Journal for the History of Science* **24**: 133–57.

SMITH, B. and NEVILLE GEORGE, T., 1961 *British Regional Geology: North Wales*. London. Pp 96.

SPEAKMAN, C., 1982 *Adam Sedgwick, geologist and dalesman.* Heathfield. Pp 145.
STODDART, D.R., 1995 Darwin and the seeing eye. *Earth Sciences History* **14**: 3–22.
THESIGER, W., 1987 *The Life of My Choice.* London. Pp 459.

(Accepted 10 February 1997.)

*Archives of natural history* **29** (1): 1–26. 2002 © The Society for the History of Natural History

# "A most glorious country": Charles Darwin and North Wales, especially his 1831 geological tour

PETER LUCAS

Pwllymarch, Cwm Nantcol, Llanbedr, Gwynedd, Wales, LL45 2PL.

ABSTRACT: Darwin's tour with Adam Sedgwick in 1831, the last of some 14 Welsh visits before the *Beagle* voyage, divides into three periods: a week, mostly with Sedgwick, from 5 August; a middle period ending by 20 August, when Sedgwick left Anglesey; and a final period during which Darwin spent some days in Barmouth, reaching Shrewsbury on 29 August. His activities are well documented, for the first period, through both men's geological notes and, for the last, in the journal of the Lowe brothers (showing Darwin reaching Barmouth from Ffestiniog on 23 August and parting from Robert Lowe on 29 August). For the middle period the circumstantial evidence points to Anglesey: whether Darwin's writings show any first hand knowledge of the island needs further examination. Robert Lowe was one of Darwin's most gifted contemporaries; his "early hero-worship" enhances the conventional picture of Darwin on the eve of the voyage. After his return to North Wales in 1842, to investigate the effects of glacial action, Darwin saw the tour as illustrating the futility of observations outside of any adequate theoretical framework.

KEY WORDS: Adam Sedgwick – Robert Lowe – Robert Dawson – W. E. Jelf – Anglesey – glaciers.

> "You must have enjoyed N. Wales, I sh$^d$ think; it is to me a most glorious country."
>
> C. E. Darwin to C. Lyell, 14 August 1863 (Burkhardt *et alii*, 1999: 590).

## INTRODUCTION

It was after "wandering about North Wales on a geological tour with Professor Sedgwick" that Darwin "arrived home on Monday 29th August" 1831 to the offer of a place on the *Beagle* (Keynes, 1988: 3). The tour was first considered in Barrett (1974). His account has been supplemented in some respects in Secord (1991) and Roberts (1996, 1998, 2000), but many gaps remain. This paper seeks to fill some of the gaps and, for the first time, to set the tour in the context of Darwin's many other visits to North Wales (listed in the appendix). And, at a defining moment in the "transformation ... into a scientific giant ... of an amiable but rather aimless young man", or so he has been described (Browne, 1995: xiii), we can ponder his impact on one of the most gifted of his contemporaries, his companion at The Cross Foxes, Mallwyd, on 28 August.

## PRELIMINARIES

### Cambridge: Lent and Easter terms 1831

"I find in Geology a never failing interest", Darwin wrote to Charles Whitley on 23 July 1834, "as [it] has been remarked, it creates the same gran[d] ideas respecting this world, which Astronomy do[es] for the universe" (Burkhardt and Smith, 1985: 397), seemingly an allusion to a passage in Herschel's *A preliminary discourse on the study of natural philosophy* (1831: 287) marked in the margin of Darwin's own copy.[1] Free at last of examinations, he had been reading the book in February 1831 (Burkhardt and Smith, 1985: 118), one of two which in his last Cambridge terms influenced him above all others (Barlow, 1958: 67–68).

What Darwin described as "the strongest desire ... from my early youth ... to understand or explain whatever I observed" (Barlow, 1958: 141), reveals itself in the interest in theory shown in his geological notes from the start of the *Beagle* voyage (noted as remarkable by Secord (1991: 153) and Herbert (1991: 168–169)). This interest can be glimpsed in the Darwin teased for his "philosophy" in the spring of 1831 and indulging in extravagant "hypotheses" in July (Burkhardt and Smith, 1985: 119, 125) and it surely brings us closer to the Darwin who made such an impression on Robert Lowe during a few days in late August.

Darwin's friendship with J. S. Henslow during his last two Cambridge terms and, through Henslow, his introduction to Adam Sedgwick (Browne, 1995: 123, 136) need no rehearsal here. Among the topics which Darwin and Henslow discussed at their many meetings must have been Henslow's (1822) geological essay on Anglesey, the "outstanding memoir" (Secord, 1991: 142) which both Sedgwick and Darwin were to praise highly (Clark and Hughes, 1890: **I**, 378; Burkhardt *et alii*, 1994: 461–462). It could have been at this time, and from Henslow, that Darwin acquired his map of Anglesey[2], the last of four hand-drawn cloth maps considered in detail in Roberts (2000).

Little or nothing is known of Darwin's contacts with Sedgwick, then planning his usual geological expedition after the Easter term. No doubt Sedgwick consulted Henslow as well as John Hailstone, his predecessor in the Woodwardian chair. "To enter N. Wales from Chester", Hailstone began his letter to Sedgwick of 6 June, "there are two roads one to the N. & the other to the S. & in the angle betwixt the two roads the first scene of your operations will lie." The letter assumed a progress west and south from Chester to the "chain of Limestone hills", between Clwyd and Conwy, "thro the heart of the Vale of Clwyd by Denbigh ... into Shropshire & even Montgomeryshire where the great Limeworks of Llanymynech are furnished by it." In the early days of their tour Sedgwick and Darwin would follow the limestone in the opposite direction. "Some part of your tour", Hailstone continued, "will most likely fix you at Shrewsbury ... I think you ought not to omit Llanymynach as the extensive Limeworks there have laid bare the beds to a great extent."[3]

If Sedgwick discussed his plans with Darwin after he received Hailstone's letter, it may have been then that he arranged to rendezvous with Darwin in Shrewsbury. The letter may also have provided the motive for Darwin's visit to the quarry at Llanymynech (Roberts, 1996)) and for a second map[4], showing the roads which reach Llanymynech from north, east and south (Roberts, 2000: 74–76).

Darwin may have remained in Cambridge until mid-June, it being only around 19 June that he reached Shrewsbury, expecting that Sedgwick would soon follow.

### Shrewsbury: mid-June to 4 August
Back in Shrewsbury waiting for Sedgwick, Darwin found himself in limbo. "I arrived at this stupid place about three weeks ago", he grumbled on 9 July, "I am trying to make a map of Shrops: but dont find it so easy as I expected" (Burkhardt and Smith, 1985: 124). "On my return to Shropshire", he later remembered, "I examined sections and coloured a map of parts round Shrewsbury" (Barlow, 1958: 68–69), no doubt the "map of Shrops". The "sections" and "map" can be identified with the geological notes which take up nine

pages at the back of the vellum-covered notebook containing Darwin's observations on his children[5] and with the two remaining maps[6], one of the Shrewsbury district and the other a westward extension of its north-western quarter (Roberts, 2000: 71–74). The notes, transcribed in Herbert and Roberts (2002: this issue), describe the four areas marked "A" to "D" on the Shrewsbury map (Roberts, 2000: 73) and show Darwin exploring the coal and gravel pits around the Meole brook south of Shrewsbury and the sandstone hills of Nescliffe to the north-west near the Holyhead road, "the general red colour, abrupt escarpment & wooded top of these hills gives much picturesque beauty to that part of Shropshire." The map making has been seen as involving an "abortive expedition" which, in his letter of 11 July (Burkhardt and Smith, 1985: 125–126), Darwin "concealed" from Henslow (Secord, 1991: 144) but it is not obvious that any relevant excursion preceded the same letter's projected "first expedition", (newly arrived) "clinometer & hammer in hand".

In mid-April Sedgwick had "set to work" on the revision of his and R. I. Murchison's joint paper on the Alps, hoping "to get this work rapidly off his hands, and to be ready to start for Wales at the beginning of the Long Vacation" but, as he explained to Murchison on 13 September, "the Alpine paper was infinitely more troublesome to reduce than I expected" (Clark and Hughes, 1890: I, 376–377). Sedgwick "did not get from Cambridge before the 1st of August", spending "one day at Dudley and two days at Shrewsbury" (Clark and Hughes, 1890: I, 378). On the two Shrewsbury days his journal shows him on 3 August noting "six miles on Welshpool road a mag. conglome at Cardeston" which he traced again "at 9th milestone", more than half way to Llanymynech, and, on 4 August, "ascending the Pontesford Hill".[7]

That Darwin was with Sedgwick on 3 August (as one would expect) is suggested by his notes for 6 August on a quarry near Ruthin, "the rock is spotted with brown, like the stone at Cardeston" (Barrett, 1974: 156; Roberts, 2000: 74), leaving 4 August as a possible occasion for his visit to Llanymynech. If accompanied by Darwin on 3 August, Sedgwick no doubt stayed with the Darwins at The Mount throughout his time in Shrewsbury though Darwin's two recorded memories, from more than 40 years later, each encompasses only a single evening (Clark and Hughes, 1890: I, 380; Barlow, 1958: 69).

## CHRONOLOGY OF TOUR

A threefold division emerges from Barrett (1974): 5–11 August, Shrewsbury to Bethesda, when Sedgwick and Darwin were mostly together and their geological notes[8] for the most part correspond; a middle period, for Darwin apparently undocumented, which ended when he "left Sedgwick no later than August 20" (Barrett, 1974: 149), the day Sedgwick reached Caernarfon from Anglesey[9]; and a final period during which Darwin "visited Barmouth to see some Cambridge friends who were reading there" (Barlow, 1958: 71).

Darwin's activities during this final period are well documented in the Barmouth journal of the Lowe brothers, Henry Porter and Robert[10], then being tutored in mathematics by Darwin's Cambridge friend, Charles Whitley. Henry was one of Darwin's few fellow members of a short-lived Cambridge dining club; his younger brother, the Oxonian Robert, would be Chancellor of the Exchequer and then Home Secretary in Gladstone's first administration of 1868–1874. In the pages of their journal Darwin reaches Barmouth from Ffestiniog on the evening of 23 August and sets out for Shrewsbury from Mallwyd, 22 miles from Barmouth, on the morning of 29 August.

## SHREWSBURY TO BETHESDA: 5 TO 11 AUGUST

### 5 August

Sedgwick's journal has just a brief entry. For him, and no doubt Darwin (theoretically he could have gone to Llanymynech), this was not a day for geologising: it was only on 6 August that Sedgwick would see his labours as commencing (Clark and Hughes, 1890: **I**, 378). One may assume that the two men took the direct route to Llangollen, a leisurely 30 miles along Telford's Holyhead road.

In 1836 Darwin would praise a road of the New South Wales colonists as "worthy of a road in England, even that of Holyhead" (Keynes, 1988: 400). Its many associations went back at least to July 1821. Prompted perhaps by a bequest in 1817 of "two shares in the Capel cerig road"[11], his father had encouraged his two sons in "a little excursion ... about a week or 10 days if it answers ... to see the new road as far as Bangor Ferry, where the new Bridge is building, & so come back some other way."[12] "[Erasmus] & Charles rode 255 miles in Wales", the Doctor reported, "& were gratified by what they saw", adding with his usual sagacity, "I believe it is useful for them to act a little for themselves."[13] On his way home from Dublin in 1827 (Burkhardt and Smith 1985: 538–539), Darwin probably travelled the whole length of the road to Shrewsbury from Holyhead and across the Menai suspension bridge (opened the previous year).

### 6–7 August

6 August, Sedgwick's journal is explicit, began with the meeting with Robert Dawson of the Board of Ordnance ascribed to 5 August in Barrett (1974: 147) and Roberts (1996: 473). Dawson had wide experience of North Wales, "the high point" of his work perhaps "his depiction of Snowdonia, on which he worked from 1816 to 1820, in the interval of teaching his cadets" (Harley and Oliver, 1992: xi) whom he took through North Wales travelling "sometimes on horseback, sometime on foot, and sometimes by chaise" (Close, 1969: 80). A reference in the journal for 24 August at Llanberis, "a series of notes from Mr Dawson's MS"[7], points to the meeting's possibly wider scope.

From Llangollen Sedgwick and Darwin headed north to the Vale of Clwyd, reaching Ruthin on 6 August and Denbigh on the Sabbath. So far at least, Darwin was within his father's medical territory. It was at Llangollen that the Doctor prescribed green nettle whippings for Mrs Preston (Mavor, 1973: 172), an acquaintance of the "ladies of Llangollen", one of them at least his patient[14]; and it was to Denbigh that he had come in 1817, "to see a patient who is very dangerously ill" (Wedgwood and Wedgwood, 1980: 181).

### 8 August

"Descended to St Asaph", Sedgwick reported to Murchison on 13 September, "thence in my gig to Conway" (Clark and Hughes, 1890: **I**, 378). It is at this point that Darwin's geological notes, hitherto using the plural "we", change to the singular "I" –

> We were struck the whole way from Llangollen to this place [St Asaph] by the almost entire absence of turf pits. St Asaph to *Abegele*. by Bettys. at the point where the road to Bettys divides from that to Abegele. a shaft was sunk for lead ore in the Limestone ... A little further on road to Bettys, I crossed a great bed of diluvium (Barrett, 1974: 147, 157).

Barrett's transcription conceals a transition. The second sentence begins a new paragraph, in a lighter ink. The first, "We were struck the whole way ... *to this place*" (present author's emphasis), evokes Darwin writing up his notes at St Asaph, perhaps during a break for lunch with Sedgwick before they separated.

The point of separation may have been a few miles further on, west of St Asaph at The Cross Foxes Inn, Glascoed.[16] Beyond the inn, the road to Bettws-yn-Rhos (today's B5381) branched off from the turnpike road from St Asaph to Abergele, making The Cross Foxes, "where the road to Bettys divides from that to Abegele", the obvious place for the two men to separate, Sedgwick continuing in the gig through St George to Abergele and Conwy.

On what the Lowe brothers in Barmouth noted as "a tolerably fine day, with a little wind"[10], Darwin continued parallel to the coast for five miles to Rhos, then north-east to Abergele where he probably spent the night. One tends to assume he was on foot but a horse was as easy to hire then as a car today. It could have been on one of the "little horses" that Wyndham had found so serviceable in the early days of Welsh tourism[16] that Darwin made his way on the expedition which has been seen as a major episode in the making of his vocation as a geologist (Secord, 1991: 148).

## 9 August

From Abergele to Conwy was 11½ miles by the turnpike road. Darwin's geological notes, terse and impersonal though they are, offer some clues to what must have been a nostalgic journey. "From *Colwyn* to L Ormes head", he noted, "broad valley the bottom with bed of Clay slate diluvium about 40 feet thick" (Barrett, 1974: 158), a reference to the old estuary of the Conwy which entered the sea at Penrhyn Bay between Rhos-on-Sea and the Little Orme.[17] The condensed note conceals a return to the scene of some of Darwin's earliest memories, from his first Welsh holiday at "Gros near Abegele" in 1813 (Burkhardt and Smith, 1985: 537). On the afternoon of 28 August, he and Robert Lowe would pass up the Mawddach below the site of his last Welsh holiday, at Caerdeon in 1869, shared with Emma Darwin just as the first had been 56 years before (Wedgwood and Wedgwood, 1980: 165).

It is not clear from his notes that Darwin went beyond the valley to either the Little or the Great Ormes Head[18] (Figures 1 & 2), the latter visited in 1827 during his stay with John Price at Bodnant, some ten miles to the south (Lucas, in press).

## 10–11 August

"Early in the morning" Sedgwick and Darwin left Conwy, Sedgwick "for the first two or three miles ... gloomy, and hardly spoke a word" (Clark and Hughes, 1890: I, 381), the episode of Sedgwick and the "d—d" waiter, suspected of failing to pass on a sixpenny tip to the chambermaid, which was evidently one of Darwin's favourite Sedgwick stories, recalled by both Francis Darwin (1887: I, 56 n.) and his brother George.[19] After leaving the Conwy valley they came over the top of Penmaenmawr and down to Aber. From there the next day they crossed the mountains to the Penrhyn slate quarries at Bethesda where Darwin's observations, which close the third sheet of his geological notes, end apparently in mid-sentence. Barrett supplies the absent punctuation and, with it, his transcription reads well enough (Barrett, 1974: 159).

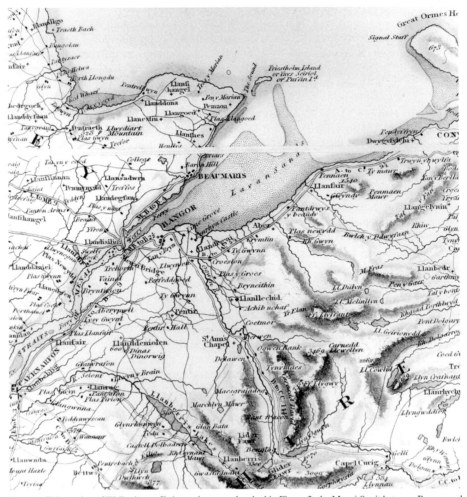

Figure 1. This section of Walker's map[28] shows the coast sketched in Figure 2; the Menai Strait between Bangor and Caernarfon; and the Holyhead road from Bangor through Tynymaes and "Nant fracon" to Capel Curig. The Bethesda quarry is opposite Ogwen Bank. Llyn Idwal is the unnamed lake below "Glider Fawr" and opposite Llyn Ogwen. (Reproduced by permission of Sedgwick Museum, Cambridge.)

Figure 2. 'We have got the prettiest room in the world here [Bangor], looking over Port Penrhyn upon Anglesey just where the steamer [*Rothsay Castle*] was wrecked with Penmaenmawr to the right & the Ormsheads.' Hensleigh Wedgwood's letter of 16 September 1831[68] describes his sister Fanny's sketch[54]. (Reproduced by courtesy of the Trustees of the Wedgwood Museum, Barlaston, Stoke-on-Trent, Staffordshire, England.)

## NINE MISSING DAYS AND CWM IDWAL: 12–20 AUGUST

Sedgwick's movements from Bethesda are quite clear, taking him down the Holyhead road to Bangor on 11 August and on 12 August across Anglesey to Holyhead to which he returned on 15 August after a weekend in Dublin.[20] On the four following days he explored Anglesey in the footsteps of Henslow (1822) of which his journal is full ("visit Holyhead Mount find it as in Henslow", "all as in Henslow", "well stated by Henslow", "perfectly described by Henslow"[7]). On 20 August he re-crossed the Menai on his way to Caernarfon. He had gone to Anglesey, he explained to Murchison on 13 September, "in the hopes of learning my lesson for Snowdonia. Henslow's paper is excellent, but the lesson is worth next to nothing" (Clark and Hughes, 1890: **I**, 378).

### Darwin's missing days

Darwin's movements after Bethesda are obscure until he re-emerges at Ffestiniog on 22 August. His geological notes consist of five separate sheets each folded in two and written on both sides, the 20 folios of Barrett's transcription. The first three sheets take him from Llangollen to Bethesda. The remaining two relate to Cwm Idwal, between Bethesda and Capel Curig (ff 13–14), and to points, including Moel Siabod and Ffestiniog, between Capel Curig and Barmouth (ff 15–20). They read like a description of a continuous geological progress from Llangollen to Barmouth but 11 days separate the visit to Bethesda on 11 August (f. 12) and the passage from Ffestiniog to Barmouth on 23 August (ff 18–20). From these dates, he could have parted from Sedgwick as early as 11, or as late as 20, August, exploring Cwm Idwal and reaching Capel Curig as early as 12 August ("a very splendid day, & very hot" according to the Lowes in Barmouth[10]) or as late as 21 August ("another splendid day"[10]). Up to nine days are quite unaccounted for.

### The evidence for Anglesey

The visit to Bethesda completed the seventh day of the tour, only the sixth of active geologising even if the Sabbath is included. Anglesey would provide the opportunity for detailed consideration of Henslow (1822) which Darwin must have been discussing with its author earlier in the summer, a further week with Sedgwick still allowing him the "few days" in Barmouth he had promised Whitley in July (Burkhardt and Smith, 1991: 466) and a return home for "the sacred 1st of September" (Darwin, 1887: **I**, 190). With the start of the expedition delayed by Sedgwick's late departure from Cambridge, a premature separation would be a fact in want of an explanation. Put another way, if not with Sedgwick, where was he?

A missing letter, perhaps also some missing geological notes, contribute to the lack of evidence. Darwin's journal entry, made retrospectively in 1838, records simply that he "went on Geological tour by Llangollen, Ruthven Conway, Bangor & Kapel Curig. where I left the Professor [Sedgwick] & crossed the mountains to Barmouth" (Burkhardt and Smith, 1985: 537, 540) (the mistaken belief that it was at Capel Curig that he left Sedgwick complicates Darwin's later accounts of this period). His letter to Henslow of 30 August (Burkhardt and Smith, 1985: 131) has not even a passing reference to Anglesey. Nor apparently are there any geological notes. Their absence could be explained if, after the Bethesda notes which end apparently in mid-sentence, there was at least one sheet which Darwin later removed[21] and gave to Henslow when they met on 3 September, basing the later journal entry on his surviving notes. But this on its own is less to explain than to explain away.

The nearest Darwin himself comes to providing a clue is in the brief account of his "short geological tour" in the letter-memoir of 24 May 1875 –

> We spent nearly a whole day in Cwm Idwal examining the rocks carefully, as he [Sedgwick] was desirous of finding fossils ... Shortly afterwards I left Professor Sedgwick, and struck across the country in another direction, and reported by letter what I saw. In his answer he discussed my ignorant remarks in his usual generous and frank manner (Clark and Hughes, 1890: I, 381).

Darwin's letter, so crucial to the reconstruction of his movements, is listed in a catalogue of letters in Sedgwick's own hand as "Mr Darwin on the fossils in Cwm Idwal" but, in another hand, is noted as "missing".[22]

There are some clues to its content in Sedgwick's "answer", his letter of 4 September. This begins with a response to Darwin's geological "information" on Cwm Idwal and Moel Siabod from which it is apparent that Darwin had been searching for fossils not with Sedgwick but on his behalf, providing him with the information from which the negligence of the "stupid red nosed waiter" prevented Sedgwick from benefiting during his stay at Capel Curig (Burkhardt and Smith, 1985: 137). Darwin's recollection of having parted from Sedgwick at Capel Curig may have led him to suppose that it was shortly *after* Cwm Idwal that he "struck off in another direction".

If he did so shortly *before* Cwm Idwal, this would equally fit a visit there as late as 21 August, after parting from Sedgwick the previous day, and one as early as 12 August, after the visit to Bethesda on 11 August. However a later date fits better with the "skeleton of my walk" beginning "21st ... from Carnarvon to Dolbadarn" described in Sedgwick's letter of 4 September (Burkhardt and Smith, 1985: 138) which led Barrett (1974: 149) to conclude that "for whatever reason, Darwin left Sedgwick not later than August 20", a date consistent with his arrival at Ffestiniog by 22 August.

That Darwin was familar with Anglesey at second hand through Henslow (1822), and through his many talks with Henslow, is not in doubt. During the voyage he used his own copy of the memoir in the *Beagle*'s library, "with numerous annotations" and "frequently cited" (Burkhardt and Smith, 1985: 557, 560), for the detailed comparison between the geology of Anglesey and the Falkland Islands.[23] Whether his observations necessarily reflect a knowledge of Anglesey at first hand is something which needs closer examination. One instance of what looks like first hand knowledge, from Quail Island in the Cape Verde Islands in January 1832 (within five months of Sedgwick's visit to Anglesey), has already been noted: "I could have scarcely credited that rock, nearly as hard as the conglomerates of older formation (viz of red-sandstone formation Anglesey) could daily be increasing under my own eyes"[24] (Secord, 1991: 152, n. 58; Roberts, 1998: 61).

**Darwin on Anglesey?**

If Darwin went to Anglesey, how long was he there? Henslow had noted six outcrops of old red sandstone (O.1 to O.6 in Henslow (1822: 388–393)). Sedgwick crossed the formations on 12 August on his outward journey to Holyhead and, on his return, on 16 August at Bodedern, "so on over the O.R. to Llanerchymedd ... North and East of the town fine quartzose Old Red".[7] Roberts concluded that Darwin came with Sedgwick until the first sighting on 12 August, separating "sometime before Sedgwick set sail from Holyhead on 13 August" (Roberts, 1998: 59). But, if Darwin came to Anglesey at all, he could have remained until 20 August before making a dash for Barmouth over the next three days.

Sedgwick stayed at Llanerchymedd on 16, and probably also 17, August. On his way to Beaumaris on 18 August he heard "of the loss of the Rothsay Castle"[7], the boat on which the Lowe brothers had been due to travel[25], "wrecked off Beaumaris in circumstances of the utmost horror" (Usher, 1955: 4). The next two days he worked his way down the Menai Strait to Caernarfon, crossing by the Garth ferry "to the other side"[7] near Port Penrhyn (some 15 miles from Capel Curig) on 19 August and, on 20 August, visiting the Moelydon ferry and Port Dinorwig, between Bangor and Caernarfon, the harbour for the Llanberis slate quarries which he explored on 23 August. It could have been from either harbour, or from the Menai bridge, that Darwin "struck across the country in another direction", his obvious destination Ty'n-y-maes above Bethesda in Nant Ffrancon, "a small Inn" (now a motel) "where the Mail changes horses" (Burkhardt and Smith, 1986: 323) and where Sedgwick stayed on 25 and 26 August (Burkhardt and Smith, 1985: 138).

One may suppose that before they parted the two men made plans for the days ahead, Darwin arranging to report to Sedgwick at Capel Curig on what he had seen. Perhaps it was because he reported there that he came to remember Capel Curig as where they separated.

## Cwm Idwal

Not later than 21 August, Darwin set out along Nant Ffrancon for what he described (in 1842) as "the wild amphitheatre in which Lake Idwell lies" (Barrett, 1977: **1**, 163), surrounded (as he noted in 1831) by its "circle of steep rocks" (Barrett, 1974: 159–160), at their heart Twll Du, the Devil's Kitchen of generations of climbers. In 1842 he would marvel at the glacial phenomena invisible to him in 1831 ("eleven years ago, spent day here saw nothing. nor sh[d] have if not pointed out"[26]), his attention fixed then on fossils (Barrett, 1974: 160).

From Cwm Idwal he reached Capel Curig and the familiar setting of Lord Penrhyn's Inn, "a sort of pivot on which the intercourse between London and Dublin is constantly turning ... sixty beds and all other accommodation in proportion are found here" (Jones, 1952: 99). Darwin knew the Inn from his expedition with Erasmus in 1821, probably also from the walking tour of 1826 and his journey home from Dublin in 1827, perhaps from the Slaney fishing expedition of 1828 (Burkhardt and Smith, 1985: 64). In 1830 he was based at the Inn for around a week from 12 August before reaching Barmouth on 20 August (Burkhardt and Smith, 1985: 105–106). Capel Curig, like Barmouth, was familiar territory.

The hotel visitors' book, "a large folio book to amuse the stranger" (Pugh, 1816: 111), has no entry for 1831[27] and so can cast no light on Sedgwick's allusion to the red-nosed waiter of Darwin's missing letter as "proving beyond a doubt that Darwin returned to the Inn after climbing Moel Siabod" (Roberts, 1998: 64). The allusion is irretrievable without the letter but Darwin's geological notes, it will be suggested, point in another direction.

## CAPEL CURIG TO BARMOUTH: 22?–23 AUGUST

Darwin set out from the Inn for Ffestiniog and Barmouth not later than 22 August, his journey taking him, as he remembered 45 years later, "in a straight line by compass and map across the mountains to Barmouth, never following any track unless it coincided with

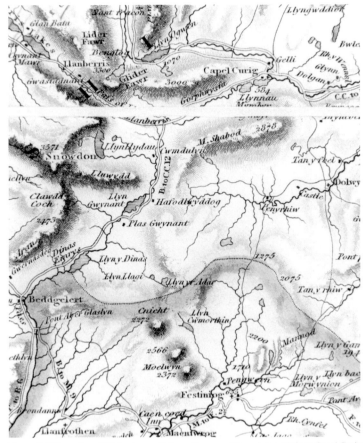

Figure 3. This section of Walker's map[28] shows the area of Darwin's walk from Capel Curig not later than 22 August 1831. Note how straight is the line south over "M Shabod" down to "Festiniog". (Reproduced by permission of Sedgwick Museum, Cambridge.)

my course", thus coming "on some strange wild places" (Barlow, 1958: 70–71). Such places notwithstanding, this was less a venture into an unknown interior than a valedictory passage through familiar landscapes.

Much has been made of the improbably "difficult terrain ... following a bearing from Capel Curig to Barmouth" (Roberts, 1998: 59). But, once adjusted for Ffestiniog, the obvious route is straightforward and can be followed on a map in something not too far from a straight line. The first leg, to Ffestiniog, is within the area reproduced in Figure 3 from Sedgwick's copy of J. & A. Walker's map of North Wales[28] (Barrett, 1974: 156 n. 35). Darwin used an edition of the map in 1842[29] and most probably in 1831 as well.

### Siabod

Darwin's journey began with the ascent of Moel Siabod (Figure 4), quite possibly not his first.[30] From the summit in reasonable weather, as on 22 August ("a very fine day indeed"[10] at Barmouth), he would have little needed either "compass or map". The plunging line

Figure 4. Fanny Wedgwood's sketch[53] looks up the Lledr valley to Moel Siabod, climbed by her Darwin cousin a few weeks before, Holyhead road visible on right. (Reproduced by courtesy of the Trustees of the Wedgwood Museum, Barlaston, Stoke-on-Trent, Staffordshire, England.)

of the col of Bwlch y Gorddinan directs the walker's eye to the shortest and easiest way south to Ffestiniog and, far beyond across a patchwork of fields and moor, to the crests of the Ardudwy mountains, the core of the Cambrian outcrop which became known as the Harlech Dome. There, if the day was clear enough, Darwin would have been able to make out, directly aligned with the col, the familiar summit of Diffwys[31] on the long ridge which goes down to Barmouth. East of the Ardudwy mountains the southern skyline is closed by the unmistakable outline of Cadair Idris, climbed by Darwin from Barmouth in 1828.[32]

### South of Siabod

For the ascent of Siabod, Sedgwick's and Darwin's notes correspond quite closely but Darwin has no counterpart to Sedgwick's notes on his steep descent to Llyn y Foel, "y[e] great Cwm with y[e] small lake on y[e] East side" (Burkhardt and Smith, 1985: 137) and back to the Inn. After the summit, his obligations to Sedgwick discharged, Darwin's brief and hurried notes point to the forward momentum of his journey[33]: "*South of Shiabed*. Trap, slate & a Feldspathic rock alternating ... for about 2 miles S of Shiabod ... *Dolyddelan*. A slate quarry apparently situated between the Feldspatic altered rock, with cleavage W 1 &½ S." (Barrett, 1974: 161).

Can the "slate quarry" be positively identified? Roberts (1998: 62) has Darwin proceeding from the summit of Siabod steeply down "in a south-south-easterly direction" to "the Dolwyddelan quarries", grid reference 743521, "to the south-east of the village", close to the alternative route to Ffestiniog by Sarn Helen. Other possibilities, more nearly approximating to "2 miles S of Shiabod", include Chwarel Ddu (SH 721521), "late 18th C. sporadic working until 1860s" (Richards, 1991: 68), and Hendre (SH 698512), "small pit working, circa 1840" (Richards, 1991: 75). The Hendre quarry is some two miles south of Siabod on the floor of the Lledr valley in Blaenau Dolwyddelan, close to Gorddinan and Hen-Gorddinan (marked on OS 1:25000 map sheet 17), from where a "rough track"[34] began its climb out of the Lledr valley up Bwlch y Gorddinan. From its summit, Walker's 1275 feet, the track went through to Ffestiniog. This was Darwin's obvious route but did he take it?

### "Northern Moelwyn"
The difficulty comes with the next, very condensed, passage in his notes: "From *this* [the slate quarry] *to Moelwyn* generally slate occasionally beds of Trap. *Northern Moelwyn.* Eastern side of it a blue slate. cleaving & line of violence SW. Dip N. Carreg y fran" (Barrett, 1974: 161).

If taken literally, this brief description covers nearly ten miles as the crow flies, south-west from the Dolwyddelan quarry to the summit of Moelwyn Mawr, well to the west of Bwlch y Gorddinan, then a "five miles ... side jaunt" (Barrett, 1974: 152, 154) east to Carreg y Fran across difficult terrain. Matters are simplified if the hypothesis of a single continuous journey is abandoned and Darwin allowed to visit Carreg y Fran from Ffestiniog on the following day but this seems psychologically unlikely, at odds with the headlong impression left by both his geological notes and his recollections of 45 years later.

The puzzle is the detour to Carreg y Fran (Rock of the Crow, SH 737449), away to the east of Bwlch y Gorddinan, separated from it by the mass of Manod Bach and Manod Mawr, and close to a different route from Dolwyddelan to Ffestiniog, "the Roman road called the Sarn Helen ... well known all over the district" (Thompson, 1983: 248). Given that Walker's map has no named summit in a wide area between Siabod and the Moelwynion, Darwin's "*Northern Moelwyn. Eastern side*" might include Allt-fawr, north east of Moelwyn Mawr (and at 2287 feet only 240 feet the lower), or perhaps even Bwlch y Gorddinan, a half hour's scramble down to the east from where access east again to Carreg y Fran is easy enough.[35] This would make for a reasonably coherent combination of the two routes but without helping to explain the reason for the detour.[36]

Another alternative is provided by the ingenuity of Roberts who has Darwin proceeding from Dolwyddelan along Sarn Helen, for this purpose converting *"Northern Moelwyn"* into a "Western Moelwynion" and contriving an "Eastern Moelwynion" in the area north of Walker's "Mannod" (Roberts, 1998: 63–64). With this bold stroke he completes the possible permutations: the direct route south by Bwlch y Gorddinan; the more easterly route along Sarn Helen; or some variation or combination of the two. Whether from Darwin's meagre notes his exact route through these "strange wild places" will ever be established with any degree of probability remains to be seen.

### Carreg y Fran
Either of two well used pre-turnpike trackways, from Penmachno to Ffestiniog and Trawsfynydd respectively (Bowen and Gresham, 1967: I, 246–248), could have taken

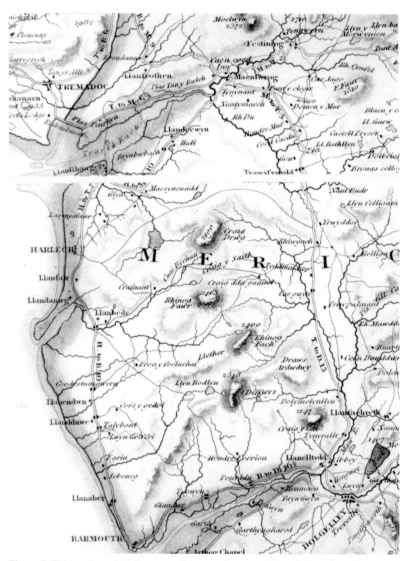

Figure 5. This section of Walker's map[28] shows the area of Darwin's walk on 23 August 1831 from "Festiniog" to Barmouth (including Rhinogs Fawr and Fach, (misplaced) "Draws Ardwdwy", Diffwys, Corsygedol and Llyn Bodlyn); also his route east with Robert Lowe through "Dolgelley" towards Mallwyd on the afternoon of 28 August. Reproduced by permission of Sedgwick Museum, Cambridge.)

Darwin from Carreg y Fran on the last lap of the journey to Ffestiniog. The more direct, the "tarmac lane" which Roberts (1998: 64) confuses with Sarn Helen, comes down Cwm Teigl. Sarn Helen itself, by then overlaid by the trackway to Trawsfynydd, passed further east to intersect the Bala–Ffestiniog turnpike above Ffestiniog.

## The Pengwern Arms, Ffestiniog

For Darwin's stay in Ffestiniog, three sources point to the same location. Samuel Holland –
later to purchase Plas Caerdeon from W. E. Jelf (Owen, 1960: 337), Darwin's landlord there
in 1869 – was in 1821 shown "the Hotel, the Pengwern Arms", where he was greeted
by "the Landlady of the Hotel, Martha Owen" (Davies, 1952: 6)). "In the evening",
Sedgwick noted for 12 June 1832, "drive to mine Hostess Martha Owen of Festiniog".[37]
And it was with "my kind old friend, Martha Owen the landlady" that Robert Lowe
stayed while tutoring at Ffestiniog in 1834 (Martin, 1893: **1**, 26). Most probably, it was
at The Pengwern Arms that Darwin stayed on 22 August and from which he set out for
Barmouth on 23 August.

## Bwlch Drws Ardudwy

Darwin's remaining geological notes relate to "*Drus Ardidy*", "W end of valley" (i.e.
Cwm Nantcol), and "road to Barmouth" (Barrett, 1974: 161–162). On the map in Roberts
(1998: 63), and also in Figure 5, a straight line from Ffestiniog to Barmouth bisects Bwlch
Drws Ardudwy. Once the detour to the coast road is eliminated[38], any likely route fits
well enough with the combination of convenient track and "straight line by compass and
map" of Darwin's later memory.

The notes confirm that he was approaching familiar territory. "One very long walk",
John Maurice Herbert recalled in his memoir of the Barmouth summer of 1828, took him,
Darwin and Thomas Butler "by Corsygedol, through Bwlch drws Ardudwy ... across a
boggy moor to ... Rhaidr du Waterfall", by which time Herbert "was so lame and dead-beat,
that Darwin and Butler had to support and almost carry me" the last six miles[39], "the
*dead march* to Dolgelley", as, far off in Rio, Darwin remembered it in 1832 (Burkhardt
and Smith, 1985: 241, and 67, 376).

Once across the Cynfal gorge from Ffestiniog, and up the other side to the turnpike road
(today's A470), Darwin could continue directly across the moorlands (as yet unencumbered
by reservoir, conifers or nuclear power station) to the gap which marked the entrance to
Bwlch Drws Ardudwy, the "wild & craggy pass between the Rhinogs Fach and Fawr"[10]
through which Robert Lowe came with Georgiana Orred on 3 September. "The rocks
that compose the hill through which the road passes", Darwin noted, "are of a most
complicated nature" (Barrett, 1974: 161). Along this "road", still today recognisably
the "narrow horse-path" of Pennant (1783: **2**, 113), Darwin came into Cwm Nantcol,
Sedgwick's "great cwm", "the scenery very beautiful".[40] From here on he would be
among scenes where he had walked, entomologised and fished in 1828 and on shorter
visits in 1829 and 1830.

## The "road to Barmouth"

"At W end of valley", Darwin noted the "remarkable regularity" with which the rock
was jointed, before, in the very last sentence of his notes, contrasting this with some
puzzling features on the "road to Barmouth"[41] (Barrett, 1974: 162). Probably he left Cwm
Nantcol by the road (now signposted to Dyffryn Ardudwy) which winds rounds the side
of Moelfre[42] with views across Cardigan Bay where in 1828 with Herbert he had landed
at low water on Sarn Badrig and, on another occasion, "meditated taking some long sea
voyage".[43] Soon there was a choice of routes, one downhill "by Corsygedol", another,
remoter and more tempting perhaps, across Pont Scethin near two lakes, Bodlyn upstream
steep under Diffwys and Erddyn (or Irrddyn) downstream.

## A clue from the Falklands?

On 16 March 1834 Darwin set out on a four-day excursion in the East Falkland Island (Keynes, 1988: 229). "In many parts of the island", he noted, "the valleys & hillsides are covered with an astonishing number of great angular fragments of quartz rocks ... spread out into level sheets or great streams ... the slope would not check the speed of an English mail coach."[44] Intrigued by the phenomenon he reminded himself: "Mem: ask Prof. Sedgwick. In N. Wales. near Barmouth. I remember being struck with a great sheet of angular fragments inclined at a small angle: it was near the borders of a lake. not far from a country house of Sir T. Mostyn."[45]

The "country house" was Corsygedol (Lloyd, 1977: 27); "Sir T. Mostyn" was Sir Thomas (1776–1831); and the "lake" was one of the "Corsygedol lakes" where Darwin and Herbert used to go "fly-fishing" in 1828[46], probably Erddyn set under Llawlech "in a wild scatter of rocks" (Condry, 1970: 120) in the bare, ice-swept valley of the Afon Ysgethin, within a mile of Pont Scethin across which coaches must have passed until the coast road to Barmouth was blasted through at the end of the eighteenth century. In the late afternoon did Darwin come this way, on the last lap of his walk to Barmouth, for a glimpse of his old haunts?

Whatever his exact route, as the sun sank across the Bay towards the hills of Llyn, Darwin came down to Barmouth, right on schedule for the "few days" there he had promised Whitley in July, his arrival noted by the Lowe brothers: "H [Henry] read the paper, & read from 6 till 7 ½ when he saw Darwin who had just arrived. He had walked from Festiniog *we hope he will stay till 1st September* ... Another splendid day."[10]

## BARMOUTH TO SHREWSBURY: 24–29 AUGUST

### 24–26 August

Darwin had reached Barmouth just in time. "Melancholy change in the weather", Sedgwick weather-bound at Llanberis, noted in his journal for 24 August.[7] Darwin comes into sight in the afternoon: "H on the hill above the town with Darwin and Whitley, rain came on, got rather wet ... sat awhile with Darwin at the Cors.y.gedol"[10], the first clue that he was staying at The Corsygedol Arms, "the town's leading hostelry" (Figure 6), its landlord William Barnett (Lloyd, 1993: 40, 43–46).[47]

"Barnett gave me a bad dinner", Herbert would complain in 1834 (Burkhardt and Smith, 1985: 376). One reason perhaps why, at least after Wednesday, Darwin took his evening meals with the Lowe brothers at their lodgings, enjoying the fine view of the Mawddach estuary and the cooking of their landlady, the bible-loving Mrs Jones (not to be confused with "Squinny ... Miss Jones ... 'as hot as love'", Darwin and Herbert's landlady in 1828 (Burkhardt and Smith, 1985: 67)). On Thursday, when Sedgwick first reached Capel Curig[48] and Darwin returned "much pleased" from Cwm Bychan[49], the brothers "dined 6 ¼, having waited for Darwin ... we gave him a quarter of an hour, but he arrived just as we finished, & put an end to a duck which, with some mutton chops, mushrooms, & apple pudding, had formed our repast."[10] On Friday, "Darwin dined with us at 6 on a leg of mutton, stewed cucumber, & a plum tart which set our teeth on edge considerably ... Afterwards ... Darwin and R discussed a bottle of Port."[10]

Figure 6. The "Inn" and the "hill". This print of Barmouth c. 1840[69] shows The Corsygedol Arms – its "imposing facade situated left of centre" above the beach with "the horse-drawn bathing machines ... an innovation of William Barnett" (Lloyd, 1974: 43) – and Cadair Idris across the estuary. (Reproduced by permission of National Library of Wales, Aberystwyth.)

## 27 August

On his last full day at Barmouth Darwin is in sight from "9 ½", when the brothers "breakfasted (Darwin with us)", to "10" in the evening when, after "some tea at the Cors.y.gedol ... H left Darwin".[10] After breakfast, Robert, Darwin and Beadon[50] "had a long converse on politics, in which Beadon was a good deal beat ... About 11 ... Darwin & R had a game of billiards; then strolled to the pierhead".[10] A planned ascent of Cadair Idris, as the weather cleared from the west, "the Carnarvonshire hills ... very clear", was deferred by a storm and "sandwiches prepared for the walk" were taken in the lodgings.[10] Frustrated by the weather from a further attempt on the mountain, "they played billiards till 4 oclock, when Darwin, Beadon, & H, strolled on the Sand Hills, & found some fungus, of a peculiar sort, which much pleased Darwin".[10] The sand hills were one of Darwin's old haunts, evoked in two of his *Beagle* letters (Burkhardt and Smith, 1985: 276, 349). Why Darwin was "much pleased" emerges from his letter to Whitley of 9 September: "there will be a paper published about the Fungus, all my conjectures were right. – If any more can be got, & put in gin, & sent to Shrewsbury: it will be capital" (Burkhardt and Smith, 1985: 151). Darwin, it seems, had already acquired his enduring talent for enlisting helpers: 12 September, "H found some more of Darwin's peculiar fungus which he put in spirits"; 20 September, "Sent a hamper to Darwin with some fungus in it"[10], the "Fungi" for which, in a postscript, Darwin thanked Whitley on 23 September (Burkhardt and Smith,

1985: 168; also 151 n. 3, 170).

Darwin's last evening meal in Barmouth "at 6 ¼" consisted of "some hashed mutton, stewed cucumber, & apple tart, after the discussion of w, Darwin & H strolled on the Harlech sands".[10] "R", for romantic reasons which will become apparent, was absent.

## 28 August

Darwin was "late" when the brothers "breakfasted ... at 9 ¼". Later, after "church", Henry found his brother and Darwin "at luncheon".[10]

Some at least of Darwin's late August departures from Barmouth were governed by the timetable for the Duke of Wellington coach which reached Shrewsbury through Mallwyd and Welshpool, starting from Barmouth on Wednesdays and from Aberystwyth on Mondays and Fridays.[51] Henry's entry shows the pattern holding good for 1831: "Saw Darwin & R start in a boat up the river on their way to Mallwydd *22 miles from here, 12 beyond Dolgelley, on the Shrewsbury road* Darwin to meet Aber.ys.twith coach tomorrow, & R to keep him company & return in the morning ... Rather fine in the morning, but clouded over again, & very windy."[10]

Darwin was travelling light: 30 August, "H ... dispatched a box to Darwin with his things"[10] – among them Saturday's gratifying fungus, no doubt also his geological hammer, in Robert's later memory weighing "14lbs." (Barrett, 1974: 149) – whose safe arrival Darwin reported to Whitley on 10 September (Burkhardt and Smith, 1985: 151).

After passing below the site of Caerdeon, where Darwin would stay in 1869, he and Lowe left the river and continued on foot over the Bwlch Oerddrws, the Pass of the Cold Gate, their passage described in the next day's entry: "The walk from Dolgelley to Dinas Mowddy is wild, & the vale about 6 miles from the former place particularly striking. Mallwydd is a small village with a good Inn, where Darwin and R arrived soon after 8, had some tea, & went to bed."[10] The "Inn" was The Cross Foxes, now The Brigands' Inn.

We come tantalisingly close to the conversation on the Bwlch Oerddrws. His father, Francis Darwin recalled, had "one or two stories about talks with strangers which I cannot remember distinctly however"; in particular "he used to speak of a long and pleasant talk he had with Bob Lowe on a coach in N Wales when they were both young men"[52] (the expedition to The Cross Foxes evidently conflated, in the memory of father or son, with the subsequent coach journey). One likely topic would have been the failings of a classical education. Darwin's dislike for the classics is well known, while Lowe –

> although an outstanding classical scholar himself ... came to detest classical studies in the schools. Dead languages, he believed, were less important than living modern languages, and living modern languages were less important than science. "To acquire the scientific habit of mind", he said "is the one invaluable thing in life" (Briggs, 1954: 267).

Was this a legacy of the journey to The Cross Foxes?

## 29 August

"This day last year", Darwin noted in his diary for 29 August 1832, "I arrived home from N. Wales & first heard of this Voyage ... it is amusing to imagine my surprise, if anybody on the mountains of Wales had whispered to me, this day next year you will be beating off the coast of Patagonia" (Keynes, 1988: 97). Henry sets the scene for that day: "R returned at 1 ½ ... In the morning they started 8 ½, on their different routes, Darwin towards Shrewsbury

& R back here ... A windy disagreeable day with rain occasionally".[10]

The coach times confirm the inference that Darwin started his journey on foot, boarding the coach somewhere before Welshpool, with its cluster of Clives at the vicarage and Powis Castle (where Fanny Wedgwood had sketched the previous August[53]). Two of the Powis Castle Clives would be among Dr Darwin's financial clients (Burkhardt and Smith, 1986: 84 n. 3). Another client, Mrs Winder[54], lived nearby at Vaynor Park[55], where Darwin and his sister, Caroline, stayed on a riding tour in 1826 before their wild passage over the Berwyns to Bala (Burkhardt and Smith, 1985: 538, 372). Seven years before that, when the Duke of Wellington coach was already running thrice weekly[56], he had taken the "stage Coach" on his way to the family holiday at Plas Edwards, Tywyn (Lucas, in press). The journey home along the Welshpool road must have been as full of memories as the outward journey along the Holyhead road 24 days before.

## LOWE'S DARWIN

To Darwin's habitually deprecating self-portrait – as in his summary of his "life at Cambridge, my time was sadly wasted there and worse than wasted" (Barlow, 1958: 60) – two of his Barmouth companions provide a salutary corrective. Herbert predicted his friend's "future fame" as early as April 1832: "you are about to couple your name, already intimately connected with science, with those of a Cuvier and a Humboldt. Don't think me guilty of Flattery – I know you will do great things, as it is impossible that your assiduity and talents should not succeed" (Burkhardt and Smith, 1985: 224). His testimony gains added weight from a more compelling witness.

Still not 20 in August 1831, Robert Lowe (Figure 7) had already spoken on the losing side in a debate in conservative Oxford dominated by Gladstone in his "first speech of compelling power" (Jenkins, 1995: 23).[57] In his Australian years Lowe can be glimpsed through Huxley's eyes in the house high above the Pacific, where "Georgiana, handsome in a velvet gown of deep red, presided over the long oak table" (Knight, 1966: 140). "I have met with many of the best men of my time since", Huxley told Lowe's biographer 45 years later, "but I have never listened to better talk than at that table" (Desmond, 1998: 703 n. 30). Though he held two of the great offices of state in Gladstone's first administration of 1868–1874, the high point of Lowe's political life had come earlier, as the dominating figure among the "Adullamites" who in 1866 frustrated the second Whig Reform Bill: "over several months Lowe put up an inspired performance against any advance to democracy" (Jenkins, 1995: 443) in "a debate which for sheer intellectual quality is generally agreed to have been unrivalled in modern times" (Blake, 1966: 441).[58]

When Darwin, his brother's friend, reached Barmouth, Lowe's attentions were fixed in quite another direction. It had been only a fortnight before, at a ball at The Corsygedol Arms, "rather a sticky affair ... got to bed before 2"[10], that he met the girl he was to marry. On Darwin's last day in Barmouth, when he and Henry dined at "6 ¼", they had "waited for R, who never came till 7 ¼, having walked out with the Miss Orreds at 4 *to Llanaber where the ladies sketched*"[59]; later that evening "R returned from the Miss Orreds, where he drank tea, at 11 ¼".[10] While Georgiana sketched, Robert composed the verses on "Llanaber Churchyard"[10] which he recited 40 years later, driving past with his Cabinet colleague, the First Lord of the Admiralty (Martin, 1893: **II**, 399). Yet the day after the romantic interlude in the churchyard he broke off his pursuit to accompany Darwin

Figure 7. From Darwin's tomb in Westminster Abbey to the marble bust of Robert Lowe, Viscount Sherbrooke, in the east porch of St Margaret's, Westminster is a walk of only a minute or so. So short a distance separates memorials of two famous dead whose paths had crossed at Barmouth in the days of their youth. (Author's photograph.)

to The Cross Foxes –

> I am proud to remember that though quite ignorant of physical science, I saw a something in him which marked him out as superior to anyone I ever met; the proof of which I gave of this was somewhat canine in nature, I followed him. I walked twenty-two miles with him when he went away, a thing which I never did for anyone else before or since (Barrett, 1974: 149).

This "early hero-worship", as it was described to Huxley[60], can be glimpsed in Whitley's report to Darwin of a visit to the Lowes in 1834 (Burkhardt and Smith, 1985: 428). The spell was to be lifelong. "Of Charles Darwin", Jowett wrote in 1893 in his memoir of Lowe, "he spoke in a strain of respect which he would not have employed towards any other living person" (Martin, 1893: **II**, 497).

## THE TOUR IN RETROSPECT: 1842

"How odd it is", Darwin wrote to Henry Fawcett in 1861, "that anyone should not see that all observation must be for or against some view if it is to be of any service!" (Burkhardt *et alii*, 1994: 269). This was nowhere better illustrated for Darwin than by the tour with Sedgwick: this was the revelation of his return to North Wales in 1842.

It was the ice which on 18 June drew Darwin back to Capel Curig (Burkhardt and Smith, 1986: 321–322, 435). Around 22 June he made his return to Cwm Idwal, before going down Nant Ffrancon where he had ridden with Erasmus 21 years before, and on 26 June, between Caernarfon and Llanberis, he climbed Moel Tryfan, his last Welsh summit.[61] On 29 June "Ch. came fr Wales"[62], "the last time I was ever strong enough to climb mountains or to take long walks, such as are necessary for geological work" (Barlow, 1958: 99).

Darwin's lasting fascination with the legacy of the ice, "it was a kind of geological novel" (Burkhardt and Smith, 1986: 387), comes out in his account of the return to Cwm Idwal[63] –

> neither of us[64] saw a trace of the wonderful, glacial phenomena all around us; we did not notice the plainly scored rocks, the lateral and terminal moraines. Yet these phenomena are so conspicuous that ... a house burnt down by fire did not tell its story more plainly than did this valley.[65] If it had still been filled by a glacier, the phenomena would have been less distinct than they are now (Barlow, 1958: 70).

George Darwin remembered the grip of the ice on his father's imagination –

> Speaking of ... the neighbourhood of Moel Trifaen[66] I think, he said that they [his father and Sedgwick] were surrounded by every mark of glaciation – "There were the moraines, the scored rocks & perched boulders, and I declare to heaven – if the glaciers had been there the evidence would have been less complete. And yet we never saw anything of it".[67]

"By this same tour", George Darwin added, "he used to illustrate the uselessness of observations without some framework of theory on which to hang them."

## CAERDEON: 1869

Darwin's last visit to North Wales, at Plas Caerdeon above the Mawddach estuary, verges on the tragic. "My father was ill and somewhat depressed throughout this visit", Francis Darwin remembered, "and I think felt saddened at being imprisoned by his want of strength, and unable even to reach the hills over which he had once wandered for days together" (Darwin, 1887: **III**, 106). Darwin's letter to J. D. Hooker of 24 July 1869 marks his farewell to so many of the scenes of his youth: "we shall be at home this day week ... & right glad I shall be ... I loathe the place, with all its beauty" (Darwin and Seward, 1903: **I**, 313).

## CONCLUSION

The first and last weeks of Darwin's 25 day absence from Shrewsbury, which began with the tour with Sedgwick, are quite fully documented. It is for the intermediate period, including the visit to Anglesey to which the circumstantial evidence points, that further consideration is most needed. The obvious starting point is Darwin's annotations in his copy of Henslow (1822) and his manuscript geological notes from the years of the *Beagle* voyage.

The tour with Sedgwick was just one of many visits which Darwin made to North Wales. Through archival records of whose richness these pages can offer only glimpses, the visits give us snapshots of Darwin over more than half a century.

## ACKNOWLEDGEMENTS

Quotations are by permission: of the Syndics of Cambridge University Library for the Darwin and Sedgwick papers in the Library; of the Sedgwick Museum, Cambridge for Sedgwick's journal and transcript; of the

Principal Archivist, Nottinghamshire Archives for the Lowe brothers' journal; and of the Archives of Imperial College of Science, Technology and Medicine, London for the T. H. Huxley papers. Quotations from the Wedgwood archives in Keele University Library are by courtesy of the Trustees of the Wedgwood Museum, Barleston, Stoke-on-Trent, Staffordshire, England. Particular thanks are due to Geoffrey Waller, also to Perry O'Donovan and John Wells, at Cambridge University Library; Rod Long at the Sedgwick Museum; and Martin Hughes and Helen Burton at Keele.

## NOTES

[1] Darwin's copy is in the Darwin Library, Cambridge University Library (CUL).

[2] CUL, Darwin Papers 265: 1, 5 (tracing of 1); most of the place names are inscribed in a hand, as yet unidentified, other than Darwin's.

[3] CUL, Sedgwick Papers, Add. 7652: 1A/56 1, 4–6.

[4] CUL, DAR 265: 3.

[5] CUL, DAR 210:11.37.

[6] CUL, DAR 265: 2, 4.

[7] Quotations from Sedgwick's journal during the tour are from Professor Owen T. Jones's transcript (it gives out on 15 September) of Journals XXI and XXII, 1831 nos I (2–29 August) and II (30 August–7 October); transcript and journals both in Sedgwick Museum, Cambridge.

[8] Darwin's notes, CUL, DAR: B5–14, are transcribed in Barrett (1974: 155–162). B3–B4, identified by Roberts (2001: 36) as a list of rocks on Angelsey, may – like ff B1–B2 (Llanymynech) and the four cloth maps – rank among the preliminaries to the tour.

[9] Near the end of his life Sedgwick (1873: xv) recalled the early days of the tour and the establishment of a "provisional base-line on the Caernarvon side of the Straits".

[10] "Journal kept jointly ... during 3 months of the summer 1831. at Barmouth, North Wales", Nottinghamshire Record Office, DDSK 218/1. The daily (unpaginated) entries are mostly the work of the better sighted Henry. Italicised quotations denote interpolations in Robert's large round hand.

[11] Dr R. W. Darwin's account books 1815–1831, CUL, DAR 227. 5: 82.26 and 52; 1831–1848, DAR 265: 9.11; also Browne (1995: 8).

[12] Dr R. W. Darwin to Josiah Wedgwood, letter of early July, Keele University Library, Wedgwood Archive E35–26594; henceforth cited as KUL, WA.

[13] Dr R. W. Darwin to Josiah Wedgwood, letter of 19 July, KUL, WA E35–26597.

[14] On 7 September 1822 Sarah Ponsonby lamented that the Doctor "had a second time passed through Llangollen, without bestowing even a few moments upon us", CUL, DAR 227.6: 370. Another chance reference shows the Doctor about to set off "to see a Lady ill by Bangor", letter to Josiah Wedgwood, 20 July 1828, KUL, WA E35–26686.

[15] Glascoed is marked on modern OS maps, 'Cross Foxes' on the OS one-inch first edition (sheet 25 of the David & Charles reprint).

[16] Wyndham "traversed a mountainous country for the space of 167 miles upon Welsh hackneys, hired from place to place ... The little horses ... were exceedingly hardy, and ... would constantly travel with heavy burdens forty miles a day, even without ... a single feed of corn ... [They] are no sooner disengaged from their saddles, than they are turned into a common pasture for the night ... Every inn in the country is provided with a paddock for this purpose" (Wyndham, 1781: 140–141).

[17] See Trueman (1971: 261, figure 34, "the geology of the Conway region").

[18] The map in Secord (1991: 146) shows him reaching both limestone promontories.

[19] Recollections of his father, CUL, DAR 112: B7 (and 46).

[20] Some of Sedgwick's Dublin contacts can be glimpsed in the letter of 25 November 1831 from Humphrey Lloyd (Trinity College, Dublin) about "a Geological Society in Dublin", CUL, Sedgwick papers, Add. 7652: IA/46.

[21] Roberts (1998: 61, 72 n. 1) sees "clear evidence that one page has been removed ... the excised page is visible on fol 10ii", yet removal of a separate sheet would not involve excision.

[22] CUL, Sedgwick Papers Add. 7652: listed between letters of 23 June and 10 September 1831 in "list of letters 1–145" in folder 1A/1–25. The archive was transferred from the Department of Geology in 1964, by which time most of the letters from Darwin, Lyell and Murchison had been removed.

[23] Diary of observations on the geology of the places visited during the voyage, CUL, DAR 33: 217–222; Barratt (1977: **I**, 211 n. 7).

[24] Diary of observations on the geology of the places visited during the voyage, CUL, DAR 32: 1.19; further evidence noted by Roberts, 2001.

[25] The *Rothsay Castle* went down in the early hours of 18 August (Usher, 1955: 5–9). News of the disaster reached the Lowes at Barmouth the next day: "We heard today of the loss of the Rothsay castle, steamer betn Liverpool & Bangor, off the Orme Head; out of 150 persons, only 15 escaped ... It is the same packet we were to have come in; but she was out of order."

[26] Darwin's 1842 geological notes, CUL, DAR 27: 1. B7 (verso).

[27] Copy in Gwynedd Archives, Caernarfon, XM/5171.

[28] Sedgwick Museum, Cambridge. Other useful maps include Dawson's "ordnance" maps provided for Sedgwick in 1832, marked "Beddgelert", "Tremadoc, Harlech Barmouth" and "Drws Ardudwy to Barmouth" (Sedgwick Museum, Cambridge); sheets 24 and 31 of the David & Charles reprint of the first edition of the "Old Series" one-inch Ordnance Survey of England and Wales, for which Dawson was chief surveyor in North Wales; and the modern OS 1:25 000 "Snowdonia outdoor leisure" sheets 17 and 18.

[29] CUL, DAR 27: 1. B2 (verso) and 21.

[30] Siabod he had very likely climbed on the walking tour in 1826. Snowdon, its immediate western neighbour (which Sedgwick was ascending on 22 August), he had climbed in 1826 (Barlow, 1958: 53–54) and August 1830 (Burkhardt and Smith, 1985: 106). In 1833 in South America he would be reminded of the summit cairns "so common in the Welsh Mountains" (Keynes, 1988: 159).

[31] Darwin's copy (Darwin Library, CUL) of Stephens (1829), listing some of his entomological trophies, has for 1829 four references to "Dyffous" (or some cognate spelling), two of them in combination with "Cader Idris".

[32] J. M. Herbert's memoir, CUL, DAR 112: B60.

[33] After the notes on the summit, there is just one sentence on "the beds composing N 1 of mountain", read as "NE" in Roberts (1998: 62), before Darwin continues south (Barrett, 1974: 161).

[34] "Pedestrians and pack animals made use of a rough track through Bwlch Gorddinan ... until 1846–7 when the present road up the Lledr valley was built" (Pritchard, 1961: 78).

[35] For example by the quarry track going off to the east on Dawson's "Beddgelert" map.

[36] To speculate would be only too easy. If, for example, it was on horseback, by Bwlch y Gorddinan or Sarn Helen, that Darwin went from Capel Curig to Barmouth the previous August, this might have influenced his route a year later, as might his visit to Ffestiniog in 1826 (Lucas, in press).

[37] Sedgwick Museum, Sedgwick's journal, XXIII, 1832 No. I, 48.

[38] See note 42 below.

[39] Herbert's memoir, CUL, DAR 112: B60–61.

[40] Sedgwick Museum, Sedgwick's journal, 13 July 1832, XXIV, 1832 No. II.

[41] These are the last of the as yet unlocated geological clues to Darwin's route from Capel Curig to Barmouth.

[42] Roberts (1998: 65) has Darwin descending the valley by the (not yet built) "Pont Cwmnantcol" down to the coast road at Llanbedr, an unnecessary detour.

[43] Herbert's memoir, CUL, DAR 112: B58, 62–63.

[44] Diary of observations on the geology of the places visited during the voyage, CUL, DAR 33: 203–205.

[45] CUL, DAR 33: 209 (verso), drawn to my attention by Sandra Herbert.

[46] Herbert's memoir, CUL, DAR 112: B58.

[47] This was the "Inn", and "hill", of Darwin's nostalgic letter to Herbert of 2 June 1833 (Burkhardt and Smith, 1985: 319–320).

[48] Further illustrating that it cannot have been at Capel Curig that he and Darwin separated.

[49] Cwm Nantcol's northern neighbour which Darwin visited with Herbert in 1828 on one of those Saturdays on which they "usually took longer excursions", CUL, DAR 112: B60.

[50] Identified in Burkhardt and Smith (1985: 151 n. 4, 613).

[51] Most obviously in 1828 when Darwin fixed his departure on Wednesday 27 August well in advance (Burkhardt and Smith, 1985: 63; Burkhardt and Smith, 1991: 466). The coach details can be extracted from Pigot (1828–1829) and from the *Salopian journal*.

[52] Recollections of his father, CUL, DAR 140(3): 69. This time was surely also in Darwin's mind when, in the context of the American Cross Fox, he made the marginal annotation "on Cross Foxes" noted in di Gregorio and Gill (1990: xvii).

[53] KUL, WA, W/M 1165.

[54] Dr R. W. Darwin's Accounts Books 1790–1815, CUL, DAR 227.5: 38.25; 1815–1831, DAR 265: 82.14.

[55] Where in Pigot (1828–1829: 1176) she is listed among the "nobility, gentry and clergy" of "Welchpool".

[56] *Salopian journal* (9 June 1819).

[57] Another who spoke for the victorious Tories was W. E. Jelf, Darwin's landlord at Caerdeon in 1869, House of Lords Record Office Office, Lowe papers: 386/106(b).

[58] Gladstone did not forget Lowe's opposition to him in 1866. "So effective were his speeches that, during this year only, he had such a command of the House as had never in my recollection been surpassed ... his position was one for the moment of personal supremacy ... I pressed his viscountcy on the Sovereign ... a man who had once soared to those heights trodden by so few ought not to be lost in the common ruck of official barons" (Brooke and Sorensen, 1971: 92).

[59] I have been unable to trace any of Georgiana's Barmouth sketches: some of her Australian sketches, from among the happier years of her marriage, are in the Mitchell Library, Sydney, Australia.

[60] Letter from A. P. Martin, 13 November 1893, Huxley Papers, Imperial College, London, 22.174.

[61] The tour can be followed in Darwin's (not fully dated) geological notes, CUL, DAR 27:1.B1–21.

[62] Emma Darwin's pocket diary, CUL, DAR 242: 8.

[63] Similarly in the letter memoir of 24 May 1875 (Clark and Hughes, 1890: I, 381).

[64] "Us" including Sedgwick, mistakenly remembered as present.

[65] An image used also in Darwin (1859: 366).

[66] This must be Tryfan east of Cwm Idwal, not the lesser summit which Darwin climbed on 26 June 1842.

[67] CUL, DAR 112: B7.

[68] To Fanny Mackintosh, KUL, WA, W/M 194. "We" are the "Welsh party" of Charlotte Wedgwood's letter to Darwin of 22 September 1831 (Burkhardt and Smith, 1985: 166); it included her three youngest siblings – Hensleigh, Fanny and Emma – and their mother, Bessy. The party's route overlapped that taken by Sedgwick and Darwin the previous month. Hensleigh's letters offer glimpses of Darwin, and of his future wife.

[69] National Library of Wales, Aberystwyth, P4335: Top. B9: B. 19.

## REFERENCES

BARLOW, N. (editor), 1958 *The autobiography of Charles Darwin 1809–1882.* London: Collins. Pp 253.

BARRETT, P. H., 1974 The Sedgwick–Darwin geologic tour of North Wales. *Proceedings of the American Philosophical Society* **118**: 146–164.

BARRETT, P. H. (editor), 1977 *The collected papers of Charles Darwin.* Chicago: University of Chicago Press. 2 volumes.

BLAKE, R., 1966 *Disraeli.* London: Methuen. Pp 819.

BOWEN, E. G. and GRESHAM, C. A., 1967 *History of Merioneth.* Volume **1** *From the earliest times to the age of the native princes.* Dolgellau: Merioneth Historical and Record Society. Pp 298.

BRIGGS, A., 1954 *Victorian people: some reassessments of people, institutions, ideas and events, 1851–1867.* London: Oldhams Press. Pp 317.

BROOKE, J. and SORENSEN, M. (editors), 1971 *The Prime Ministers' papers: W. E. Gladstone* **1**: *autobiographical.* London: Her Majesty's Stationery Office. Pp 263.

BROWNE, J., 1995 *Charles Darwin, voyaging.* London: Jonathan Cape. Pp 605.

BURKHARDT, F. H. and SMITH, S. (editors), 1985 *The correspondence of Charles Darwin*, **1** (1821–1836). Cambridge: Cambridge University Press. Pp 702.

BURKHARDT, F. H. and SMITH, S. (editors), 1986 *The correspondence of Charles Darwin*, **2** (1837–1843). Cambridge: Cambridge University Press. Pp 603.

BURKHARDT, F. H. and SMITH, S. (editors), 1991 *The correspondence of Charles Darwin*, **7** (1858–1859, supplement 1821–1857). Cambridge: Cambridge University Press. Pp 671.

BURKHARDT, F. H. *et alii* (editors), 1994 *The correspondence of Charles Darwin*, **9** (1861). Cambridge: Cambridge University Press. Pp 609.

BURKHARDT, F. H. *et alii* (editors), 1999 *The correspondence of Charles Darwin*, **11** (1863). Cambridge: Cambridge University Press. Pp 1080.

CLARK, J. W. and HUGHES, T. M., 1890 *The life and letters of the reverend Adam Sedgwick.* Cambridge: Cambridge University Press. 2 volumes.

CLOSE, C., 1969 *The early years of the ordnance survey.* New York: Augustus M. Kelly. Pp 164.

CONDRY, W. M., 1972 *Exploring Wales.* London: Faber & Faber. Pp 304.

DARWIN, C., 1859 *On the origin of species by means of natural selection, or the preservation of favoured races in the struggle for life.* London: John Murray. Pp 502.

DARWIN, F., 1887 *The life and letters of Charles Darwin.* London: John Murray. 3 volumes.

DARWIN, F, and SEWARD, A. C. (editors), 1903 *More letters of Charles Darwin: a record of his work in a series of hitherto unpublished letters.* London: John Murray. 2 volumes.

DAVIES, W. L. (editor), 1952 *The memoirs of Samuel Holland one of the pioneers of the North Wales slate industry.* The Merioneth Historical and Record Society Extra Publications **1**, 1. Bala: R. Evans & Son. Pp 32.

DESMOND, A., 1998 *Huxley: from devil's disciple to evolution's high priest.* London: Penguin Group. Pp 820.

DI GREGORIO, M. A. and GILL, N. (editors), 1995 *Charles Darwin's marginalia.* Volume **1**. New York: Garland. Pp 895.

FISHER, J. (editor), 1917 Richard Fenton Tours in Wales (1804–1813). *Archaeologia Cambrensis* **17** (sixth series) supplement: 1–371.

HARLEY, J. B. and OLIVER, R. R., 1992 *The old series ordnance survey maps of England and Wales Scale: 1 inch to 1 mile Vol VI Wales.* Lympne Castle, Kent: H. Margary.

HENSLOW, J. S., 1822 Geological description of Anglesea. *Transactions of the Cambridge Philosophical Society* **1**: 359–452.

HERBERT, S., 1991 Charles Darwin as a prospective geological author. *British journal for the history of science* **24**: 159–192.

HERBERT, S. and ROBERTS, M. B., 2002 Charles Darwin's notes on his 1831 geological map of Shrewsbury. *Archives of natural history* 29: 27–29.

HERSCHEL, J. F. W., 1831 *A preliminary discourse on the study of natural philosophy.* London: Longman, Rees, Orme, Brown, & Green; & John Taylor. Pp 372.

JENKINS, R., 1995 *Gladstone.* London: Macmillan. Pp 698.

JONES, E. G. (editor), 1952 *A description of Carnarvonshire (1809–11) by Edmund Hyde Hall.* Caernarvon: G. Evans. Pp 383.

KEYNES, R. D. (editor), 1988 *Charles Darwin's Beagle diary.* Cambridge: Cambridge University Press. Pp 464.

KNIGHT, R., 1966 *Illiberal Liberal: Robert Lowe in New South Wales, 1842–1850.* Carlton, Victoria: Melbourne University Press. Pp 299.

LLOYD, L., 1974 *Maritime Merioneth: the town and port of Barmouth (1565–1973).* Llanfair: L. Lloyd. Pp 124.

LLOYD, L., 1977 Corsygedol, Ardudwy's principal estate. *Journal of the Merioneth Historical and Record Society* **8**: 27–60.

LLOYD, L., 1993 *Wherever freights may offer: the maritime community of Abermaw/Barmouth 1565 to 1920.* Caernarfon: Gwasg Pentecelyn. Pp 384.

LUCAS, P., 2002 (in press) "Three weeks which now appears like three months": Charles Darwin at Plas Edwards, 1819. *National Library of Wales journal* 33.

LUCAS, P. (in press) Jigsaw with pieces missing: Charles Darwin with John Price at Bodnant, the walking tour of 1826 and the expeditions of 1827.

MARTIN, A. P., 1893 *Life and letters of the Rt. Hon. Robert Lowe, Viscount Sherbrooke.* London: Longmans, Green. 2 volumes.

MAVOR, E., 1973 *The ladies of Llangollen: a study in romantic friendship.* Harmondsworth: Penguin Books. Pp 242.

OWEN, H. J., 1960 Caerdeon chapel. *Journal of the Merioneth Historical and Record Society* 3: 331–342.

PENNANT, T., 1783( "1781") *The journey to Snowdon*, Book **2** in *A tour in Wales*, Volume **2**. London: H. Hughes. (3 books in 2 volumes.)

PIGOT AND CO., 1828–1829 *Pigot and Co.'s national commercial directory for 1828–9; comprising a directory and classification of the merchants, bankers, professional gentlemen, manufactures and traders, in [specified counties in England] And in the whole of North Wales.* London & Manchester: Pigot & Co. Pp 1180.

PRITCHARD, R. T., 1961 Caernarvonshire turnpike trusts: Ysbyty and Penmachno Trusts. *Transactions of the Caernarfonshire Historical Society* 22: 76–80.

PUGH, E., 1816 *Cambria depicta.* London: W. Clowes. Pp 476.

RICHARDS, A. J., 1991 *A gazeteer of the Welsh slate industry.* Capel Garmon, Llanrwst: Gwasg Carreg Gwalch. Pp 239.

ROBERTS, M. B., 1996 Darwin at Llanymynech: the evolution of a geologist. *British journal for the history of science* **24:** 133–157.

ROBERTS, M. B., 1998 Darwin's dog-leg: the last stage of Darwin's Welsh field trip of 1831. *Archives of natural history* **25:** 59–73.

ROBERTS, M. B., 2000 I coloured a map; Darwin's attempts at geological mapping in 1831. *Archives of natural history* **27:** 69–79.

ROBERTS, M. B., 2001 Just before the *Beagle*: Charles Darwin's fieldwork in Wales, summer 1831. *Endeavour* 25: 33–37.

SECORD, J. A., 1991 The discovery of a vocation: Darwin's early geology. *British journal for the history of science* **24:** 133–157.

SEDGWICK, A., 1873 Preface, pp i–xxxiii, in SALTER, J. W. *Catalogue of Cambrian and Silurian fossils.* Cambridge: Cambridge University Press.

STEPHENS, J. F., 1829 *A systematic catalogue of British insects: being an attempt to arrange all the hitherto discovered indigeneous insects in accordance with their natural affinities.* London: Baldwin & Cradock. 2 parts in 2 volumes.

THOMPSON, M. W. (editor), 1983 *The journeys of Sir Richard Colt Hoare through Wales and England 1793–1810.* Gloucester: Alan Sutton. Pp 288.

TRUEMAN, A. E., 1971 *Geology and scenery in England and Wales.* Harmondsworth: Penguin Books. Pp 400.

USHER, G., 1955 An Anglesey disaster. *Transactions of the Anglesey Antiquarian Society and Field Club* *1955*: 2–11.

WEDGWOOD, B., and WEDGWOOD, H., 1980 *The Wedgwood circle 1730–1897: four generations of a family and their friends.* London: Studio Vista. Pp 386.

WYNDHAM, H. P., 1781 *A tour through Monmouthshire and Wales, made in the months of June, and July, 1774. And in the months of June, July, and August, 1777.* 2nd ed. Salisbury: E. Easton. Pp 214.

Received: 11 April 2000. Accepted: 10 July 2000.

## APPENDIX

DARWIN'S KNOWN AND PROBABLE VISITS TO WALES

| | |
|---|---|
| 1813 | Family seaside holiday at Rhos-on-Sea (Barlow, 1958: 21–22; Burkhardt and Smith, 1986: 438) |
| 1819 | Family seaside holiday at Plas Edwards, Tywyn (south of Barmouth) (Lucas, 2002) |
| 1820 | To Pistyll Rhaiadr (on south-eastern slopes of Berwyns) on "old Dobbin" with Erasmus (Burkhardt and Smith, 1986: 441) |
| 1821 | To Bangor Ferry on horseback with Erasmus (notes 12 and 13 above) |
| 1822 | Riding tour to Montgomery (eight miles from Welshpool) with "Elizabeth" (Burkhardt and Smith, 1985: 538) |
| 1826 | 1. Walking tour through North Wales including Bodnat, Ffestiniog and Snowden (Lucas, in press) |
| | 2. Riding tour to Vaynor Park and Bala with sister, Caroline (Barlow, 1958: 54; Burkhardt and Smith, 1985: 538) |
| 1827 | 1. Holyhead to Shrewsbury on return from Dublin (Burkhardt and Smith, 1985: 538–539) |
| | 2. To John Price at Bodnant with Erasmus, expeditions to Great Ormes Head and Carneddau (Lucas, in press). |
| 1828 | Summer vacation reading party at Barmouth (Barlow, 1958: 58; Burkhardt and Smith, 1985: 539) |
| 1829 | 1. Abortive visit to Barmouth (Burkhardt and Smith, 1985: 87–88) |
| | 2. Barmouth with sisters (Burkhardt and Smith, 1985: 91) |
| 1830 | Capel Curig and Barmouth (Burkhardt and Smith, 1985: 105–106) |
| 1831 | The tour with Sedgwick (Barlow, 1958: 69–71; Burkhardt and Smith, 1985: 540) |
| 1842 | Ten days based at Capel Curig (Barlow, 1958: 99; Burkhardt and Smith, 1986: 435) |
| 1869 | Family holiday at Plas Caerdeon near Barmouth (Darwin, 1887: **III**, 106) |

*J. Soc. Biblphy nat. Hist.* (1980) 9 (4): 515–525

# Charles Darwin's plant collections from the voyage of the *Beagle*

By DUNCAN M. PORTER

Department of Biology,
Virginia Polytechnic Institute & State University,
Blacksburg,
Virginia 24061, U.S.A.

The much-celebrated voyage of H M S *Beagle* began at Devonport, after several false starts, on 27 December 1831. The *Beagle* returned to England at Falmouth on 2 October 1836, having been gone four years and 279 days, a long trip even for those times. During this extraordinary surveying voyage, Charles Darwin (1809–1882), as the *Beagle's* naturalist, made many geological, zoological, and botanical collections and observations.

Darwin had received training in the natural sciences during his two years at Edinburgh University (1825–1827), and his interest therein lay mainly in zoology, particularly with marine invertebrates. This training continued at Cambridge University (1828–1831), where one professor stands out in his influence on the young Darwin: the Reverend John Stevens Henslow (1796–1861), Professor of Botany. Darwin attended Henslow's lectures on botany, 'and liked them much for their extreme clearness, and the admirable illustrations; but I did not study botany.'[1]

Henslow's influence was not in teaching Darwin to be a botanist, but rather as a friend who introduced him to the wider world of science and scientists. Following his graduation from Cambridge,[2] it was Henslow who persuaded him to study geology, and who asked Adam Sedgwick (1785–1873), the Woodwardian Professor of Geology, if Darwin might accompany him on a geological field trip to North Wales. Darwin had not attended Sedgwick's 'eloquent and interesting' lectures because of his experience with the 'incredibly dull' lectures on geology and zoology of Robert Jameson (1774–1854), Regius Professor of Natural History at Edinburgh.[3]

Darwin undertook his field studies in Wales with Sedgwick just prior to embarking on the *Beagle,* and throughout the voyage (and afterward) referred to himself as a geologist.[4] His primary interest in geology on the voyage is borne out by the voluminous geological notes and large number of collections he made. The geological notes total 1383 pages, almost four times as many as those on zoology (368 pages).[5] Reading Lyell's *Principles of geology*[6] had an acknowledged profound effect on Darwin during the voyage. The first volume was given to him prior to their leaving by Captain Robert Fitz-Roy (1805–1865), the second was sent to him by Henslow in 1832.

When one reads Darwin's notes, letters, and diary written on the voyage, which are variously at the Cambridge University Library, Down House, and the Royal Botanic Gardens, Kew, one cannot help but gain the impression that he was most interested in geology, somewhat so in zoology, and least so in plants. In a letter to Henslow from Rio de Janeiro early in the voyage (18 May 1832), he writes: 'Geology & the invertebrate animals will be my chief object of pursuit through the whole voyage.'[7]   Indeed, in his autobiography[8] Darwin stresses investigating the geology of all places visited and of collecting animals, but he is silent on the subject of botany.

There are no known separate series of notes on plants like those that he made for geology and zoology now at the Cambridge University Library. The only notations concerning his plant collections, besides remarks in his diary and letters, are to be found in his field note-books, where they are intermingled serially with the notations regarding zoological and geo-logical specimens also collected on his journeys. These small pocket notebooks, which Darwin carried with him on his travels to record specimens collected and observations, are now at Down House.

'The note-books contain mainly geological notes, varying from about half to as much as nine-tenths of the entries in different books.'[9] There are twenty-two notebooks in total.[10] Fourteen deal with inland travels, two are rough drafts of geological papers and miscellaneous notes, and six are catalogues of specimens collected during the voyage. Three or four addit-ional notebooks, including those from Australia and New Zealand, appear to have been lost.[11]

The field notebooks served as the source materials for Darwin's more expanded geological and zoological notes, discussed above. These notes, in turn, served as the bases for the five volumes published on the zoology of the *Beagle,*[12] and the three volumes on the geology.[13] More recently, they have provided us with Darwin's ornithological notes.[14]

Lest the reader gain the idea from the foregoing that Darwin entirely neglected botany on the voyage of the *Beagle*, rest assured that he did not. In the beginning it may be said that he did not pay especial attention to plant collecting. Yet even at two of the first landfalls, Santiago in the Cape Verde Islands and the Abrolhos Archipelago off Brazil, he estimated that he collected nearly all the flowering plants of these islands coincident with his geologiz-ing.[15] Darwin began to collect plants in ernest only following a letter from Henslow of 15 January 1833,[16] which was sent after his first plant specimens and seeds reached Cambridge. In this letter, Henslow gave Darwin some sound advice on collecting and pressing plant speci-mens.

Darwin previously wrote (15 August 1832): 'As for my plants, 'pudet pigetque mihi [it shames and disgusts me] .' All I can say is that when objects are present which I can observe & particularize about, I cannot summon resolution to collect where I know nothing.'[17] In answer, Henslow stated in his letter of 15 January: 'Most of the plants are very desirable to *me.*'[18] Darwin, however, did not receive this answer until July 1834, a year-and-a-half after it was written.

In a later letter (31 August 1833), following the arrival of further specimens, Henslow wrote: 'The plants delight me exceedingly, tho' I have not yet made them out — but with Hooker's work and help I hope to do so before long'.[19] Hooker was William Jackson Hooker (1785–1865), at that time Regius Professor of Botany at Glasgow University, who, as we shall see, played a vital role in our story. To this, Darwin answered (March 1834): 'I am very glad the plants gave you any pleasure, I do assure you I was so ashamed of them, I had a great mind to throw them away; but if they give you any pleasure I am indeed bound, & will pledge myself to collect whenever we are in parts not often visited by Ships & Collec-tors.'[20]

The majority of the plant collections appears to have been made after this letter was penned, particularly in southernmost South America and the Galápagos Islands. Two months before arriving at these islands, Darwin wrote to his cousin William Darwin Fox (*ca* 1805–1880) from Lima, Peru (July 1835): 'I look forward to the Galapagos with more interest than any other part of the voyage.'[21] Although this enthusiasm was for geological and not biological reasons, Darwin certainly is better known today for his biological observations on the archipelago, rather than the geological.

In January 1836 he wrote Henslow from Sydney: 'I last wrote to you from Lima, since which time I have done disgracefully little in Nat. History; or rather I should say the Galapagos Islands, where I worked hard. – Amongst other things, I collected every plant, which I could see in flower, & as it was the flowering season I hope my collection may be of some interest to you.'[22] Darwin did an admirable job of collecting in these islands, particularly considering that it was *not* the best time of year for flowers. In six weeks, he gathered 173 taxa, twenty-four per cent of the presently-known flora.

Prior to the voyage, Henslow agreed to receive Darwin's collections as they were sent back from the *Beagle*. Darwin wrote him on 18 October 1831: 'I have talked to everybody: & you are my only resourse; if you will take charge, it will be doing me the greatest kindness. – The land carriage to Cambridge, will be as nothing compared to having some safe place to stow them; & what is more having somebody to see that they are safe. – I suppose plants & Birds-skins are the only things that will give trouble: but I know you will do what is proper for them.'[23] So, even before embarking, Darwin was anticipating the collection of plants. As was true for Darwin's other expenses on the voyage, payment for shipment of the specimens from their point of arrival in England (Falmouth, Liverpool, or Portsmouth) to Cambridge was borne by his father, Dr Robert Darwin (1766–1848) of Shrewsbury.

Henslow's parting advice to Charles Darwin was, 'With a little self denial on your part I am quite satisfied you must reap an abundant harvest of future satisfaction.'[24] Prophetic words, indeed!

During the voyage, Darwin shipped to Henslow nine or ten consignments of specimens from South America, including over a thousand (mostly animals) preserved in 'spirits of wine'. Collections made since the last shipment, sent from Chile in June 1835, including those from the Galápagos Islands, were taken home on the *Beagle*.

Not all of Darwin's carefully gathered plant specimens reached Cambridge, however. Of a gale off Cape Horn in 1833, when the *Beagle* was unsuccessfully attempting to enter the Pacific Ocean, Darwin wrote to Henslow (11 April 1833): 'A sea stove one of the boats & there was so much water on the decks, that every place was afloat; nearly all the paper for drying plants is spoiled & half of this cruizes collection'.[25]

It is not known for sure whether Henslow agreed to identify Darwin's plants prior to the *Beagle's* departure. He may have done so simply because he had urged Darwin especially to collect plants. Identification proved a formidable task for one conversant only with the British flora, so as we have seen, Henslow sought help from his friend Prof. William Hooker. This is not surprising, as at this time Hooker and George Walker Arnott (1799–1868) were publishing their series of papers on 'Contributions towards a flora of South America and the Islands of the Pacific'. A few of Darwin's collections found their way into this series, including *Senecio darwinii* Hook. & Arn., described by them from Patagonia in 1841.[26]

Although Henslow wrote Darwin on 31 August 1833 that Hooker would help him with the identifications, over two years later (24 November 1835) he wrote Hooker: 'So soon as I have done with proof sheets of my little vol. in Lardner & have looked over & distributed my annual aquisitions in British Botan. I mean to have a regular attack upon Darwin's plants, & will send you specimens of all that I can.'[27]

By 21 January 1836 Henslow had prepared a package of specimens for Hooker: 'If you will have the goodness to name for me those which I now send I shall be able to get on rapidly with the collection – . . . I should like to get my list complete by Darwin's return if possible as he will then I doubt not begin to think of publishing his voyage & if there is any thing new among his plants would like to mention it – Your experience will enable you at a glance to suggest the specimens which are probably new – . . . I wish he had put up more duplicates than I find he has – but as his chief pursuit was Zoology & Geology I must be satisfied with what he has sent me.'[28]  After preparing a second package for Hooker, Henslow wrote again nine days later (30 January 1836) that, 'The public will have far greater confidence in your remarks & descriptions than in any attempt of mine – Darwin's letters contain very little Botanical allusion, as he is no Botanist – His collections were made to please me—'.[29]

With Hooker's help, Henslow did publish two papers on Darwin's collections: 'Descriptions of two new species of *Opuntia;* with remarks on the structure of the fruit of *Rhipsalis.* '[30] on a few of Darwin's cacti, and 'Flora Keelingensis. An account of the native plants of the Keeling Islands.'[31]  Of the latter, Henslow wrote to Hooker on 9 March 1838: 'I have now sent you merely the *rough draft* which you will be so good as to return for me to correct & *condense* - . . . Be so good as to let me have what you can *speedily* as I have only 6 weeks before lectures begin, & if I do not get my M.S. in time I must again lay it aside till they are over -'.[32]  As we shall see, the tone of this letter is very much like those that Darwin was writing to Henslow at this time.

Upon returning to England, Darwin sent Henslow his Galápagos Islands plants, which with the Keeling Islands specimens appear to be almost the only botanical collections made after leaving Peru. His failure to do  any more plant collecting than this may have been due to the limited space for specimen storage on the *Beagle* being almost filled up by his Galápagos gatherings,[33] but more likely was due to the areas next visited (Tahiti, New Zealand, Australia, Tasmania, Mauritius, Cape of Good Hope) already being well-collected.

As we have seen, Darwin paid particular attention to the collecting of his Galápagos plants, and he was especially interested in their identification. In conjunction with his request to the Government for a grant to publish the *Zoology of the Beagle,* Darwin wrote to Henslow on 18 May 1837: 'I forgot to ask you: if I succeed with government & *if* afterwards it appears advisable, should you object to publish the botany of the Galápagos in it, as a part of the fauna? – I certainly should like, if possible, some part of the botany kept together when there are materials for any general results.'[34]

At this time, Darwin was writing up his journal for publication, and he needed a number of botanical questions answered. In every letter written to Henslow from May 1837 until a few days before the manuscript was sent to press in August, Darwin queried Henslow on his plants. In May he wrote: 'There are about half a dozen plants, of which if I do not know the names of genus or something about them, I must strike out long passages in my journal.'[35] His last letter in August implored: 'I send my MSS to the press the day after tomorrow . . . Pray write *soon* & tell me whether you can answer me any of the questions, so that I may know. – I should want first the two or three about *America.* '[36]  Apparently the questions went unanswered. The first edition of the *Journal of Researches*[37] is singularly sparse in botanical information, although Henslow did read the proofs for Darwin.[38]

Darwin's *Autobiography*[39] indicates that his interest in the variation of Galápagos species included *both* plants and animals, not just the tortoises and finches which are the usual examples given. This may be the primary reason that Darwin kept pumping Henslow for information on his Galápagos plants.

In spite of his apparently minimal assistance to Darwin in identifying the collections, Henslow should not be judged too harshly. This was a very busy period for him. Besides being Professor of Botany, Henslow also was Director of the University Botanic Garden; planning for the new garden, which opened in 1846, probably took more time than his professorial duties. In addition, he was curating the Herbarium, planning museums in Ipswich and Cambridge, and pursuing his own research on the British flora and fossil plants.

In 1837 Henslow was presented with 'the valuable but exacting living of Hitcham in Suffolk'.[40] In 1839 he gave up his residence in Cambridge and moved to Hitcham. He wrote to Sir William Hooker, who had been knighted in 1836, (17 December 1839): '. . . & now reside here continuously excepting for the month in spring when I have to deliver my lectures – I am not 40 miles from Cambridge & have ready communication enough, but miss the public library, & can have only a part of my Herbarium about me - However, I am in a very retired spot & have much more leisure for my pursuit of Botany, notwithstanding my parish avocations, which I no way neglect, than I had in Cambridge.'[41]

However much time he had for botany in the beginning of his stay in Hitcham, it soon vanished. Henslow applied himself conscientiously to his pastoral duties and did less academic botany. He never applied himself more than superficially to Darwin's plants, many of which remain unidentified to this day. Henslow performed admirable service in getting Darwin onto the *Beagle* and in receiving his collections as they dribbled back from across the world, but he failed in his good-hearted attempts to work up the plants.

If the issue had ended here, almost all of Darwin's collections still might remain unidentified. However, Dr Joseph Dalton Hooker (1817–1911) now entered the scene. J. D. Hooker was the son of Sir William Hooker, and he later became Henslow's son-in-law. Hooker met Darwin briefly in London on two occasions in 1839.[42]

On 30 September 1839, Hooker left England on his own voyage, accompanying Captain Sir James Clark Ross (1800–1862) as botanist and assistant surgeon with HMS *Erebus* and *Terror* to the Antarctic. Hooker's voyage in many ways parallels that of Darwin's on the *Beagle*. Darwin, in fact, served as an inspiration to him: his enthusiasm was fueled by reading proof-sheets of Darwin's *Journal of Researches*, sent to him by his father just prior to his departure. He returned to England on 4 September 1843, fired with enthusiasm to publish a *Flora Antarctica*,[43] which his father had urged him to do.

Even before Hooker returned, on 12 March 1843 Darwin wrote to Sir William, now Director of the Royal Botanic Gardens, Kew: 'I am very glad to hear you talk of inducing your son to publish an Antarctic Flora. I have long felt much curiosity for some discussion on the general character of the flora of Tierra del Fuego, that part of the globe farthest removed in latitude from us. How interesting will be a strict comparison between the plants of these regions & of Scotland or Shetland. I am sure I may speak on the part of Prof. Henslow that all my collection (which gives a fair representation of the Alpine flora of Tierra del Fuego & of Southern Patagonia) will be joyfully laid at his disposal.'[44]

In his first letter to J. D. Hooker (21 November 1843), Darwin wrote:[45]

But I have run from the subject, which made me write, of expressing my pleasure that Henslow (as he informed me a few days since by letter) has sent to you my small collection of plants. You cannot think how much pleased I am, as I feared they would have been all lost, & few as they are they cost me a good deal of trouble. There are a very few notes, which I believe Henslow has got, describing the habitats &c. of some few of the more remarkable plants. I paid particular attention to the Alpine flowers of Tierra Del., & I am sure I got every plant which was in flower in Patagonia at the seasons, when we were there. . . . I hope Henslow will send you my Galapagos Plants (about which Humboldt even expressed to me considerable curiosity). I took much pains in collecting all I could. A Flora of this Archipelago would, I suspect, offer a nearly parallel case to that of St. Helena, which has so long excited interest.

Hooker answered (28 November 1843): 'Many thanks for your kind letter of congratulations & also for your offer of assistance in examining the plants you collected, of which I shall most thankfully avail myself. It is very liberal of you to place them so at my disposal & I do hope that I shall show myself not to be altogether unworthy of the trust. . . . Professor Henslow has kindly promised to send your Galapagos plants as well as the Antarctic, I am not aware of any collector having been there but yourself and Douglas.'[46] This was David Douglas (1798−1834), collector for the Horticultural Society, who visited the archipelago briefly in January 1825. Actually, Darwin was preceded not only by Douglas, but also by Dr John Scouler (1804−1871), who was ship's-surgeon and naturalist with Douglas; by James McRae (?−1830), another collector for the Horticultural Society, who followed Douglas and Scouler by two months; by Hugh Cuming (1791−1865), sailmaker and naturalist, who collected in the islands in 1829; and by Archibald Menzies (1754−1842), naturalist and surgeon with Captain George Vancouver (1758−1798) on HMS *Discovery,* who made the first known Galápagos plant collections in 1795.[47]

Henslow delivered Darwin's collections and notes to Hooker over a period of months, beginning in late 1843. On 21 November 1843, Henslow wrote Hooker: 'I shall be delighted to place Darwin's plants in your hands − & beg you will make just whatever use of them you please − not forgetting to give me a rap on the knuckles for having done so little with them − Some few I have named & incorporated with the herbarium, but I can give you the list of these − My duties here [Hitcham] quite debar me from pursuing Botany with any vigor now. . . . I have an interesting set of plants from the Galapagos from Darwin − & have been intending again & again to set to work at them − they are there − Would these be wanted by you?'[48]

Henslow seems continually to have been discovering plants that had been overlooked and not forwarded with the bulk of the collection. As late as 28 February 1846 he wrote to Hooker: 'I have stumbled upon a few of Darwin's ferns which have not been sent to you −'[49] On 1 July 1846 he wrote to Sir William: 'I have stumbled upon 2 or 3 more of Darwin's *S. W.* American plants, but I believe these must be the *last* I had overlooked in picking them out of the packets for Dr Jos.'[50] In November of that year (6 November 1846), however, he wrote again to Sir William: 'Tell Dr Jos. I will bring with me on the 19th (as far as London) the 2 bottles he alluded to − but one of them contains a Calceolaria which I suppose Darwin fancied was an orchid.'[51]

Besides the Darwin plants, Henslow also forwarded a Galápagos collection made by James McRae. He wrote to Hooker on 10 December 1843:[52]

I shall leave at 13 Clements Green [his brother's house] the Galapagos plants, in my way thru Town tomorrow − You will find a few with them from *Macrae* which belong to the Horticultural Soc − I had begun by a few random notes to examine them before I left Cambridge, & have left them just as

I inserted them at the time – Since I came here [Hitcham] I have had no time for them – always intending to recommence – but never being able to do so among my numerous engagements & duties – You will find an interesting set of plants – Pray publish them in any way you prefer. I am too happy to see justice done to Darwin's exertions to think of making stipulations of any sort – Do just as you please – giving him due credit for collecting in a branch of science which formed no part of his studies, & solely to oblige me.

In Hooker, Darwin had found someone unencumbered by too many other duties, someone who could devote more time than Henslow to his plant collections. Hooker began with the Falkland Island and Tierra del Fuego specimens, identifying them and citing them in his *Flora Antarctica*. True to form, Darwin immediately began peppering him with questions about his plants, many of them the same questions on distribution and relationships he had asked Henslow six years previously. This time, however, he received the answers. These answers were quite important for Darwin in writing the second edition of his *Journal of Researches*.[53] Although this edition is more concise than the first, the information on plants in it is much greater, particularly in the chapter on the Galápagos Islands, which Hooker read for him in proof. Hooker rapidly became Darwin's source for botanical knowledge, his confidant, his best friend.

Darwin and McRae's Galápagos specimens, along with those few of Douglas, Scouler, Cuming, and some others, served Hooker as the basis for two papers on the Galápagos flora. These were worked up from 1843 to 1846 and published in the *Transactions of the Linnean Society* in 1847. The first[54] was the initial attempt to produce a flora of these interesting islands. The second[55] was the first of many papers that Hooker was to devote to plant geography. They are both classics in their fields, the one because it is so important for the typification of many Galápagos endemics, the other because it was the first to hypothesize the various dispersal mechanisms by which the plant colonizers arrived in the islands and from whence they came.

A few of Darwin's other specimens served as types for new species described in various publications by Hooker, Sir William, and others. Nevertheless, the majority of his plant collections remain unidentified in the Cambridge University Herbarium, a situation I hope to rectify in the near future.

Because Hooker accomplished his work on the Galápagos plants mainly while he was at Kew, and because there is a partial set of these plants there, many taxonomists have assumed that the types of the many new species he described must be there as well. However, I have just completed a study on Darwin's Galápagos specimens and find that those at Kew are duplicates, the main set being at Cambridge.[56] A few of Darwin's *Beagle* specimens are to be found in other herbaria (i.e., the Gray Herbarium, Harvard University; the herbarium of the Missouri Botanical Garden; the Manchester Museum). These were sent to their curators by Hooker or the Kew botanist George Bentham (1800–1884), and are duplicates of specimens at Cambridge or Kew.

What happened to Darwin's "memoranda" on his plants[57] and Henslow's notes on the Galápagos specimens? In November 1845, Hooker wrote to Darwin implying that he was returning both specimens and notes to Henslow.[58] The whereabouts of the latter, however, is unknown. Perhaps Hooker discarded them, rather than returning them to Henslow, although this is unlikely. Henslow himself may have discarded or misplaced them. They are still being sought; hopefully, some day they will be discovered in the archives of Cambridge or Kew or at Down House. It may be that, rather than being a separate set of notes, Darwin's "memoranda" are the comments on plants that he made in his field notebooks.

With the publication of the two Galápagos papers by Hooker, work on the *Beagle* plant collections stopped.[59] Henslow was immersed in his parish work; Hooker early in 1846 took a position with the Geological Survey to study Britain's fossil flora, and in 1847 left to spend four years in India to collect the Himalayan flora and later to go on to other botanical interests. Both Henslow and Hooker continued to correspond with Darwin. Henslow's correspondence becomes less botanical and more general, however, while Hooker's retains much botanical discussion to the end.

Francis Darwin and A. C. Seward, writing in 1903, characterized Charles Darwin and Joseph Hooker's relationship thus:[60]

> The close intercourse that sprang up between them was largely carried on by correspondence, and Mr. Darwin's letters to Sir Joseph have supplied most valuable biographical material. But it should not be forgotten that, quite apart from this, science owes much to this memorable friendship, since without Hooker's aid Darwin's great work would hardly have been carried out on the botanical side. And Sir Joseph did far more than supply knowledge and guidance in technical matters: Darwin owed to him a sympathetic and inspiring comradeship which cheered and refreshed him to the end of his life.

By supplying answers to Darwin's botanical questions which Henslow could not (or would not) answer, Hooker started Darwin down the parth which culminated in the many botanical works he was to publish after 1859.

ACKNOWLEDGEMENTS

Research for this paper was made possibly by a grant from the Penrose Fund of the American Philosophical Society and travel funds from the Virginia Polytechnic Institute & State University Education Foundation. I am indebted to the directors and staffs of the Cambridge University Library, Cambridge University Herbarium, Royal Botanic Gardens, Kew, and Down House for the opportunity of examining their collections of Darwiniana, Kew allowing me to quote from unpublished letters; Peter Gautrey, Peter Sell, Miss E. Smith, Grenville Lucas, and Philip Titheradge have been especially helpful. I owe special thanks to Miss Mea Allan for sharing some unpublished correspondence between Henslow and the Hookers with me, which I quote with her permission. Indeed, all biologists are indebted to her for showing us[16] the influence of botany on Darwin. I am grateful to my wife Sarah and to my colleagues Professors Albert Moyer and David West for their comments on drafts of this paper, and to Dr Sydney Smith, St Catharine's College, Cambridge for sharing his profound knowledge of Darwin and for his continuing encouragement of my studies. None of this would have been possible without the assistance of Mrs Shirley Lucas, of which I am most grateful.

SOURCES

Except for Darwin's letters to William Darwin Fox (note 21) and Sir William Hooker (note 44), I have seen the originals or facsimiles of all cited correspondence. However, if a given letter has been published, as most have, a reference to the place of publication is given. The provenance of the letters is as follows: Darwin to Henslow(Original Letters, from Chas. Darwin, to Professor Henslow. 1831–1837. Royal Botanic Gardens, Kew), Darwin to J. D. Hooker (Darwin Papers, Vol 114. Cambridge University Library), Henslow to Darwin (Original Letters. . . . Kew; Darwin Papers, Vol. 97. CUL), Henslow to J. D. Hooker (English Letters. . . .Kew; R. A. Hooker Letters, photocopies held by Miss Mea Allan), Henslow to W. J. Hooker (English Letters. . . . Kew; R. A. Hooker Letters), and J. D. Hooker to Darwin (Darwin Papers, Vol. 100. CUL). Original punctuation and spelling have been retained in all cases.

NOTES AND REFERENCES

1 BARLOW, N. (ed.), 1958. *The autobiography of Charles Darwin, 1809–1882.* Harcourt, Brace, & World, New York. p. 60.

2 In later life, Darwin played down the role of Cambridge in his education; in fact he stood tenth in his class.

3 The quotations are from BARLOW, *Autobiography* pp. 60, 52.

4 He did so many times in correspondence, and once in print (DARWIN, C., 1855. Does sea-water kill seeds? *Gard. Chron.* no. 21, 26 May, p. 356). The leading botanist of the time in the United States, Asa Gray (1810–1888), also called him a geologist (GRAY, A., 1876. *Darwiniana: essays and reviews pertaining to Darwinism.* Appleton, New York; reprint ed. by DUPREE, A. H., 1963. Belknap Press of Harvard University Press, Cambridge, Massachusetts. p. 132). In spite of his many botanical papers and books, Darwin continually referred to himself as a "Botanical Ignoramus" (e.g. in a letter of 21 November 1843 to J. D. Hooker, printed in DARWIN, F. (ed.), 1888. *The life and letters of Charles Darwin.* Appleton, New York. 2 vols. 1: 382). However, it must be remembered that this was mainly to persons like Gray and Hooker, the leading botanists of the day.

5 GRUBER, H. E. and GRUBER, V., 1962. The eye of reason: Darwin's development during the *Beagle* voyage. *Isis* 53: 186–200.

6 LYELL, C., 1830–1832. *Principles of geology.* vols. 1 and 2. John Murray, London.

7 BARLOW, N. (ed.), 1967. *Darwin and Henslow, the growth of an idea. Letters 1831–1860.* Bentham-Moxon Trust and John Murray, London. Letter 19: 51.

8 BARLOW, *Autobiography.* pp. 77ff.

9 BARLOW, N. (ed.), 1946. *Charles Darwin and the voyage of the Beagle.* Philosophical Library, New York. p. 150. Herein are printed a number of mainly zoological extracts from the notebooks.

10 BARLOW, *op. cit.,* indicates that there are 24 notebooks, but this is incorrect.

11 BARLOW, *op. cit.*

12 DARWIN, C. (ed.), 1839–1843. *Zoology of the voyage of H.M.S. Beagle.* 5 vols. Smith Elder, London.

13 DARWIN, C. 1842. *The structure and distribution of coral reefs.* 1844. *Geological observations on the volcanic islands.* 1846. *Geological observations on South America.* Smith Elder, London.

14 BARLOW, N. (ed.), 1963. Darwin's ornithological notes. *Bull. Brit. Mus. (Nat. Hist.) hist. Ser.* 2 (7): 201–278.

15 BARLOW, *Darwin and Henslow.* Letter 20: 58.

16 BARLOW, *op. cit.* Letter 22: 66. ALLAN, M. 1977. *Darwin and his flowers: The key to natural selection.* Taplinger, New York. pp. 84–85,

17 BARLOW, *op. cit.* Letter 20: 58.

18 BARLOW, *op. cit.* Letter 22: 66. ALLAN, *op. cit.:* 84.

19 BARLOW, *op. cit.* Letter 28: 78. ALLAN, *op. cit.:* 85.

20 BARLOW, *op. cit.* Letter 31: 84. ALLAN, *loc. cit.*

21 DARWIN, *Life and Letters.* 1: 234.

22 BARLOW, *Darwin and Henslow.* Letter 41: 113.

23 BARLOW, *op. cit.* Letter 13: 43.

24 BARLOW, *op. cit.* Letter 17: 50.

25 BARLOW, *op. cit.* Letter 26: 71.

26 In *Hook. J. Bot.* 3: 333. Darwin specimens continue to be used to typify new names, the latest being *Spilanthes darwinii* D. M. Porter, *Madroño* 25: 58, 1978, from the Galápagos Islands.

27 *English Letters/1832–35/H–L/Vol. V,* Royal Botanic Gardens, Kew. Letter 117. His 'little vol. in Lardner' was *Principles of descriptive and physiological botany* in Lardner's *Cabinet Cyclopedia* London, 1836.

28 *English Letters/1835–36/H–W/Vol. VIII,* Royal Botanic Gardens, Kew. Letter 5.

29 *Op. cit.* Letter 6.

30 HENSLOW, J. S., 1837. *Mag. Zool. Bot.* 1: 466–469.

31  HENSLOW, J. S., 1838. *Ann. Nat. Hist.* **1**: 337–347.

32  *English Letters/H–Z/1838/Vol. XI*, Royal Botanic Gardens, Kew. Letter 7.

33  The small amount of space available to Darwin on the *Beagle* is discussed by SMITH, S., 1960. The origin of 'The Origin.' *Adv. Sci.* **64**: 396.

34  BARLOW, *Darwin and Henslow.* Letter 47: 129.

35  BARLOW, *op. cit.* Letter 47: 128.

36  BARLOW, *op. cit.* Letter 50: 133. The word which I have deciphered from Darwin's handwriting as 'want', Barlow gives as 'count'.

37  DARWIN, C., 1839. *Narrative of the surveying voyages of His Majesty's ships* Adventure *and* Beagle, *between the years 1826 and 1836, describing their examination of the southern shores of South America, and the* Beagle's *circumnavigation of the globe. Vol. 3. Journal and remarks. 1832–1836.* Henry Colburn, London.

38  BARLOW, *Darwin and Henslow:* 14.

39  BARLOW, *Autobiography:* 80.

40  BARLOW, *Darwin and Henslow:* 16.

41  *English Letters/H–Z/1839/Vol. XIII,* Royal Botanic Gardens, Kew. Letter 7.

42  The first meeting was in the company of Asa Gray at the Hunterian Museum, Royal College of Surgeons (DUPREE, A. H., 1959. *Asa Gray, 1910–1888.* Belknap Press of Harvard University Press, Cambridge, Massachusetts. p. 81), in January, a meeting that Hooker apparently forgot. In later life, he wrote to Darwin's son Francis that, 'My first meeting with Mr. Darwin was in 1839, in Trafalgar Square. I was walking with an officer who had been his shipmate for a short time in the *Beagle* seven years before, but who had not, I believe, since met him.' (DARWIN, *Life and Letters,* 1: 380). Because of Hooker's statement, this meeting invariably is cited in the literature as their first. Sidney Smith has pointed out to me that the unidentified officer can be none other than Robert McCormick (1800–1890), surgeon on the *Beagle* who returned to England early in the voyage because of differences with Captain Fitz-Roy and other of the officers. He was soon to be surgeon, and nominally Hooker's superior, on the *Erebus* and *Terror* voyage.

43  HOOKER, J. D., 1844–1847. *The botany of the Antarctic voyage of H. M. discovery ships* Erebus *and* Terror *in the years 1839–1843, under the command of Captain Sir James Clark Ross. Part I. Flora Antarctica.* 2 vols. London.

44  DARWIN, F. and SEWARD, A. C. (eds.), 1903. *More letters of Charles Darwin.* Appleton, New York. 2 vols. **2**: 242–243, Letter 575.

45  DARWIN, *Life and Letters,* 1: 382–383; much in ALLAN, *Darwin and his flowers:* 134–135. On 18 September 1839, referring to Darwin's *Journal of Researches,* the great German scientist Alexander von Humboldt (1769–1859) had written to him: 'How much I regret that Mr. Henslow could not finish the examination of your curious collection (pp. 460, 537, 541), or at least the keying of the families containing some known species.' BARRETT, P. H. and CORCAS, A. F., 1972. A letter from Alexander Humboldt to Charles Darwin. *Jour. Hist. Med.* **27**: 169.

46  ALLEN, *op. cit.,* p. 135.

47  The early history of botanical collecting in the Galápagos Islands is discussed in PORTER, D. M. (in press). The vascular plants of Joseph Dalton Hooker's "An Enumeration of the plants of the Galápagos Archipelago; with descriptions of those which are new." *Botanical Journal of the Linnean Society.*

48  *R. A. Hooker Letters.* Letter 1.

49  *R. A. Hooker Letters.* Letter 9.

50  *English Letters/1846/Vol. XXIV,* Royal Botanic Gardens, Kew. Letter 258.

51  *Op. cit.* Letter 265.

52  *R. A. Hooker Letters.* Letter 2.

53  DARWIN, C., 1845. *Journal of researches into the Natural history and geology of the countries visited during the voyage of H. M. S.* Beagle *round the world, under the command of Capt. Fitz-Roy, R. N.* ed. 2. John Murray, London.

54  HOOKER, J. D., 1847. An enumeration of the plants of the Galapagos Archipelago; with descriptions of those which are new. *Trans. Linn. Soc. Bot.* **20**: 163–233.

55 HOOKER, J. D., 1847. On the vegetation of the Galapagos Archipelago, as compared with that of some other tropical islands and of the continent of America. *op. cit.* **20**: 235–262.

56 Problems of typification with Hooker's Galapagos plant names are taken up in PORTER, *op. cit.*

57 'I have Darwin's *memoranda* to bring up with me – & if I can I will bring also the Galapagos plants.' Henslow to J. D. Hooker, 9 September 1843. *English Letters/A–H/1843/Vol. XIX,* Royal Botanic Gardens, Kew. Letter 326.

58 'I have often tried to make your notes [illegible word] on to the species. I wish you would come & take a look at them before I return them to Henslowe.' *Darwin Papers,* Vol. 100, Cambridge University Library. Letter 15.

59 The fungi were published upon by the Reverend Miles Joseph Berkeley (1803–1889): BERKELEY, M. J. 1839. Notice of some fungi collected by C. Darwin, Esq., during the expedition of H. M. Ship *Beagle. Ann. nat. Hist.* **4**: 291–293. 1842. Notice of some fungi collected by C. Darwin, Esq., in South American and the Islands of the Pacific. *Ann. Mag. nat. Hist.* **9**: 443–448. 1845. On an edible fungus from Tierra del Fuego, and an allied Chilean species. *Trans. Linn. Soc.* **19**: 37–43.

60 DARWIN and SEWARD, *More Letters.* **1**: 38–39.

*Archives of Natural History* (1988) **15** (2): 197–231

# Charles Darwin's *Beagle* Collections in the Oxford University Museum

By GORDON CHANCELLOR,

City Museum & Art Gallery,
Priestgate, Peterborough
PE1 1LF

ANGELO DiMAURO,

Department of Biology,
University of Connecticut,
Torrington Campus,
Torrington Connecticut 06790, USA

RAY INGLE,

Department of Zoology,
British Museum (Natural
History), Cromwell Road,
London SW7 5BD

GILLIAN KING,

Oxford University Museum,
Parks Road,
Oxford OX1 3PW.

SUMMARY

One hundred and ten numbered lots of zoological specimens and several specimens of rocks collected by Charles Darwin during the second voyage of H.M.S. *Beagle* (1831–1836) are contained in the collections of the Oxford University Museum. The circumstances which led to their acquisition are outlined. Most of this material comprises "higher" crustaceans but a few other invertebrate groups are also represented. A systematic identification has been made of these specimens and their condition is briefly described.

INTRODUCTION

Following the return of H.M.S. *Beagle* from her second hydrographic commission in October 1836, much of the zoological material collected on the voyage by Charles Darwin was subsequently entrusted for scientific description to Thomas Bell (1792–1880). In 1862 Bell's zoological collections (including invertebrates sent to him by Darwin) were purchased by John Obadiah Westwood (1805–1893), the first Hope Professor of Zoology at Oxford University, to enrich the Hope Entomological Collection in his care. In 1889 a greater part of this Bell Collection was transferred to the Oxford University Museum's Zoological Collections (Davies & Hull, 1976).

In 1936 some *Beagle* specimens were sent by G. D. Hale Carpenter (Hope Professor of Zoology (Entomology) from 1922 to 1948) to W. T. Calman of the British Museum (Natural History) for examination. We must presume that this material was returned to Oxford, although there is no record in the Hope Department loan catalogue of the transaction and the only documentation of its having occurred is a letter from Carpenter to Dr Isabella Gordon of the British Museum (Natural History) dated 1945 asking her to look for the material sent to Calman. In 1962 and 1975 further zoological specimens were transferred from the Hope Collections to the Zoological Collections of the Oxford University Museum, and studies of these collections since 1976 have revealed that they contain many of the crustaceans described by Bell, including type specimens, many of which were thought to have been lost (see Di Mauro, 1982). It has been possible to establish that 110 numbered lots of these specimens originated from the *Beagle* collections. These have been identified by us as far as it has been possible from current literature (see p. 201), and they have been correlated with Darwin's own lists of collections and other notes kept by him from the *Beagle* voyage.

In addition to the material mentioned above, there are a number of insects in the Hope Entomological Collections (mainly from Australia) labelled "C. Darwin" in his own handwriting which are also part of the *Beagle* collections. The acquisition of this material is explained by Poulton (1909: 202–3) and formed the subject of the paper on Darwin's insect collection by K. G. V. Smith (1987).

Finally, we have identified several specimens of rocks collected by Darwin during the *Beagle* voyage and now in the Geological Collections of the Oxford University Museum. These Darwin specimens form part of the Charles Daubeny (1795–1867) collection (see Gunther, 1904:188; 1925:244). These rocks, all labelled in Darwin's handwriting, comprise two specimens (Darwin's dry specimen catalogue numbers 2248 and 2256) labelled "Andesitic trachyte. Baths of Cauquenes, described p. 175 in my Geolog. Observ. on S. America . . . " collected in September 1834; two specimens labelled "Andesitic syenite; larger specimen from T. del Fuego; smaller one, Cordillera of Chile . . . " (the latter bears the corner of one of Darwin's green printed number labels); two specimens of obsidian labelled "From St. Helena" (these would have been numbered between 3700 and 3728); two pieces of fossil wood, one labelled "Van Diemen's Land. Lower part of the tree", the other "Peru. Petrified wood".

There may be other Darwin specimens in the Daubeny collection, notwithstanding that the *Beagle* rock collection in the Mineralogy and the Petrology Museum in Cambridge is almost complete, but since Darwin's name appears nowhere else in Daubeny's manuscript catalogue to his collection, such specimens will be discovered only when Darwin's numbers or handwriting are recognised.

## PRESENT STATE OF THE COLLECTION

The Darwin zoological material discussed here comprises chiefly crustaceans and a few other invertebrates. Some of these specimens are preserved dry whereas others are in alcohol, but it is clear from Darwin's notes that orginally all the material was preserved in spirits. The dry specimens are stored in cabinet drawers containing the Bell collection and comprise mainly brachyuran crabs filling two drawers, one of which is shown in Figure 1. There are three additional specimens among other non-Darwin material in three other drawers. Several dry specimens are associated with labels in Darwin's handwriting and/or still bear the metal tags with which he numbered them while on the *Beagle*. Other specimens are labelled in Bell's handwriting and bear his manuscript names (i.e 'darwinii' or 'darwinii n. sp.'); some of these

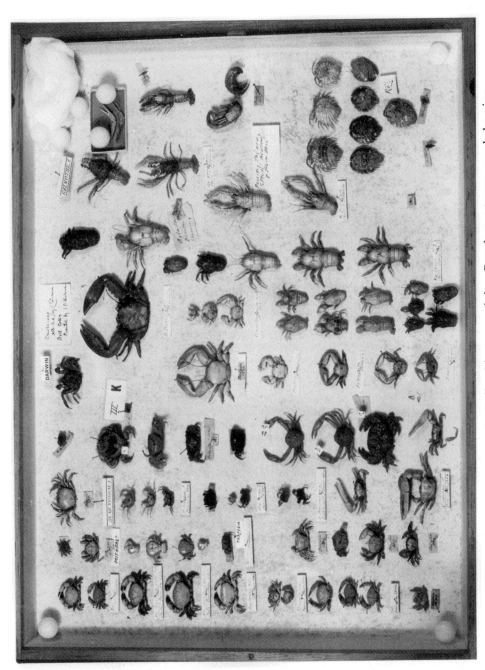

Figure 1. A drawer from the Bell cabinet containing *Beagle* crustaceans and showing the present condition of this dry material.

are visible in Figure 1. Six specimens are associated with brown paper labels (in addition to a metal tag) bearing a specimen number in a handwriting which appears to be that of Darwin's servant and amanuensis, Syms Covington. The spirit material is now all contained in special 'Darwin' jars (see below) (jars 1–9 for the Crustacea, with separate jars for Myriapoda, Arachnida and Annelida respectively); the specimens are individually numbered in small tubes except for a few large specimens in their own jars. Most of this spirit material can be clearly recognised as *Beagle* specimens. Unfortunately there remain many specimens, dry and in spirit, which are probably part of the *Beagle* collection but which bear no number or label of any sort and are therefore not included in this paper.

## HISTORICAL BACKGROUND

In recent years there has been a growth of interest in Darwin's *Beagle* collections and the present whereabouts of most of these collections have been established (see Porter, 1985, for an authoritative overview). The bulk of the animals, plants, rocks and fossils which Darwin collected was well cared for when it reached England and, by the standards of many early nineteenth century voyages of discovery, Darwin was well served by the mineralogists, palaeontologists, botanists and zoologists he recruited to describe those specimens which he did not wish to deal with himself, such as the barnacles and arrow worms. One of the few groups of animals from the voyage never properly described was the Crustacea Darwin had entrusted to Thomas Bell (an acknowledged expert on the group) for inclusion in a projected complete *Zoology of the Beagle*. In the event, only the vertebrates from the voyage were described, of which Bell dealt with the reptiles (see Bell, 1842–43), while the Crustacea sank into oblivion.

It is clear that all the specimens given to Bell were originally preserved in spirits by Darwin, and nearly all the extant specimens have numbers that agree with those listed in the "Catalogue for specimens in spirits of wine" (referred to hereafter as the "Catalogue") and which is today displayed in Down House, Darwin's home in Kent. This "Catalogue" comprises three red pocket notebooks with soft covers and was briefly described by Barlow (1945;265). It is a chronological listing of specimens collected throughout the voyage and the numbers are divided as follows: 1–660 (first notebook), 661–1346 (second notebook), 1347–1529 (third notebook with this last number representing the last specimen entry).

All the crustaceans listed in this "Catalogue" are marked in pencil with a letter "C" (Herbert, 1974); this was to enable Syms Covington (see above) to copy out a separate list to accompany the Crustacea collection for the use of the zoologist who intended to study the collection. Our attempts to locate this list have been unsuccessful. It is possible that Covington never prepared the list of Crustacea before he departed for Australia in 1839 but this seems unlikely if the majority of the lists were prepared on the *Beagle* before she reached England.

There are about 230 lots marked with a "C" in the "Catalogue" but we have only succeeded in identifying 110 of these lots in the Oxford University Museum collections, so that it would appear that much material still remains to be discovered. In this paper, with a very few exceptions, only the 110 lots we have identified are listed, so that for a complete list of Darwin's *Beagle* Crustacea the reader must consult Darwin's original "Catalogue".

Another source of information concerning Darwin's zoological collections is the manuscript "Diary of observations on zoology of the places visited during the voyage (of H.M.S. *Beagle*)", referred to in this paper as the "Zoology Diary" which Darwin wrote concurrently with his more general "Personal Diary" of day-to-day affairs (see Barlow, 1933; Porter, 1985). The "Zoology Diary" is today in the Darwin archive at Cambridge University Library. With the kind assistance of Mr Peter Gautrey it has been possible to examine it and abstract all data relevant to the Oxford University Museum specimens. Since this "Zoology Diary" was used by Darwin in the writing of his official *Journal of Researches* (first published in 1839) some of its contents are of particular interest. Occasional cross-references to the "Zoology Diary" occur in the "Catalogue"; these are given, for example, as "V. 19(b)" (i.e. *vide* page 19, note (b)). The text of the "Zoology Diary" today comprises four bound volumes:, DAR 30(i); DAR 30(ii); DAR 31(i); DAR 31(ii); although Darwin's illustrations and index are bound in DAR 29(iii). (See *Handlist of Darwin Papers at the University Library Cambridge*, 1960. p. 10)

There are two references to Crustacea in a manuscript entitled "Notes on the preservation of specimens", which is bound as the last part of DAR 29(iii) in the Cambridge University Library. These references read as follows: "Not to keep Crustacea with other things owing to the black tinge they give to the spirits". "Crustacea ought to have proof spirits and the Brachyuri well emptied of water & incission in the abdomen". This manuscript would appear to have been written by Darwin before the *Beagle* set sail, as instructions to himself on how to preserve the specimens he intended to collect.

Finally, a manuscript catalogue compiled by Thomas Bell and in the Hope Library contains Darwinian "Catalogue" numbers entered against Bell's own numbers along with tentative generic identifications.

## INTERPRETATION OF THE "CATALOGUE" AND REAPPRAISAL OF THE *BEAGLE* SPECIMENS

Darwin's "Catalogue" is written in ink, except for entries 1474 to 1529 which are in pencil (see Herbert, 1974:239). Information is inscribed on each right-hand page of the notebook. At the top of each page appears (nearly always) the date and locality followed by a vertically arranged sequence of numbers and tentative identifications with occasional further information. Sometimes several localities are given below the main entry at the page heading, and occasionally additional information appears on the left-hand page opposite the main right-hand entry. These additional comments are merged with the main entries in the list of *Beagle* specimens given here (pp 202–226), although enclosed within square brackets. Since these comments were numbered by Darwin, they can be tied unambiguously to, and were almost certainly written at the same time as, the main entries.

A taxonomic reappraisal of the *Beagle* specimens has been made with the help of study reference collections in the British Museum (Natural History) and the following literature: Abele (1972), Barnard (1950), Bollman (1889), Bott (1969, 1973), Bousfield (1982), Campbell & Griffin (1966), Chace (1956, 1968), Crane (1975), Fize & Serene

(1955), Forest & Saint Laurent (1967), Garth (1946, 1958), Haig (1955, 1960, 1964), Holthuis (1952, 1954), Lucas (1980), Manning (1969, 1971), Manning & Holthuis (1981), Menzies (1961), van Name (1936), Pretzmann (1972), Rathbun (1911, 1918, 1925, 1930, 1937), Sakai (1976), Schmitt (1942), Stebbing (1914), Stephenson & Hudson (1957.

The following entry (taken from page 227) of this *Beagle* specimen list is given as an interpretive example of the format used. All editorial matter is indented, or italicised within brackets if it interrupts the ''Catalogue'' text.

1833 March Tierra del Fuego ----------------- top of page entry in ''Catalogue''

551 Sphaeromidae from stones & a Crust.     ''Catalogue'' number and Darwin's
Macrouri. from stomach of a Gadus           identification
G[*ood*] S[*uccess*] Bay

‖----------‖--------------------------------------------- current editorial comment (in this
                                                         instance to explain abbreviated locality)

        6/12924 ------------------------------------- Oxford University Museum registration
                                                       number

        Decapoda: Galatheidae -------------------- order and family names

        *Munida subrugosa* (White) -------------- generic name, specific name, author of
                                                   species

Remarks: relate to the present condition of specimens and relevant information.

In interpreting Darwin's notes it is useful to know that he took his names for colours from Patrick Syme's edition of *Werner's nomenclature of colours* (Edinburgh, 1821) and that he also had available on the *Beagle* Jean Baptiste Bory de Saint-Vincent's edition of the seventeen volume *Dictionnaire classique d'histoire naturelle* (Paris, 1822–1831) (see under specimen 1098). Darwin also refers (under specimen 1057) to Juan Ignacio Molina's *Compendio de la historia geografica natural y civil del Reyno de Chile* . . . (Madrid, 1794–1795). A complete list of the books available to Darwin on board the *Beagle* is provided in an appendix to Burkhardt & Smith (1985).

LIST OF *BEAGLE* SPECIMENS

January 1832 Cape Verdes
12 Jan. 17 & 18.: 5 small Crustacea from Quail island
    4/12897
    Decapoda: Paguridae
    *Clibanarius* sp.
    Remarks: fragmented.
    4/12898
    Decapoda: Xanthidae
    *Epixanthus helleri* A.Milne Edwards

4/12899
Decapoda: Majidae
*Micropisa ? ovata* Stimpson

4/12899
Decapoda: Majidae
*? Herbstia* sp.
Remarks: condition poor.

4/12900
Decapoda: Porcellanidae
*Petrolisthes ? cessacii* (A.Milne Edwards)
Remarks: condition very poor.

[Jan. 22nd]
40 Centipedes Arachnida. NE of Port Praya
4/12901–2
Isopoda: Idoteidae
*?Idotea* sp.

59 Containing Crustacea. Echiura. Siphunculus & white animal allied to it: Actinea.
Fisurella. Chiton. W of Quail Island

Dry 2/14516
Decapoda: Calappidae
*Cycloes bairdii* Stimpson
Remarks: this may be a new record for the Cape Verde Islands. Chace (1968)
gives the range as St Helena and the S. Atlantic. The specimen was labelled by
Bell "Cycloes. Cy: Darwinii B. Darwin 59".

91 Crustacea. Chiton. Bulla [Doris *deleted?*] (see 1477)


1832 St Jago to Fernando Noronha

108 Porpita. v. 19(b) Crustacea [Biphora *deleted*] Salpa

4/12903
Amphipoda: Gammaridea
Remarks: condition poor.

March
[144.145.146.148.150.151.152.153 Caught at Bahia from Feb 29th to March 17th—]

144 Crab

4/12904
Decapoda: Grapsidae
*Sesarma (Holometopus) angustipes* Dana

145 Shells. Crustacea & Fish

4/12905
Decapoda: Majidae
*Mithrax (Mithraculus) coryphe* (Herbst)

4/12906
Decapoda: Xanthidae
Remarks: condition too poor to allow further identification.

4/12907
Isopoda: Valvifera
Remarks: condition very poor but telson distinctive.

[160. . .177 Abrohlos. March 29th]

172 Crustacea

4/12908
Decapoda: Dromiidae
*Dromidea antillensis* Stimpson

4/12909
Decapoda: Dorippidae
*Ethusa* sp.
Remarks: condition poor.

1832 May. Rio de Janeiro

206 Julus (May 5th)

Myriapod jar/12910
Diplopoda: Spirostreptidae
Remarks: fragmented.

222 Salt Water Crab: Julus: Lepisma: Worms. Wood lice. Acari

Dry 105/14517
Decapoda: Portunidae
*Ovalipes punctatus* (de Haan)
Remarks: the metal tag numbered 223 with this specimen is anomalous since 223
is not listed as a crustacean in the "Catalogue".

1832 Monte Video: August

347 Fish. Coast of Patagonia Latitude 38° 20′ August 26th—Soundings 14 fathoms.
Caught by hook & Line V. 77.

Remarks: the fish referred to is *Percophis brasilianus* Cuvier described by Jenyns
(1840–1842:23). This fish was "caught by hook and line in fourteen fathoms water
on the coast of Patagonia. in lat. 38° 20′" and was noted by Darwin as "when
cooked, was good eating". This specimen is no longer in the Zoological Museum
Cambridge and is presumably not extant.

351 Isopod (Bopyrus ?) on fish: & [curious Decapod *deleted*: Porcellana *inserted*].
can swim tail first: & Amphipode &c: [Habitats on Corallina. same as (347) *deleted
in pencil*: 14 fathoms coast of Patagonia *inserted in pencil*].

4/12911
Decapoda: Porcellanidae
*Petrolisthes* sp.
Remarks: condition very poor.

4/12912
Isopoda: Cirolanidae
*Cirolana ? urostylis* Menzies

1832 Sept: Bahia Blanca
407 Pagurus in a Buccinum

Dry 16/14518
Decapoda: Majidae
*Mithrax (Mithraculus) nodosus* Bell
Remarks: this specimen appears to have been misnumbered as it is clearly a spider crab and not a pagurid. This species occurs on the Pacific side of S. America far removed from the Argentine locality of 407.

414 Plagusia: body pale.

4/12913
Decapoda: Grapsidae
*Cyrtograpsus altimanus* Rathbun

436 Plagusia (two species)

Separate jar/12914
Decapoda: Grapsidae
*Cyrtograpsus angulatus* Dana

1832
Octob: Bahia Blanca

448 Small Crustacea, from Coralline &c i.e. not pelagic.—

4/12915
Isopoda: Idoteidae
*Idotea ? metallica* Bosc
Remarks: condition rather poor.

4/12916
Isopoda: Idoteidae
*Idotea ? metallica* Bosc

4/12917
Isopoda: Idoteidae

4/12918
Isopoda: Idoteidae
Remarks: condition poor, only anterior half present.

4/12919
Amphipoda: Caprellidae
Remarks: condition poor.

6/12920
Amphipoda: Isaeidae
*?Megamphopus* sp.
Remarks: condition poor.

6/12921
Amphipoda: ?Aoridae
Remarks: condition poor.

1832 November. & Dec. Monte Video

475 Polydesmus, dusky red: & Scutigera: found in ship (British?).

Myriapod jar/12922
Diplopoda: Polydesmidea
Remarks: badly broken.

1832 Dec:

491 Crust. Mac V. 98(b) (its young ?) & Crust. Amphipus-off San Blas. Dec. 4th:
Remarks: there are long descriptions of entry 491 and of other Crustacea collected
on Dec. 4th, in DAR 30 (ii). The following entry appears in Darwin's "Personal
Diary" for December 5th–6th while at sea off Patagonia. "During these two
delightful days we have been gliding onwards: but at a slow pace. I have been
employed in examining some small Crustacea; most of which are not only of new
genera, but very extraordinary ones" (see Barlow, 1933). The crustaceans mentioned are probably those described in Darwin's *Journal* on page 189 (see Darwin,
1839; Burkhardt & Smith, 1985: 307).

1/9854
Amphipoda: Gammaridea
Remarks: condition very poor.

2/11955
Mysidacea: Mysida
Remarks: fragmentary.

1832 Dec. 20th Good Success Bay:

507 Crust: (Sphaeromidae) very fine *on* Fucus. Hermit Island

1/9852
Amphipoda: Gammaridea
Remarks: condition poor.

2/11939
Amphipoda: Gammaridea
Remarks: several specimens in one tube with one reference number.
*? Dryopoides* sp.
*Kuria* sp., several reasonably good specimens. Gammarideans—one single damaged
and one very small specimen.

2/11949
Isopoda: Sphaeromidae
*Cassidinopsis emarginata* (Guérin-Méneville)
*Exosphaeroma ? lanceolata* (White)
*Amphoroidea typa* H. Milne Edwards
*?Dynamenella* sp.
Remarks: assorted genera all in one tube, the *?Dynamenella* and *?Exosphaeroma*
specimens are damaged.

1/9893
Isopoda: Sphaeromidae
*Amphoroidea typa* H. Milne Edwards

510 Pagurus. Goree Sound

6/12923
Decapoda: Paguridae
*Pagurus comptus* White
Remarks: two fragments only.

1833 March Tierra del Fuego

551 Sphaeromida from stones & a Crust. Macrouri. from stomach of a Gadus.—
G[*ood*] S[*uccess*] Bay

  6/12924
  Decapoda: Galatheidae
  *Munida subrugosa* (White)

  6/12925
  Isopoda: Sphaeromidae
  *Exosphaeroma gigas* (Leach)

1833 March Falkland Islands

556 Lithobius: [Gonoleptes *deleted*]: Arachnida: Oniscus. Lumbricus: Falkland
Islands

  Remarks: there is a dry amphipod (Dry 90) in the collection which is not associated
  with any Darwin number, but which is pinned to a small paper label inscribed
  "East Falkland" in Darwin's handwriting. We have identified this as Amphipoda:
  Paramphithoidae.

  2/11951
  Isopoda: Sphaeromidae
  *Cassidinopsis emarginata* (Guérin-Méneville)

560 Sphaeromida & genus closely allied to Atylus; all under stones: Falkland Isd:

  1/9889
  Isopoda: Sphaeromidae
  *Exosphaeroma lanceolata* (White)

  1/9850
  Isopoda: Sphaeromidae
  *Exosphaeroma studeri* Vanhöffen

  1/9856
  Amphipoda: Eusiridae
  *Bovallia ? regis* Stebbing

589 Crustacea: crawling on Corallines; The long one is of a very curious structure;—

  6/12926
  Tanaidacea
  Remarks: condition rather poor.

  6/12927
  Amphipoda: Eusiridae
  *?Pontogeneia* sp.

  6/12928
  Isopoda: Sphaeromidae
  *Isocladus calcarea* (Dana)

1833 E. Falkland Isd:

601 Cancer.

  Dry 96/14519
  Decapoda: Atelecyclidae

*Peltarion spinulosum* (White)

Remarks: there is a large brown paper label associated with this specimen with the number "601" inscribed in Covington's handwriting. Another specimen of this species is represented in the Bell collection labelled "Tierra del Fuego 72" but which does not obviously belong to any *Beagle* material collected from that region.

603 2 Macrouri. Amphipoda. Sphaeroma.

6/12929
Decapoda: Paguridae
*Pagurus comptus* White
Remarks: condition poor, hind-end damaged.

6/12932
Amphipoda: Ampithoidae
*Ampithoe ? brevipes* (Dana)
Remarks: conditions rather poor.

6/12930
Isopoda: Sphaeromidae
*Exosphaeroma lanceolata* (White)

6/12931
Decapoda: Caridea
Remarks: condition poor.


1833 May Maldonado

611 Fresh W Crust. Amphipod.—colour coppery & metallic lustre.
1/9881
Amphipoda: Gammaridae

617 Lithobius & Julus. Sierra las Animas

Myriapod jar/12933
Chilopoda: Scolopendromorpha

Myriapod jar/12934
Diplopoda: Chilognatha
Remarks: broken, only females present.

622 Oniscus. Lithobius. Scolopendra. rocky hills.—

6/12935
Isopoda: Sphaeromidae
*Isocladus calcarea* (Dana)
Remarks: very poor condition.

Myriapod jar/12936
Chilopoda: Scolopendromorpha


1833 June. Maldonado

667 Julus (2 species) Lithobius

Myriapod jar/12937
Diplopoda: Chilognatha
Remarks: condition very poor.

737 Crab. caught in dry hole in one of the low islands of the R. Parana. above Rosario:—

Dry 26/14521
Decapoda: Trichodactylidae
*Dilocarcinus (Dilocarcinus) pagei cristatus* Bott

753 Crustacea, inhabiting the above [mouth of R. Negro Coral *inserted in pencil*]

1/9848
Amphipoda: Caprellidae
*?Caprella* sp.

1834 Jan: 8th:—

782 Various marine productions. 4 or 5 miles from shore; 19 fathoms; Lat. 48° 56'.

786 Crustacea found with (782)

Arachnid jar/12940
Pycnogonida

6/12938
Isopoda: Sphaeromidae
*Isocladus* sp.

6/12939
Decapoda: Hymenosomatidae
*Halicarcinus planatus* (Fabricius)
Remarks: specimen very small.

1834 Jan:

800 Crust. sea beach. [P. Desire]

6/12941
Isopoda: Sphaeromidae
*Exosphaeroma* sp.

6/12942
Isopoda: Idoteidae
*Edotea* sp.

801 Crustacea. pelagic. Watchmen Cape L. 48°.18'. [Caught at night. could not catch any by day. under similar circumstances: small white Entom. [numerous *inserted*] creeping (?). Small white Ento: with long antennae very numerous at night Lat 51°. 53' Long 68° 11'.—]

6/12943
Amphipoda: Gammaridea
Remarks: headless and breaking up.

6/12944
Cumacea: Diastylidae
*Diastylis* sp.

802 Crust. Isopoda.—I believe certainly was on the body of a large dog fish. colour above. mottled greenish grey & tile red, edge [dark brown *inserted*]. [Same locality as last.—]

7/12945
Isopoda: Cirolanidae
*?Cirolana* sp.

1834 Jan: Sts. Magellan.—

810 3 Crustacea; Amphipod, with three spines, mottled pink & white; Hab:
[St Gregory's Bay. Sts of Magellan 15 fathoms.]

7/12946
Amphipoda: Eusiridae
*?Bovallia* sp.
Remarks: condition poor.

7/12947
Amphipoda: Caprellidae
Remarks: condition poor.

814 Crab. white "above tile red" with pimples of "art. blood do." eggs color of
yoke of egg.—[Near Elizabeth. 5 fathoms.]

Dry 99/14522
Decapoda: Lithodidae
*Paralomis granulosa* (Jacquinot)
Remarks: Elizabeth Island is in the Straits of Magellan. The specimen is associated
with a brown paper label in Covington's hand and inscribed "814 Spirits".

815 Crab. mud colour: eggs bright "scarlet red". [Near Elizabeth. 5 fathoms.]

Dry 100/14523
Decapoda: Majidae
*Eurypodius latreillei* Guérin-Méneville

820 Crab. Macr. [Port Famine]

Dry 104/14524
Decapoda: Galatheidae
*Munida subrugosa* (White)
Remarks: see Barlow (1965: 231–2, 250) for transcriptions of DAR 29 (ii). MS
pp 30, 32, 56 which refer to Crustacea in the stomachs of birds. This specimen
(820) is specifically referred to as from the much distended stomach of a "Puffinus".

822 Crustacea. Cape Negro

Annelida jar/12948
Polychaeta: ? Amphinomidae
Remarks: condition rather poor.

1834 Port Famine

830 Crab, black "brownish orange with purple," legs. mottled "orpiment orange"

Dry 108/14525
Decapoda: Majidae
*Eurypodius latreillei* Guérin-Méneville

833 Crust: Schizopod. St Sebastian Bay. *Vast* numbers. 12 Fathoms snow white
except black eyes. [Caught at night:—Here there were very many Whales:—5 miles
out at sea.—]

834 Crustacea. S of C. Penas. 11 fathoms. 3 miles out at sea; caught at night. [Largest & most abundant specimen color [pale *deleted*] red. like half-boiled crab; *excessively* numerous: (833) & 2nd sized amphipod (with dark blue eyes & back) also very numerous.]

> 7/12949
> Amphipoda: Hyperiidae
> *Parathemisto* sp.

> 7/12950
> Mysidacea: Mysida
> Remarks: fragments only.

1834 Feb: T. del Fuego

837 Crustacea, 13 Fath. 2 miles from shore, caught at night.—Cape Ines. Feb. 19th.—

839 Crust.—some miles to South of (837) under similar circumstances. (ship at anchor at night). [The Amphipod (largest & most specimen) excessively numerous. in different places, different sorts appear predominant.—Viz this species here.— a Shizopod [*?*] near St Sebastian (833.834) & other at C. Penas.]

> 2/11952
> Amphipoda: Hyperiidae
> *Parathemisto* sp.

> 1/9875
> Decapoda: Galatheidae
> *Munida gregaria* (Fabricius)

841 Crust: one mile from shore, caught by night. East of Wollaston Isd.

> 7/12952
> Amphipoda: Aoridae
> *Lembos ? fuegiensis* (Dana)
> Remarks: condition poor.

> 7/12953
> Amphipoda: ? Hyperiidae
> Remarks: very small and fragile.

> 7/12954
> Amphipoda: Hyperiidae
> *Parathemisto* sp.
> Remarks: condition poor.

> 7/12951
> Isopoda: Sphaeromidae
> *Dynamenella ? eatoni* (Miers)

842 Crust. on Corallines. low water mark on Corallines. Wollaston Isd.

> Arachnid jar/9858
> Pycnogonida: Nymphonidae
> *?Nymphon* sp.

> 1/9844
> Amphipoda: Gammaridea
> Remarks: condition poor.

1834 Feb: Tierra del Fuego

850 Crust: 1 mile from shore: 16 F[*athoms*] caught at night. NE end of Navarin Isd.
[Is the small & most numerous specimen with rudimentary legs the young of (834).]

1/9846
Copepoda: Calanoida (2 specimens)
Mysidacea (1 specimen)
Stomatopoda (pseudozoe larva, 1 specimen)
Decapoda (larvae, several specimens)

2/11960
Decapoda: Galatheidae
*?Munida* sp.
Remarks: condition rather poor.

851 Crust: from sea-weed &c at bottom. 16 Fathoms. NE end of Navarin Isd.

Arachnid jar/12960
Pycnogonida: Pycnogonidae
*?Pycnogonum* sp.

7/12958
Decapoda: Caridea
Remarks: condition poor.

7/12955
Decapoda: Hymenosomatidae
*Halicarcinus planatus* (Fabricius)

7/12956
Isopoda: Sphaeromidae
*Isocladus calcarea* (Dana)
Remarks: specimens tightly coiled.

7/12959
Amphipoda: Gammaridea
Remarks: small and condition very poor.

860 Crust. Mac. V. 217

7/12961
Decapoda: Hippolytidae
*Nauticaris magellanica* (A. Milne Edwards)
Remarks: specimen damaged and incomplete. DAR 30 (ii) p 217 reads "March
1st. East end of Beagle Channel.—Roots of Fucus. G. Back "Hyacinth & brownish
red" with oblong. marks & spots of gem-like "ultra-marine blue". one white
transverse mark & longitudinal one on tail; 1st great legs. same colour as body
but penultimate limb centre part white edged with "do blue". antepenultimate
ringed with white. "do blue" & "do red". other legs with basal limbs faintly
ringed. but ultimate limbs orange.—Sides with oblique stripes "reddish brown".
Animal most beautiful.—"

1834 March. Tierra del Fuego

867 Crust. Brachy. & Macro. Hab: East end of Beagle Channel

1/9882
Decapoda: Palaemonidae
*?Palaemon* sp.
Remarks: damaged.

Dry 25/14530
Decapoda: Atelecyclidae
*Acanthocyclus albatrossis* Rathbun

Dry 27–9/14527–9
Decapoda: Hymenosomatidae
*Halicarcinus planatus* (Fabricius)
Remarks: a manuscript name is attached to the specimen "H. leachii (Male) Darwin
867".
*Hymenosoma leachii* Guérin-Méneville is a synonym of *Halicarcinus planatus*
(Fabricius). See also 917.

Dry 3/14526
Decapoda: Majidae
*Mithrax (?Mithrax)* sp.

868 Crust. Amphi. & Isopod. Hab: do. [*i.e. as for 867*]

2/11956
Amphipoda: Aoridae
*Lembos fuegiensis* (Dana)

1/9853
Amphipoda: Talitridae
*Transorchestia ?chiliensis* (H. Milne Edwards)

1/9845
Isopoda: Sphaeromidae
*Dynamenella ? eatoni* (Miere)

1834. March. E. Falkland Isd.

903 Crab. South Coast of E. Falkland Isd.

Dry 101/14531
Decapoda: Majidae
*Eurypodius latreillei* Guérin-Méneville

917 Crust: caught as all these marine productions by pulling up roots of kelp.

Dry 32/14534
Decapoda: Hymenosomatidae
*Halicarcinus planatus* (Fabricius)
Remarks: the specimen bears a label inscribed "H. Leachii (female) Darwin 917";
several legs are missing.

Dry 31/14533
Decapoda: Hymenosomatidae
*Halicarcinus planatus* (Fabricius)
Remarks: appears to be a juvenile; all legs missing.

Dry 30/14532
Decapoda: Hymenosomatidae
*Halicarcinus planatus* (Fabricius)
Remarks: specimen bears another label inscribed "E. Falkland I. (female) 175".
The number is probably Bell's.

1/9878
Amphipoda: Aoridae
*Lembos ? fuegiensis* (Dana)

1/9857 & 9877
Amphipoda: Eusiridae
*Bovallia ? regis* Stebbing

1/9849
Decapoda: Paguridae
*Pagurus comptus* White
Remarks: only head region represented.

1834 June Port Famine.—

980 Crustacea. Kelp Roots.

7/12964
Amphipoda: Eusiridae
*Bovallia ? regis* Stebbing

7/12965
Amphipoda: Ampithoidae
*?Ampithoe* sp.

7/12962
Isopoda: Sphaeromidae
*Exosphaeroma studeri* Vanhöffen

7/12963
Isopoda: Sphaeromidae
*Dynamenella ?eatoni* (Miers)
Remarks: fragmentary.

1834  July. Chiloe

993 Crust. Scolopendra &c &c

7/12966
Decapoda: Bresiliidae
*Bresilia* sp.

Myriapod jar/12967
Chilopoda: ?Scolopendromorpha
Remarks: condition poor.

Myriapod jar/12968
Chilopoda
Remarks: condition very poor.

997 Crab. Brachy: both species tinged with dark crimson red [Kelp roots.]

Dry 6–7/14535–6
Decapoda: Xanthidae

*Gaudichaudia gaudichaudii* (H. Milne Edwards)
Remarks: bearing a label inscribed "Xa. Gaudichaudii?
Chile? Darwin 997".

999 Crust. Kelp roots

7/12969
Decapoda: Alpheidae
*Betaeus truncatus* Dana

7/12970–1
Decapoda: Caridea
Remarks: condition very poor.

1000.1001 Squilla: often caught when fishing with nets here; given me.

Dry 102/14537
Stomatopoda: Squillidae
*Pterygosquilla armata* (H. Milne Edwards)

1005 Small Crust: at sea off Valparaiso, taken chiefly 3 or 4 feet beneath surface.
[day time *inserted*]. [some minute ones are curious]

2/11958
Mysidacea: Mysida

8/12974
Decapoda: Caridea
Remarks: condition poor.

8/12972–3
Amphipoda: Gammaridea
Remarks: condition poor.

Annelid jar/12975
Polychaeta: Nereidae
Remarks: condition poor.

2/11953
Copepoda
Branchiopoda
Conchostraca
Remarks: not identified.

Valparaiso
1027 Vaginulus V. 272 [*see also 1180*]

1028 Crust: adhering in numbers on under side of Asterias (1031)

1/9904
Isopoda: Idoteidae
*Edotea magellanica* Cunningham
Remarks: the association with *Asterias* may be fortuitous.

1834 Aug: Valparaiso
1029 Crab, above dark "Cochi. R" legs "Hyacinth & tile R" front pincers purplish.

Dry 107/14538
Decapoda: Grapsidae
*Grapsus grapsus* (Linnaeus)

1030 Crab, whole body mottled with "Carm. & purplish R"

94/14539
Decapoda: Cancridae
*Cancer polyodon* Poeppig

1057 Fresh-water Crust. —Mac. —[Mentioned by Molina as the builder. The mud which it brings up in making its burrows is placed so as to form a circular wall several inches high round the edge of mouth of burrow. —Burrow in marshy field generally near a brook.]

Dry 56–73/14540–57
Decapoda: Aeglidae
*Aegla laevis talcahuano* Schmitt
Remarks: this material includes both Darwin and Hugh Cuming specimens and it has not been possible to attribute specimens to the relevant collectors. The specimens are all associated with the Catalogue number 1103 (q.v.).

1058 2 Scorpions. Gonlipla. Scolopendra Julus. under stones. Mountains [the largest Julus emits much yellow fluid with very pungent smell like mustard.—]

Myriapod jar/12976
Chilopoda: Geophilomorpha

Myriapod jar/12977
Diplopoda: ?Spirostreptidae
Remarks: fragmentary.

1834 Nov. & Dec. Archip: of Chiloe

1078 Crab

Dry 93/14558
Decapoda: Cancridae
*Cancer plebejus* Poeppig
Remarks: the specimen has a brown paper label inscribed in Covington's handwriting "1078 Spirits".

1084 Crabs in the greatest numbers under stones.

Dry 48/14559
Decapoda: Porcellanidae
*Petrolisthes tuberculatus* (Guérin-Méneville)
Remarks: labelled "P. striata Mr Darwin" and on reverse "1084".

Dry 50–1/14560–1

Decapoda: Porcellanidae
*Petrolisthes tuberculatus* (Guérin-Méneville)
Remarks: labelled "P. sulcatus n.sp. Darwin" and on reverse "1084".

1093 Onisci. F. Water Leaches &c &c

8/12982
? Copepoda
Remarks: only a fragment extant.

Arachnid jar/12978
Arachnida: Araneae
Remarks: legs detached.

Annelida jar/12981
Annelida: Hirudinea

Myriapod jar/12979
Chilopoda: ?Scolopendromorpha
Remarks: fragmenting.

Myriapod jar/12980
Diplopoda: Chilognatha
Remarks: badly broken.

1098 Crust: amphipod; burrows & feeds in the leaves of a Fucus [Growing like the Durvilleae utilis Dic. Class.—Plates.]

8/12983
Amphipoda: ?Dexaminidae
Remarks: very small.

1101 Lumbricus. Oniscus Scolopendra. C Tres Montes

Arachnid jar/12985
Arachnida: Araneae

Arachnid jar/12986
Arachnida: Araneae
Remarks: some legs missing.

Annelid jar/12984
Annelida: Oligochaeta

1103 Crabs. F. Water stream [C. Tres Montes] [*see under 1057*]

2/11835
Amphipoda: Gammaridae
*Melita* sp.

1/9860
Isopoda: Sphaeromidae
*Exosphaeroma ? lanceolata* (White)

1/9855
Decapoda: Hippolytidae
*Nauticaris magellanica* (A. Milne Edwards)

1/9880
Decapoda: Paguridae
*Pagurus edwardsi* (Dana)

1104 Crust. pelagic colourless C. Tres Montes [Caught in day time in harbour several yards beneath the surface]

1/9879
Copepoda: Calanoida
Remarks: condition very poor. In Darwin's field pocket-book (see Barlow, 1945:229) there is the following entry dated 27th November 1834: "Small crustacea, purple clouds of infinite numbers pursued by flocks of Famine Petrels" (see also Darwin, 1839:355).

1109 Crustacea pelagic; nighttime; [13 fathoms at C. Tres Montes]

1/9851
Mysidacea: Mysida

3/12055
Decapoda: Hippidae
*Emerita ?analoga* (Stimpson)
Remarks: the number assigned is uncertain. The specimen was marked "177" probably indicating its Bell Catalogue number. Number 177 in that Catalogue gives "Isopoda" and "Darwin 1109".

1/9876
Isopoda: Cirolanidae
*Excirolana ? chilensis* Richardson
Remarks: condition poor.

1110 Crabs: The Amphipoda red coloured under putrid kelp.—

1/9847
Amphipoda: Gammaridea
Remarks: condition very poor.


1834 Decemb. Isd. of Inchin. North Part of Tres Montes

1114 Crustacea (littoral)

8/12991
Amphipoda: Gammaridea
Remarks: condition poor.

8/12988
Decapoda: Majidae
*Pelia ? pacifica* A. Milne Edwards

8/12989
Decapoda: Hymenosomatidae
*Halicarcinus planatus* (Fabricius)

8/12990
Isopoda: Sphaeromidae
*Amphoroidea typa* H. Milne Edwards

8/12987
Decapoda: Porcellanidae
*Allopetrolisthes ? angulosus* (Guérin-Méneville)

1124 Crustacea. by night: pelagic in harbour several feet beneath surface. [caught in middle of cove]

2/11942
Decapoda: Alpheidae
*Synalpheus spinifrons* (H. Milne Edwards)

1127 Crust. Mac: caught in middle of cove, beneath surface, at night, must [be excessively numerous, because both great herds of seal & Tern appear to live on them; Above light purplish black, mouth, joints, rings of abdomen & all thin places, fine purplish red:—Anna Pinks [?] Harbor. Jan 4th.—]

Dry 74–5/14562–3
Decapoda: Galatheidae
*Munida gregaria* (Fabricius)
Remarks: specimen 74 has the chelipeds present but detached. The specimens are labelled "Gr. gregaria Chili? Darwin" and on the reverse "1127". Gr = (Grimothoe) was the name once given to the late larval stage of *Munida gregaria*.

Dry 80–1/14564–5
Decapoda: Diogenidae
*Clibanarius ? panamensis* Stimpson

8/12992
Cirripedia: Thoracica
Remarks: badly damaged.

1835 Jany. Chiloe

1152 Crab. above uniform dull red.

Dry 92/14566
Decapoda: Cancridae
*Cancer plebejus* Poeppig

1155 Balanus. buried in a [stick?] [& a Crust Macrouri *deleted*] [*see 1499*]

1164 Amphipoda: Crust: feeding on dead crab on sand beach at Cucao-Chiloe

8/12993
Amphipoda: Talitridae
*?Talorchestia sp.*

1166 Hermit Crab: Chonos Archipel.

5/11893
Decapoda: Paguridae
*Pagurus ? perlatus* H. Milne Edwards

1835 Feb. Valdivia

1180 Vaginulus V. 272(a)
Remarks: this entry is an anomaly because there is no reference to Crustacea.

Dry 23/14567
Decapoda: Grapsidae
*Hemigrapsus crenulatus* (H. Milne Edwards)
Remarks: unnumbered but associated with label inscribed by Darwin "Valdivia".

3/12444
Decapoda: Porcellanidae
*Petrolisthes edwardsii* (Saussure)

1195 Crab-Coquimbo

Dry 52/14568
Decapoda: Porcellanidae
*Petrolisthes desmarestii* (Guérin-Méneville)
Remarks: specimen inscribed "P. Darwinii Mr Darwin" and on the reverse "1195".

1203 Squilla Coquimbo

Separate jar/12440
Stomatopoda: Squillidae
*Pseudosquillopsis ? lessoni* (Guérin-Méneville)

1835 Coquimbo

1209: 210 Crabs

Dry 22/14569
Decapoda: Grapsidae
*Hemigrapsus crenulatus* (H. Milne Edwards)

1212 Scolopendra: Julus

Myriapod jar/12994
Chilopoda: Scolopendromorpha
Remarks: fragile.

1214 Scolopendra. Julus Valparaiso

Arachnid jar/12995
Arachnida: Scorpiones

Myriapod jar/12996
Chilopoda: Scolopendromorpha

Myriapod jar/12997
Diplopoda: ?Spirostreptidae
Remarks: fragmented.

1223 1224 Crabs. [Coquimbo]

Dry 106/14570
Decapoda: Pseudothelphusidae
*Pseudothelphusa sp.*

*The following specimens are not numbered but are almost certainly from Peru*

Dry 21/14571
Decapoda: Pinnotheridae
*Pinnotherelia laevigata* H. Milne Edwards & Lucas
Remarks: this rather poorly preserved specimen is associated with a label inscribed
by Darwin "Lima". It could therefore belong to any of the samples from Peru
known to have contained crustaceans (possible numbers are 1247, 49, 50, 56, 57,
1262). The specimen has been mistakenly placed next to specimens labelled
*Trapezia.*

Dry 9–10/14572–3
Decapoda: Xanthidae
*Pilumnoides perlatus* (Poeppig)
Remarks: labelled "Pi: perlatus Peru? Darwin".

1835 July. Lima. S. Lorenzo

1247 Crab, above purple, legs speckled.

Dry 95/14574
Decapoda: Cancridae
*Cancer plebejus* Poeppig
Remarks: bears a brown label numbered "1247" in Covington's hand.

1249 Crab. white, back with purple punctures & legs marked with paler do.

Dry 97/14575
Decapoda: Portunidae
*Ovalipes punctatus* (de Haan)
Remarks: bears an attached broken metal tag which reads ". . 49" and which we presume to have been originally 1249.

1250 Crab. beautifully marked with dark Lilac purple. in regular forms, colour brighter on legs. [Crab. very common, when taken draws all its legs close to its body & shams death.—]

Dry 109/14576
Decapoda: Calappidae
*Hepatus chiliensis* H. Milne Edwards
Remarks: an attached brown paper label is inscribed "1250 Spirits" in Covington's hand.

1835 August Lima—

1256 Crabs. Coquimbo Mr. King.

Dry 53–5/14577–9
Decapoda: Porcellanidae
*Allopetrolisthes angulosus* (Guérin-Méneville)
Remarks: labelled "P. unidentata n.s Mr. Darwin".

Dry 15/14580
Decapoda: Majidae
*Microphrys* sp.

1259 Crab. Mac. [Coquimbo]

2/11836
Decapoda: Alpheidae
Remarks: some legs missing.

2/9859
Decapoda: Bresiliidae
*?Bresilia* sp.
Remarks: slightly damaged.

1262 Decapod Notapod [Callao]

Dry 49/14581
Decapoda: Porcellanidae
*Petrolisthes* sp.
Remarks: labelled "P. elongata. Darwin".

1835 Septemb. Galapagos. Chatham Isd.

1269 Fish. See (1270)

1270 Crust Parasit on Fish (1269)

2/11950
Copepoda: ?Siphonostomata
Remarks: in fair condition. Jenyns (1840–42) remarked that all Darwin's fishes reached England intact and that each had a numbered tin label attached. Specimen 1269 was described by Jenyns as *Cossyphus darwini* (now *Pimelometopon darwini*

(Jenyns)) and is in the British Museum (Natural History); accession number 1918.1.31.11.

Dry 79/14582
Decapoda: Palaemonidae
Remarks: fragmentary, not numbered but labelled "Galapagos" in Darwin's handwriting. Possible Galapagos numbers for crustaceans are 1277-8.85. 90-1, 1316.

1835 Sept. Galapagos.—Charles Isd.

1298 Scolopendra. dark reddish brown grow to 14 inches long.—

Separate jar/12998
Myriapoda: Scolopendromorpha
*Scolopendra galapagoensis* Bollman
Remarks: it is interesting to note that this specimen was collected more than 50 years prior to the description of the species by Bollman in 1889; the holotype locality is Chatham Island (United States National Museum number 594).

1835 Otahiti

1332 Fresh water. Prawns

Separate jar/12442
Decapoda: Palaemonidae
*Macrobrachium* sp.

1333 Fresh water shrimp & one marine.

1/9890
Decapoda: ?Palaemonidae
Remarks: specimen damaged, the number appears to be incomplete (e.g. "13. .") and could refer to either Tahiti or New Zealand.

December Bay of Isds.—New Zealand

1346 Crabs—tidal rocks

Dry 47/14584
Decapoda: Porcellanidae
*Petrolisthes elongatus* (H. Milne Edwards)
Remarks: labelled "P. violacea Darwin 1346".

Dry 45/14583
Decapoda: Xanthidae
*Daira perlata* Herbst
Remarks: unnumbered and has two labels, one inscribed by Darwin "New Zealand" and the other "Mr C. Darwin" by ?Bell. It is doubtful whether this specimen is from New Zealand.

Dry 41/14585
Decapoda: Grapsidae
*Paragrapsus gaimardii* (H. Milne Edwards)
Remarks: not numbered (see below).

Dry 91/14586
Pycnogonida
Remarks: this and the previous specimen are unnumbered but both bear labels

inscribed by Darwin "Hobart Town". There are no crustaceans recorded from
this locality in the "Catalogue".

April Keeling Islands

1403 Crabs; the Decapod is nearly white runs like lightning. with erected eyes on the
white sand beaches: Hermit crab, coloured bright scarlet frequent a particular
univalve. & swarms on the coast & in all parts of the dry land far from water:
Another Hermit Crab is likewise found inland [its front legs form a most perfect
operculum to its shell.—]

Remarks: there is no hermit crab in the collection from Keeling, which is to be
regretted in view of the above observations and comments in the "Personal Diary"
for April 2nd (Barlow, 1933; see also Darwin, 1839: 544).

Dry 40/14587
Decapoda: Ocypodidae
*Uca (Planuca)* sp.
Remarks: unnumbered female but labelled "Keeling Isd" by Darwin.

Dry 38/14588
Decapoda: Xanthidae
*? Actaea* sp.
Remarks: unnumbered but labelled "Keeling Isd" by Darwin.

Dry 35/14589
Decapoda: Ocypodidae
*Austruca lactea* (de Haan)
Remarks: unnumbered but may be from Keeling or Tahiti.

8/12999
Decapoda: Ocypodidae
*Ocypode ceratophthalma* (Pallas)


1836 April. Keeling Isds.

1418 Various crabs. beneath stones.

5/13001
Decapoda: Portunidae
*Thalamita admete* (Herbst)

8/13000
Decapoda: Alpheidae
*Synalpheus* sp.

5/13003
Decapoda: Caridea
Remarks: condition very poor.

5/13004
Amphipoda and ?Caridea
Remarks: condition very poor.

5/13002
Decapoda: Porcellanidae
*Petrolisthes ? nobilii* Haig

1428 Crab. V.362

Remarks: unfortunately this famous species *Birgus latro* (Linnaeus) appears not to be represented in the Oxford *Beagle* collections, but we feel its importance merits its inclusion in the present list. An account of this species is given in the "Personal Diary" (Barlow, 1933) and in Darwin (1839).

1835 May. Mauritius.—

1450 Crab

Dry 98/14590
Decapoda: Xanthidae
*Zozymus aenaeus* (Linnaeus)
Remarks: bearing a brown paper label inscribed "1450 Spirits" in Covington's hand.

June Cape of Good Hope.

1460 Fresh Water Crust. Amphipoda.

3/11974
Amphipoda: Gammaridae
*?Melita* sp.

Decemb 1836

Remarks: the following entries are all in pencil and, according to the date, must have been made after the *Beagle* reached England (see Herbert, 1974: 239).

1477 (91) Crust. St Jago

5/13006
Stomatopoda: Gonodactylidae
*?Gonodactylus* sp.
Remarks: condition poor.

5/13005
Decapoda: Xanthidae
*Micropanope* sp.
Remarks: condition poor.

1489 number lost. Crust. Falkland Islds.?? [East S. America *inserted*].

Arachnid jar/9862
Pycnogonida: Pycnogonidae
*?Pycnogonum* sp.

1498 Crab. Keeling Isd (?)

5/13007
Decapoda: Coenobitidae
*Coenobita rugosa* H. Milne Edwards

1499 (1155). Crab.—Chiloe.—

Dry 76—7/14491—2
Decapoda: Galatheidae
*Munida subrugosa* (White)

Remarks: these specimens are labelled "M. Darwinii Chile ? Darwin" and on the reverse side "1499". Bell (1847: 196) referred to "a new species obtained by Mr Darwin and through his kindness, is in my possession" and, on page 207 of the same publication, to the "existence of a new and elegant" species which I find among the fine collection of Crustacea procured by my friend Mr. Darwin".

1500 Crust Brachyuri. Mauritius Dr Page

Dry 11—13/14593—5
Decapoda: Xanthidae
*Leptodius sanguineus* (H. Milne Edwards)
Remarks: associated with the label inscribed "Chl: Darwin 1500". The species was originally referred to the genus *Chlorodius*. Dr Page has not been identified.

1502 Crust. Brachyuri. Mauritius Dr Page

Dry 25/14596
Decapoda: Grapsidae
*Sarmatium* sp.

1503 Crust. Brachyuri. Mauritius Dr Page

Dry 14/14597
Decapoda: Majidae
*?Microphrys* sp.
Remarks: listed as "probably *Mithrax*" in Bell's MS catalogue.

1504 Crust. Brachyuri. Mauritius Dr Page

Dry 36/14598
Decapoda: Ocypodidae
*Uca (Thalassuca) ?vocans* (Linnaeus)
Remarks: labelled "G. vocans Darwin 1504" and is listed in Bell's MS catalogue as "Gelasimus vocans".

Dry 37/14599
Decapoda: Ocypodidae
*Austruca lactea annulipes* (A. Milne Edwards)
Remarks: unnumbered but next to Dry 36.

Dry 33/4/14600—1
Decapoda: Hymenosomatidae
*Halicarcinus planatus* (Fabricius)
Remarks: labelled "H. Darwinii (n.s.) Falkland Is. Darwin 1504". This is an error since the number refers to Mauritius.

2/11927
Decapoda: various families as listed below

Xanthidae
*Liomera rugata* (H. Milne Edwards)
*Trapezia ferruginea* Latreille
*Chlorodiella niger* (Forsskål)
*Atergatopsis ? signata* (Adams & White)

Goneplacidae
*Carcinoplax* sp.

Galatheidae
*Galathea* sp.

Majidae and Xanthidae undetermined.

1507 Hermit Crabs [Mauritius]

Dry 17—20/14602—5
Decapoda: Xanthidae
*Trapezia digitalis* Latreille
Remarks: labelled "Tr. Darwin 1507".

Separate jar/12443
Anomura: Coenobitidae
*Coenobita rugosa* H. Milne Edwards

1544

Dry 8/14606
Decapoda: Xanthidae
*Eurypanopeus transversus* (Stimpson)
Remarks: labelled "Xantho laevis n.s. Darwin 1544".

1616

Separate jar/12441
Decapoda: Palaemonidae
*Macrobrachium* sp.
Remarks: these last two mentioned entries must have received numbers subsequent to the return of the *Beagle* but were not entered in the "Catalogue"; the last number recorded in this is 1529.

## ACKNOWLEDGEMENTS

We are especially grateful to Mr Jimmy Hull whose impeccable curation of the Darwin spirit specimens has helped considerably towards the preparation of this paper. We also thank Dr Tom Kemp, Curator of the Zoological Collections, for giving us every facility for the study of this material and for his continual encouragement.

Thanks are also due to Mrs A. Smith of the Hope Department Library for making available the Bell archival documents and to Mr Peter Gautrey of Cambridge University Library for allowing us access to the Darwin archival material, for which he cares so expertly.

The Royal College of Surgeons of England have graciously permitted the published extracts from the Catalogue contained in the Darwiniana at Down House and we will forever remember the rainy winter afternoons spent studying this, and the particular kindness of Mr Philip Titheradge, the Custodian.

We also thank Miss Monica Price, Assistant Curator of the Mineralogical Collections at the Oxford University Museum, who drew our attention to Gunther's (1925) reference to Darwin material in the Daubeny collection, and we thank Mr Ivor Lansbury of the Entomological Collections who showed us some of the Darwin insects in his care.

## REFERENCES

ABELE, L. G., 1972 The status of *Sesarma angustipes* Dana, 1952. *S. trapezium* Dana, 1852 and *S. miersii* Rathbun, 1897 (Crustacea: Decapoda: Grapsidae) in the western Atlantic. *Caribbean Journal of Science* **12**: 165–170.

BARLOW, N. (Ed.), 1933 *Charles Darwin's Diary of the Voyage of H.M.S. Beagle.* Cambridge: Cambridge University Press.

BARLOW N. (Ed.), 1945 *Charles Darwin and the Voyage of the Beagle.* London: Pilot Press.

BARLOW N. (Ed.), 1965 *Darwin's Ornithological Notes. Bulletin of the British Museum (Natural History). Historical Series* **2**: 201–278.

BARNARD, K. H., 1950 Descriptive catalogue of South African decapod Crustacea. *Annals of the South African Museum* **38**: 1–837.

BELL, T., 1842–43 *Reptiles.* In Darwin, C. *The Zoology of the voyage of H.M.S. Beagle, under the command of Capt. Fitzroy, R.N. during the years 1832 to 1836* . . . Pt. V. i–vi, 51 pp. London: Smith, Elder & Co.

BELL, T., 1847 *A History of British Crustacea* Pt. V: 193–240. London: John van Voorst.

BOLLMAN, C. H., 1889 Myriapoda. In Scientific results of explorations by the U.S. Fish Commission Steamer Albatross. *Proceedings of the United States National Museum* **12**: 211–216.

BOTT, R., 1969 Die Susswasserkrabben Sud-Amerikas und ihre Stamnesgeschichte. *Abhandelungen hersg. von der Senckenbergischen Naturforschenden Gesellschaft* 518: 1–94.

BOTT. R., 1973 Die verwandtschaftlichen Beziehungen der *Uca*-Arten. *Senckenbergiana Biologica* **54**: 315–325.

BOUSIFIELD, E. L., 1982 The amphipod superfamily Talitrodea in the Northeastern Pacific region. I. Family Talitridae: Systematics and distributional ecology. *Publications in Biological Oceanography.* No. 11. National Museum of Canada. Ottawa i–vii. 73 pp.

BURKHARDT, F. & SMITH, S., (Eds) 1985. *The Correspondence of Charles Darwin. 1821–1836.* Volume 1. Cambridge: Cambridge University Press.

CAMPBELL, B. M. & GRIFFIN, D. J. G., 1966 The Australian Sesarminae (Crustacea: Brachyura): Genera *Helice, Helograpsus,* nov., *Cyclograpsus,* and *Paragrapsus. Memoirs of the Queensland Museum* **14**: 127–174.

CHACE, F. A., 1956 Porcellanid Crabs. In: *Résultats Scientifiques. Expédition Oceanographique Belge dans les Eaux Côtières Africaines de l'Atlantique Sud (1948–1949).* Bruxelles III 5: 1–54.

CHACE, F. A., 1968 A new crab of the genus *Cycloes* (Crustacea: Brachyura; Calappidae) from St. Helena, South Atlantic Ocean. *Proceedings of the Biological Society of Washington* **81**: 605–612.

CRANE, J., 1975 *Fiddler crabs of the world (Ocypodidae: Genus Uca).* Princeton: Princeton University Press.

DARWIN, C., 1839 *Journal of Researches into the Geology and Natural History of the various countries visited by H.M.S. Beagle etc.* London: Henry Colburn.

DAVIES, K. C. & HULL, J., 1976 *The Zoological Collections of the Oxford University Museum.* Oxford: Oxford University Press.

DIMAURO, A. A. Jr., 1982 Rediscovery of Professor Thomas Bell's type Crustacea (Brachyura) in the dry crustacean collection of the Zoological Collections, University Museum, Oxford. *Zoological Journal of the Linnean Society of London* **76**: 155–182.

FIZE A. & SERENE, R. 1955 Les Pagures du Vietnam. *Notes. Institut Oceanographique Nhatrang.* **45**: 1–228.

FOREST, A. & SAINT LAURENT, M. de., 1967 Crustacés Decapodes: Pagurides. In: Resultants Scientifique des Campagnes de la "Calypso" au large des côtes Atlantiques de l'Amerique du Sud (1961–1962) VIII 6. *Annales de l'Institut Oceanographique* **45**: 47–170.

GARTH, J. S., 1946 Littoral brachyuran fauna of the Galapagos Archipelago. *Allan Hancock Pacific Expeditions* **5** (10): 341–601.

GARTH, J. S., 1958 Brachyura of the Pacific Coast of America, Oxyrhyncha. *Allan Hancock Pacific Expeditions* **21** (1):1–499.

GUNTHER, R. T., 1904 *A History of the Daubeny Laboratory Magdalen College Oxford.* London: Henry Frowde.

GUNTHER, R. T., 1925 *Early Science in Oxford*. Vol. III. Oxford: Oxford University Press.

HAIG, J., 1955 Reports of the Lund University Chile Expedition 1948–1949. 20. The Crustacea Anomura of Chile. *Acta Universitatis Lundensis* **51** (12):1–68.

HAIG, J., 1960 The Porcellanidae (Crustacea, Anomura) of the Eastern Pacific. *Allan Hancock Pacific Expeditions* **24**: 1–440.

HAIG, J., 1964 Porcellanid Crabs from the Indo-West Pacific, Part I. in: Papers from Dr. Th. Mortensen's Pacific Expedition 1914–1916. 81. *Videnskabelige Meddelelser fra Dansk Naturhistorisk Forening i Kjøbenhavn* **126**: 355–386.

HERBERT, S., 1974 The place of Man in the development of Darwin's theory of transmutation, Part I. To July 1837. *Journal of the History of Biology* 7:217–258.

HOLTHUIS, L. B., 1952 Reports of the Lund University Chile Expedition 1948–1949. 5. The Crustacea Decapoda Macrura of Chile. *Acta Universitatis Lundensis* **47** (10): 1–110.

HOLTHUIS, L. B., 1954 On a collection of decapod Crustacea from the Republic of El Salvador (Central America). *Zoologische Verhandelingen*. **23**: 1–43.

JENYNS, L., 1840–1842 *Fish*. In Darwin, C. *The Zoology of the voyage of H.M.S. Beagle, under the command of Capt. Fitzroy, R.N. during the years 1832 to 1836* . . . Pt. IV. i–xv, 172 pp. London: Smith Elder & Co.

LUCAS, J. S., 1980 Spider Crabs of the family Hymenosomatidae (Crustacea; Brachyura) with particular reference to Australian species: Systematics and Biology. *Records of the Australian Museum* **33**: 148–247.

MANNING, R. B., 1969 Stomatopod Crustacea of the Western Atlantic. *Studies in Tropical Oceanography* **8**: 1–380.

MANNING, R. B., 1971 Eastern Pacific Expeditions of the New York Zoological Society. Stomatopod Crustacea. *Zoologica* **56**: 15–113.

MANNING, R. B. & HOLTHUIS, L. B., 1981 West African Brachyuran Crabs. *Smithsonian Contributions to Zoology* **306**: i–xii, 1–379.

MENZIES, R. J., 1961 Reports of the Lund University Chile Expedition 1948–1949. 42. The zoogeography, ecology and systematics of the Chilean marine isopods. *Acta Universitatis Lundensis*. **57** (11): 1–162.

NAME, W. G. van, 1936 The American land and freshwater isopod Crustacea. *Bulletin of the American Museum of Natural History* **71**: 1–535.

PORTER, D. M., 1985 The *Beagle* collector and his collections. In: Kohn, D. S. (Ed.), *The Darwinian Heritage: A Centennial Retrospect:* Princeton University Press: Wellington, New Zealand, Nova Pacifica.

POULTON, E. B., 1909 *Charles Darwin and the Origin of Species*. London: Longmans, Green & Co.

PRETZMANN, G., 1972 Die Pseudothelphusidae (Crustacea, Brachyura). *Zoologica* **42**: 1–180.

RATHBUN, M. J., 1911 The stalk-eyed Crustacea of Peru and the adjacent coast. *Proceedings of the United States National Museum* **38**: 531–620.

RATHBUN, M. J., 1918 The grapsoid crabs of America. *Bulletin of the United States National Museum* **97**: i–xxii, 1–461.

RATHBUN, M. J., 1925 The spider crabs of America. *Bulletin of the United States National Museum* **129**: i–xx, 1–598.

RATHBUN, M. J., 1930 The cancroid crabs of America. *Bulletin of the United States National Museum* **152**: i–xvi, 1–609.

RATHBUN, M. J., 1937 The oxystomatous and allied crabs of America. *Bulletin of the United States National Museum* **166**: i–vi, 1–278.

SAKAI, T., 1976 *Crabs of Japan and the Adjacent Seas*. Tokyo: Kodansha Ltd.

SCHMITT, W. L., 1942 The species of Aegla, endemic South American fresh water crustaceans. *Proceedings of the United States National Museum* **91**: 431–520.

SMITH, K. V. G., (ed.) 1987 Darwin's Insects. Charles Darwin's entomological notes. *Bulletin of the British Museum (Natural History)*. Historical series **14**: 1–141.

STEBBING, T. R. R., 1914 Crustaceans from the Falkland Islands collected by Mr Rupert Vallentine F.L.S. Part II. *Proceedings of the Linnean Society of London* (1914): 341–378.

STEPHENSON, W. & HUDSON, J. J., 1957 The Australian Portunids (Crustacea: Portunidae). *Australian Journal of Marine & Freshwater Research* **8**: 312–368.

**Index of identifications**

| | | | |
|---|---|---|---|
| *truncatus, Betaeus* | 999 | *Vaginula* sp. | 1180 |
| *tuberculatus, Petrolisthes* | 1084 | Valvifera (undet.) | 145 |
| *typa, Amphoroidea* | 507, 1114 | *vocans, Uca* | 1504 |
| *Uca* sp. | 1403 | Xanthidae (undet.) | 145, 1504 |
| *urostylis, Cirolana* | 351 | | |

*Archives of Natural History* (1996) **23** (2): 279–286

# Supplementary notes on Darwin's insects

By KENNETH G.V. SMITH

70 Hollickwood Avenue,
London N12 OLT.

Since the publication of my account of Darwin's insects (Smith, 1987) further specimens have been located by several workers in the collections of various institutions. A few species overlooked by me are also included. These are all reported here as far as possible and assigned to the relevant Darwin numbers and entries as given in Darwin's "Insects in Spirits of Wine" (a short manuscript list in Darwin's hand in the Cambridge University Library) and his "Insect Notes" (in Syms Covington's hand in The Natural History Museum Library) in the style of my previous paper. Throughout The Natural History Museum is referred to as BM(NH).

## "INSECTS IN SPIRITS OF WINE"

376. Pulex from hairy underside of Tatusia [=*Dasypus*, armadillo Pichiy=Pichi—local name, see Darwin, 1845: 95–96] (375) curious vagabond Ricinia. Bahia Blanca.

SIPHONAPTERA, Malacopsyllidae: *Phthiropsylla agenoris* Rothschild. Professor R.L.C. Pilgrim (1992) found one male of this species remounted from the pin labelled "Flea on *Dasypus minutus* Bahia Blanca N. Patagonia Darwin" in W.S. Macleay's handwriting in the Macleay Museum,[1] University of Sydney. The original label was attached to the slide by Ricardo Palma, National Museum of New Zealand. The species is typically found on armadillos in the Neotropical Region (Smith, 1987: 20; Smith, 1987: 43).

## "INSECT NOTES"

325. Numerous single Coleoptera. Hemiptera from Bahia Brazil.

COLEOPTERA, Eumolpidae: *Bromiodes squamosus* Bryant (1923: 262). The original description includes a *single* Darwin specimen; thus I refer this here rather than to other Bahia entries (348, 349, 3858–3864). I have examined this specimen in the BM(NH) and it is labelled '3' written in ink upon a white label and has a printed label 'Darwin Coll 1885–119'. This is not a Darwin label. On the original MS of the "Insect Notes" there is a note by G.R. Waterhouse, "Many specimens from this collection were presented by C.O. Waterhouse Reg. No. 85. 119. Some of them bear Nos 1–4, as per label:–1, Sydney 3528; 2, Van Diemen's Land; 3, Bahia; 4, King George's Sound Australia".

136

Bryant notes that this beetle attacks coconut palms, a tree which Darwin mentions as one of those forming part of the "thousand beauties" of the scenery of Bahia in his *Journal of Researches* (1845: 497).

618. Coleoptera Do. [Rio de Janeiro].

COLEOPTERA, Corylophidae: *Corylophodes glabratus* Matthews (1887: 109) "Found near Rio de Janeiro". One specimen in BM(NH) labelled "Rio" with "618" on verso. Matthews was apparently unaware that this was a Darwin specimen (and those under 1322 and 3524 below).

1321. 1322. 1323. Coleoptera [Maldonado].

COLEOPTERA, Corylophidae: *Sacium alutaceum* Matthews (1887: 106). "Found near Maldonado, in South America." One specimen in BM(NH) labelled "Maldonado" with "1322" on the verso.

3524. 3525. 3526. Insects by sweeping Do. [Hobart Town].

COLEOPTERA, Corylophidae: *Sericoderus australis* Matthews (1887: 108). Found near Hobart Town in Tasmania. One specimen in BM(NH) labelled "Hobart Town" with "3524" on the verso. This species was later synonymized by Matthews (1899: 121)[2] with *S. fulvicollis* Reitter.

Oedomeridae: *Ischomena sublineata* Waterhouse, (C.O.). One specimen in the BM(NH) collection labelled "Tasmania, Hobart, C. Darwin". The species was described by C.O. Waterhouse (1877, *Cistula Entomologica* 2: 229 [*Sessinia*]) but no Darwin material was cited in that paper.

3528. Insects sweeping near Sydney, S. Covington.

COLEOPTERA, Coccinellidae: *Scymnus cardinalis* Mulsant "1887–42". "3528". This specimen was in the Fiji Sugar Corporation's Collection which was incorporated into the collections of the University of the South Pacific, Fiji. How the specimen came to be there is not known, but it was presumably acquired by exchange. The BM(NH) Accession Number 1887–42 is for "2000 Coleoptera from various localities presented by G.R. Waterhouse Esq. Collected by Charles Darwin in the voyage of the *Beagle*." Professor R.A. Beaver has very kindly presented the specimen to the BM(NH). Two other *Scymnus* species (and many other insects) were included for this entry in my previous account (Smith, 1987).

THYSANOPTERA; Phlaeothripidae: *Idolothrips spectrum* Haliday in Walker (1852) New Holland. No captor given but according to Bagnall (1908: 207) this was a Darwin specimen, and he erected the new genus *Acanthinothrips* to receive it.

According to Mound and Palmer (1983) the type depository of this species is unknown. Although the specimens were sent to Haliday for description they have not been found in the Haliday Collection in the National Museum of Ireland (Dr J.P. O'Connor, pers. comm.). Incidentally, the puzzle of the word 'Fitans' relating to Darwin thrips sent to Haliday by Walker (Smith, 1987: 30) is explained by the late Dr M.W.R. de V. Graham (an expert on Walker) who kindly informed me (pers. comm.) that the word is really 'Titans' (Walker was stressing their unusually large size). The largest species are usually tropical (size range of thrips 0.5–*ca* 15.0 mm, *I. spectrum* being the latter). Thus, my quote should read ". . . Thrips (of which there are some Titans half an inch long)".

This entry is interesting in that Darwin's assistant and amanuensis, Syms Covington (the "Insect Notes" are in his hand, see Smith, 1987: 113, note 6), is actually mentioned by name. A revised and enlarged edition of the short biography of Covington (cited in Smith, 1987) is provided by Ferguson (1988).

3561. Small insects sweeping on coarse grass or brushwood. King George's Sound. March [1836].

HEMIPTERA, Delphacidae: *Pseudembolophora macleayi* Muir (1920: 181). Muir (1934: 575) says "Four specimens somewhat damaged . . . These specimens were taken by Charles Darwin at King George's Sound, South West Australia (1836). The genus and species were described by Muir (1920) from three specimens in the Macleay Collection, Sydney also taken at King George's Sound."

3688. 3689. 3690. 3691. Small insects sweeping in Valleys of mountains near Simons Bay. [Cape] June.

DIPTERA, Empididae: ?*Syneches* sp. Two specimens, each labelled "Cape of Good Hope. C. Darwin" (printed), "Darwin Coll. 1885–119" (printed), "3690" (handwritten) in BM(NH). These specimens differ from all species of *Syneches* included in my monograph of African Empididae (Smith, 1969b) and may prove to be not congeneric. They will be described elsewhere by Dr Bradley Sinclair who kindly drew my attention to them in the Diptera Accessions drawers in the BM(NH).

The following recently located Darwin specimens cannot be assigned to specific numbers or entries.

COLEOPTERA, Anthicidae: *Notoxus* sp. in the Lyman Entomological Museum. Labelled: "Voyage/of the/Beagle" (handwritten in blackish ink on bluish-green paper); "Ex-Musaeo/Parry" (printed in black and surrounded by a black rectangle on a pale buff label); "Muséum Paris" (printed)/'Col La Ferté" (handwritten in blackish ink on very pale blue nineteenth century Paris Museum label); "Collection of the/ Lyman Entomological Museum/and Research Laboratory/Ste-Anne-de-Bellevue/ Quebec, Canada" (printed in black on white label).

The genus *Notoxus* is widespread geographically, so a precise locality or entry cannot be deduced.

The late Professor D.K. McE. Kevan kindly sent this specimen (found by Francois Genier) for examination, but had no idea how it came to be at the Lyman Museum.[3]

The Parry referred to was certainly Frederick John Sidney Parry (1810–1885), a coleopterist whose collection and library were auctioned by Stevens on 16 May 1885 (Allingham, 1924; Chalmers-Hunt, 1976). According to McLachlan (1885) "At one time he [Parry] had a general collection of Coleoptera but latterly it was limited to Lucanidae and Cetoniidae". This suggests that his general collection of beetles, including Anthicidae, could have been privately dispersed *before* his death. This may explain how the Anthicidae could have come earlier to La Ferté [Marquis F. Thibault de la Carte de La Ferté-Senectère (1808–1886)] who had published a monograph on the family in 1848. La Ferté's collections mostly went to Paris (Muséum national d'Histoire naturelle) on his death, though the [identified?] Anthicidae supposedly went to Eugene Louis Bouvier (1856–1944). One can only assume that the Darwin specimen, unrecognized as such (and probably unidentified even to family) came to be in Canada by an exchange of material with the Paris Museum.

Two Parry specimens of *Hypaulax ampliata* Bates (F.) (Tenebrionidae), bearing typical "Voyage of the Beagle" blue labels and like the above without data, were reported on in my previous account (Smith, 1987: 100) (and tentatively referred to King George's Sound), but these came to the BM(NH) via Frederick Bates (B.M. Accession No. 1881–19) (brother of Henry Walter Bates).

EPONYMS

A further insect generic Darwin eponym has come to light.

*Darwiniphora* Schmitz, 1953, Phoridae (part). *Die Fliegen der palaearktischen Region* (Lieferung 171) (1952) **4**, (33): 280. (Diptera, Phoridae). Erected for *Conicera duplicata* Schmitz 1929. Chile. Not based on Darwin material.

Similarly one more Darwin insect eponym in a specific name has been found.

*Polybothris darwini* Théry, A., 1912, *Annales de la Société Entomologique de Belgique* **56**: 114. (Coleoptera, Buprestidae). Madagascar. Not based on Darwin material.

NEW LITERATURE

Finally, two items of new literature give interpolations of some of Darwin's insect specimens described in other works.

A magnificent work by Bouček (1988) includes an assessment and reclassification of the Hymenoptera Chalcidoidea described by Francis Walker (1839) including Darwin material collected in Australia (Sydney, King George's Sound and Tasmania (Hobart)) ("Insect Notes", entries 3524–3528 and 3561). The works of the eccentric systematist Girault who coined several Darwin eponyms (Smith, 1987: 110–111) are also included in Bouček's treatment (though Girault did not describe from Darwin's material).

Graham (1987) provides precise dates for some of the Haliday plates intended to illustrate Volume 2 of Walker's *Monographia Chalciditum* (1839) but published in the *Entomologist* (see below). Unfortunately, the only plate (P) depicting Darwin specimens was not among those specifically treated by Graham, but it can be inferred from his comments to have appeared in issue number 21 of the very rare Volume 1 of *The Entomologist* (1840–1842), probably in August 1842. Some of these figures had incidentally also been used to adorn the covers of *A Bibliography of Irish Entomology* (Ryan *et al.*, 1984) as a tribute to A.H. Haliday, Ireland's greatest entomologist. Of these, the two specimens in side view on the lower half of the cover depict Darwin specimens: *Thoracantha furcata* Walker (upper) and *Eucharis iello* Walker (lower) (compare with fig. 16, Smith, 1987).

Another recent work (Nicholas and Nicholas, 1989) reproduces some of the historic insect illustrations used in my account (Smith, 1987) with further information on Darwin's collecting during the 62 days he spent in Australia.

As appropriate collections of Darwin's entomological contemporaries are studied in the future (e.g. the Rippon Collection, *vide* Kirk Spriggs, 1995) further items collected on the *Beagle* voyage may well come to light; these are most likely to be among unidentified material.

## NOTES

[1] The Macleay Museum, University of Sydney, contains some 500,000 specimens (Hornung, 1984) mostly collected from 1756 to the mid-1800s. The collection was virtually inaccessible from 1912–1982 but the recent appointment of a curator has made the collection available for study. The Museum was founded by Alexander Macleay (1767–1848), an amateur entomologist who, apart from his own collecting, traded with most of the important entomologists of his day and purchased at auctions. Alexander's eldest son, William S. Macleay (1792–1865) worked for the British Government in Paris where his acquaintances included Cuvier and Latreille. In 1839 he joined his father in Sydney, inherited his collection and built upon it. He was a friend of Lt J.B. Emery of H.M.S. *Beagle* (third surveying voyage, 1837–43, in command of Captain Lort Stokes) in 1839, also in 1849 T.H. Huxley, assistant-surgeon of H.M.S. *Rattlesnake* (Fletcher, 1921–1929: 614–5). An excellent account of the third *Beagle* voyage, including collecting in Australia (without Darwin of course) is provided by Horden (1989).

[2] Of further interest to entomological historians is a note in the editor's preface to Matthew 1899 "the figures having been transferred to zinc by Mr J. Collin of Newmarket and I have to thank him for the trouble he has taken to make them accurate copies". The Plate A is labelled "J.E. Collin transf" and plates I–VIII are labelled "A. Matthews ad. nat. del. J.E. Collin transf." and all marked West, Newman [*sic*] imp'. J.E. Collin was a famous dipterist (obituary Smith, 1969a; bibliography Smith, Cogan and Pont, 1969) who as a young man illustrated books and papers on Diptera written by his equally eminent dipterist uncle G.H. Verrall. Apparently this aspect of Collin's work was more widely called upon than previously appreciated. An important item omitted from the Collin bibliography and of some interest to historians of evolution is "Note. Drosophila sub obscura sp. n. ♂ ♀, *Journal of Genetics* 33(1): 60". This 1936 item appears at the end of a paper by C. Gordon on *D. melanogaster* and *D. subobscura* and was overlooked. The species described became one of the mostly widely used in genetic speciation studies!

[3] The Lyman Entomological Museum contains the largest University insect collection in Canada, only exceeded in size by the National Collection at Ottawa. It was founded in 1914 when Mr H.H. Lyman, a distinguished amateur lepidopterist (with interests in Diptera and Hymenoptera) bequeathed his entomological collections and library to McGill University, with financial provision for its maintenance and expan-

sion. Exchanges with other institutions and individuals throughout the world have been a long-established policy (Vickery and Moore, 1964).

# REFERENCES

ALLINGHAM, E.G., 1924 *A romance of the rostrum*. London. Pp 333.

BAGNALL, R.S., 1908 On some new genera and species of Thysanoptera. *Transactions of the Natural History Society of Northumberland* **3**: 183–216.

BOUČEK, Z., 1988 *Australasian Chalcidoidea (Hymenoptera, a biosystematic revision of genera of fourteen families, with a reclassification of species*. Wallingford. Pp 832.

BRYANT, G.E., 1923 New injurious Phytophaga from India and Brazil. *Bulletin of Entomological Research* **13**(3): 261–265.

CHALMERS-HUNT, J.M., 1976 *Natural history auctions 1700–1972*. London. Pp xii+189.

DARWIN, C., 1845 *Journal of researches into the natural history and geology of the countries visited during the voyage of H.M.S. Beagle round the world*. London. Pp viii + 519.

FERGUSSON, B.J., 1988 *Syms Covington of Pambula, assistant to Charles Darwin on the voyage of the H.M.S. Beagle 1831–1836*. Second edition, revised and enlarged. Merimbula [Australia]. Pp 35.

FLETCHER, J.J., 1921, 1929 The Society's heritage from the Macleays. (part) I. *Proceedings of the Linnean Society of New South Wales* **45** (1920): 567–635; (part) II. *Proceedings of the Linnean Society of New South Wales* **54**: 185–272.

GRAHAM, M.W.R. de V., 1987 Some early issues of *The Entomologist* (Vol. 1, 1840–1842) which provide precise dates for Haliday plates of Chalcidoidea (Hym.). *Entomologist's Monthly Magazine* **123**: 185–189.

HORDEN, M., 1989 *Mariners be warned! John Lort Stokes and H.M.S. Beagle in Australia, 1837–1843*. Melbourne. Pp xxiv+359 [+ 24].

HORNUNG, D.S. Jr., 1984 The Macleay insect collection *Antenna* **8** (4): 172–175.

KIRK-SPRIGGS, A.H., 1995 Robert Henry Fernando Rippon (*ca* 1836–1917), naturalist and zoological illustrator. *Archives of Natural History* **22**(1): 97–118.

McLACHLAN, R., 1885 Major F.J. Sidney Parry [obituary]. *Entomologist's Monthly Magazine* **21** (1884–1885): 240.

MATTHEWS, A., 1887 New genera and species of Corylophidae in the collection of the British Museum. *Annals and Magazine of Natural History* (5) **19**: 105–116.

MATTHEWS, A., 1899 *A monograph of the Coleopterous families Corylophidae and Sphaeriidae* (edited by Philip B. Mason). London. Pp 220.

MOUND, L.A. and PALMER, J.M., 1983 The generic and tribal classification of spore-feeding Thysanoptera (Phlaeothripidae: Idolothripinae). *Bulletin of the British Museum (Natural History) Entomology Series* **46**(1): 1–174.

MUIR, F., 1920 A new genus of Australian Delphacidae. *Proceedings of the Linnean Society of New South Wales* **45**: 181–182.

MUIR, F., 1934 New and little-known Fulgoroidea (Homoptera). *Annals and Magazine of Natural History* (10) **14**: 561–584.

NICHOLAS, F.W. and NICHOLAS, J.M., 1989 *Charles Darwin in Australia, with illustrations and additional commentary from other members of the Beagle's company including Conrad Martens, Augustus Earle, Captain Fitzroy, Philip Gidley King and Syms Covington*. Cambridge. Pp xiv+175.

PILGRIM, R.L.C., 1992 An historic collection of fleas (Siphonaptera) in the Macleay Museum, Sydney, Australia. *Proceedings of the Linnean Society of New South Wales* **113**(1): 77–86.

RYAN, J.G., O'CONNOR, J.P. and BIERNE, B.P., 1984 *A bibliography of Irish entomology*. Glenageary (Ireland). Pp 363.

SCHMITZ, H., 1953 Phoridae (part). *Die Fliegen der Palaearktischen Region* (Lieferung 171) (1952) **4** (33): 273–320.

SMIT, F.G.A.M., 1987 *An illustrated catalogue of the Rothschild Collection of fleas (Siphonaptera) in the British Museum (Natural History). vii. Malacopsylloidea (Malacopsyllidae and Rhopalopsyllidae)*. London. Pp 380.

SMITH, K.G.V., 1969a James Edward Collin, F.R.E.S., 1876–1968 [obituary]. *Entomologist's Monthly Magazine* **104** (1986): 145–148.

SMITH, K.G.V., 1969b The Empididae of Southern Africa (Diptera). *Annals of the Natal Museum* **19**: 1–342.

SMITH, K.G.V., 1987 Darwin's Insects. Charles Darwin's entomological notes. *Bulletin of the British Museum (Natural History)* Historical Series **14**(1): 1–143.

SMITH, K.G.V., COGAN, B.H. and PONT, A.C., 1969 A bibliography of James Edward Collin (1876–1968). *Journal of the Society for the Bibliography of Natural History* **5**(3): 226–235.

THÉRY, A., 1912 Descriptions de Buprestides nouveaux de Madagascar. *Annales de la Société d'entomologique de Belgique* **56**: 114.

VICKERY, V.R. and MOORE, G.A., 1964 The Lyman Entomological Museum, 1914–1964. *Canadian Entomologist* **96**(12): 1489–1494.

WALKER, F., 1839 *Monographia Chalciditum. Vol. 2. Species collected by C. Darwin, Esq., London.* London. Pp 100.

WALKER, F., 1852 *List of the specimens of homopterous insects in the collection of the British Museum.* **4**: 909–1188. London.

(Accepted 11 August 1995.)

EDITORIAL NOTE.

In the original printing, page 286 was blank.

*Archives of Natural History* (1983) **11** (2): 315–316

# More Darwin *Beagle* notes resurface

By DUNCAN M. PORTER

Department of Biology,
Virginia Polytechnic Institute & State University,
Blacksburg, Virginia 24061

In May 1981, I reported on the discovery at the Cambridge University Herbarium of Charles Darwin's "Plant Notes" from the voyage of the *Beagle* (Porter, 1981). At about the same time, I made a visit to the British Museum (Natural History) to examine Darwin's "Reptile and Amphibian Notes". The latter were cited as residing in the Museum in 1903 (Woodward, 1903), and subsequent investigation showed that they were deposited in 1845. Darwin originally sent them to Thomas Bell (1792–1880), Professor of Zoology at King's College, London. Bell (1842–43) wrote the reptile and amphibian volume for the *Zoology of the Beagle*. Also, I suspected that the "Insect Notes" might be at the Museum, although they were not listed by Woodward.

George R. Waterhouse (1810–1888), Honorary Curator of the Entomological Society of London (and Curator of the Zoological Society), later Keeper of Mineralogy and Geology at the British Museum (Natural History), published a series of papers of Darwin's *Beagle* insect collections, commencing in 1838 (Waterhouse, 1838). Presumably, he had the "Insect Notes" at hand, as he mentioned "Mr. Darwin's notes" in at least one paper (Waterhouse, 1842). Later, George C. Champion (1851–1927), Librarian of the Entomological Society, referred to "the incomplete copy of Darwin's register at the Museum" (Champion, 1918).

On examining the Museum's library card catalogue, I found that no Darwin manuscript of notes on insects was listed. When I enquired of Miss Pamela Gilbert, Entomology Department Librarian, as to the existence of such a manuscript, she answered that at one time it was in the Department's library, but it was now missing. She then very kindly asked if I would like to examine the book in which it had once resided.

The table of contents of volume D of the "Index to Insect Room Lists" reads: "Page 21. Darwin. C. List of numbers referring to Insects collected by–during Voyage of Beagle." Appended to this are the notes: "list missing. 5–4. 27." and "Still missing. Nov. 1976. RS." However, on turning to the indicated page, I found the manuscript still to be there. I pointed this out to Miss Gilbert, who responded that it was a copy of the original manuscript, not being in Darwin's handwriting. As they are labelled "Copy of Darwin's notes in reference to Insects collected by him", one can understand how the significance of these "Insect Notes" was not hitherto recognized.

Like all of the other lists of *Beagle* specimens known (in addition to those on reptiles and amphibians and on insects at the Museum, notes on birds, fish, mammals, molluscs, and plants are at the Cambridge University Library), they are in the hand of Syms Covington (ca 1816–1861), Darwin's servant and amanuensis during most of the voyage and until 1839. As with the "Plant Notes", there are additions and corrections in Darwin's handwriting, and a few additional notes are added, presumably by Waterhouse. The "Insect Notes" thus are analagous to those other notes on *Beagle* specimens of which we are aware (Porter, 1982). Like the other lists, the "Insect Notes" were prepared for the specialists who were to identify the specimens for Darwin.

143

As is true for the "Plant Notes", the "Insect Notes" contain a number of interesting observations that were never published by Darwin, or when published they were much edited from the original. An example of the former is on a dipteran collected in May 1832 in Rio de Janeiro, Brazil: "570. Dipter. called sand fly, Caught whilst inflicting its painful bite on the knuckle, its favourite place." An example of the latter is on an interesting South American hemipteran (*Triatoma infestans*, the "Great Black Bug of the Pampas"), collected in July 1835: "3423. Bug, caught at Inquique. Peru. Is called in the Mendoza country. Benchuca: is mentioned by many travellers, as so great a pest and blood-sucker: inhabits crevices in old walls. This specimen, when caught, was very thin; even on showing it a finger, would, when placed on a table, immediately run at it with protruded sucker – Being allowed, sucked for 10 minutes: became bloated and globular, 5 or 6 times its original size; 18 days afterwards was again ready to suck: Being kept 4 & ½ months, became of proper proportions, as thin as at first; I then killed it. A most bold and fearless insect." This bold and fearless insect, the vector of Chagas' Disease, a protozoan infection (*Typanosoma cruzi*) of the blood, stands accused of causing Darwin's well-known later infirmities (Adler, 1959).

In this regard, it is interesting to find that this was not Darwin's first encounter with the benchuca. Earlier in 1835 he wrote: "2913. Bug. Mentioned by all Authors, as so great a pest near Mendoza & in the Traversias; sucks very much blood, frequents houses; but this was caught in sandy ravines of Cordilleras of Copiapo; called Benchuca, caught in my bed."

ACKNOWLEDGEMENTS

I am very much indebted to Miss Pamela Gilbert for bringing the "Index to Insect Room Lists" to my attention. Research on Darwin's *Beagle* specimens was funded by a grant from the National Geographic Society while I was a Visiting Fellow at Clare Hall, University of Cambridge.

REFERENCES

ADLER, S., 1959 Darwin's illness. *Nature, London* **184**: 1102–1103.

BELL, T., 1842–43 *The Zoology of the Voyage of H.M.S. Beagle, Under the Command of Captain FitzRoy, R.N., During the Years 1832 to 1836. Part V. Reptiles.* Smith, Elder, London.

CHAMPION, G. C., 1918 Notes on various South American Coleoptera collected by Charles Darwin during the voyage of the *Beagle*, with descriptions of new genera and species. *Entomologist's Monthly Magazine* **54**: 43–55.

PORTER, D. M., 1981 Darwin's missing notebooks come to light. *Nature, London* **291**: 13.

PORTER, D. M., 1982 Charles Darwin's notes on plants of the *Beagle* voyage. *Taxon* **31**: 503–506.

WATERHOUSE, G. R., 1838 Descriptions of some of the insects brought to this country by C. Darwin, Esq. *Transactions of the Royal Entomological Society of London* **2**: 131–135.

WATERHOUSE, G. R., 1842 Carabideous insects collected by Mr. Darwin during the voyage of Her Majesty's Ship Beagle. *Annals and Magazine of Natural History* **9**: 134–139.

WOODWARD, B. B., 1903 *Catalogue of the Books, Manuscripts, Maps and Drawings in the British Museum (Natural History).* vol. 1. Trustees of the British Museum, London, p. 422 cites: "Darwin (C. R.) [MS. Lists and Notes by C. R. Darwin, J. E. Gray and T. Bell relating to the Reptiles and Amphibia obtained by C. R. Darwin during the Voyage of the Beagle.] fol. [1832–45.]".

*Archives of Natural History* (1992) **19** (1): 29–37

# FitzRoy's foxes and Darwin's finches

By W.R.P. BOURNE

Department of Zoology,
Aberdeen University,
Tillydrone Avenue,
Aberdeen AB9 2TN.

There has been much speculation whether the development of Charles Darwin's ideas about the origin of species was influenced by the Captain of H.M.S. *Beagle*, Robert FitzRoy, who was both a scientist and religious fundamentalist. There is little evidence that FitzRoy ever had much influence on Darwin's ideas about the animals usually identified as the source of his views on the Galapagos, however. Little attention has been paid to the presence, at the end of the passage where Darwin first comments on their variation, of a reference to the occurrence of a similar phenomenon in the Falkland Fox *Dusicyon australis*, although the crew of the *Beagle* first encountered the fox two and a half years earlier. On examining their reports on the Falklands, which appears to be the first place where they had much time to study the animals of islands, it seems that initially it was FitzRoy who paid most attention to the fox, and speculated about its origin and variation. He commented on the similarity of the latter to that found in domestic animals, whereas Darwin thought it provided evidence for a "centre of creation" and questioned if it varied. The Falklands may also be one of the first places where Darwin encountered the co-existence of sibling species of finch.

## INTRODUCTION

An extraordinary amount has been written about the voyage of H.M. Surveying Sloop *Beagle* around the world between 1831 and 1836. I would hesitate to add more if I had not had cause to scrutinise it again from an unusual angle. Over the last seven years I have made five visits as a Royal Fleet Auxiliary ship's surgeon to the Falkland Islands, and became curious about the impact of development on their ecology (Bourne, 1988). So I compared the reports of early visitors, who included many of the greatest naturalists of their day. The parallel accounts of the visits of the *Beagle* by FitzRoy (1839) and Darwin (1839) proved particularly interesting, partly because they are among the best descriptions of the early stages of development, partly because they reveal contrasting attitudes in the observers, and partly because they collected several birds said to come from the islands which are no longer found there, including one of two sibling species of finch (Woods, 1988).

On following the information about Darwin's specimens back to his original notes,[1] I was reminded that his first suggestion that species evolve (Barlow, 1935) quoted a Falkland example in addition to the insular variation of animals in the Galapagos, which they did not reach for another eighteen months after leaving the Falklands, as follows: "The only fact of a similar kind of which I am aware, is the constant asserted difference between the wolf-like Fox of East and West Falkland Isds.—If there is the slightest foundation for these remarks the Zoology of Archipelagoes—will be well worth examining; for such facts (would *inserted*) undermine the stability of Species." It has apparently already been postulated by Robert G. Frank jr. that the Falklands may have played an important part in the development of Darwin's ideas about evolution (Broad, 1982), but since I cannot trace a statement by Frank it may be useful to provide another view.

145

## THE FALKLAND FOX

The Warrah, *Dusicyon australis*, is discussed by Gorham (1972) and Clutton-Brock (1977). It was related to a group of South American forms intermediate between the dogs and foxes which were domesticated by the Indians, but was larger and more rufous, with white peripheral markings and a more dog-like skull showing some peculiarities of the dentition. It was unusually tame, and was already vanishing from the accessible eastern part of the islands by the time the *Beagle* arrived. It was soon further reduced by fur-traders, and exterminated by shepherds in the 1870s. It remains debatable whether it arrived naturally on ice or driftwood along the Falkland Current in the remote past, and then developed some of the characteristic peculiarities of animals of oceanic islands, or was a hybrid with domestic dogs introduced by either the Fuegian Indians or Spanish authorities (Hamilton Smith, 1843: 252), who are known to have liberated dogs in the Juan Fernandez group during the eighteenth century in order to keep down the goats used as food by pirates.

Nearly all the early visitors to the Falklands describe these animals, and since both FitzRoy (1839) and Darwin (1839) include similar accounts, it may be of interest to compare them (Table I). FitzRoy, who had had a brilliant career in the Navy (Mellersh, 1968), and was still only 28, discusses it at length with extracts from all the relevant literature in the course of a history of the islands presumably intended as a reply to criticism that he had interfered in their internal affairs. Darwin, who was four years younger, merely provides a portrait less than a third as long in the course of a much more exciting account of a camping trip across East Falkland with the Gauchos during 16–19 March 1834 to the place named after him. While he is much more readable and attracted a wider audience, historians may prefer FitzRoy (compare the extracts from their journals given by Stanbury, 1977).

It would appear from their basically similar accounts that the officers of the *Beagle*, who may have been unusually interested in the large, tame Falkland Foxes because they hunted foxes at home, were originally informed that these animals

### TABLE I

Comparison of the information provided by FitzRoy (1839) and Darwin (1839) in their parallel accounts of the voyage of H.M.S. *Beagle* for the Falklands and their fox, the Warrah, *Dusciyon australis*.

| Topic | FitzRoy | Darwin |
|---|---|---|
| Account of Falklands | 61 pages. | 13 pages. |
| Account of Fox | 3½ pages; twice as large as English fox, long thick woolly fur, fierce, fearless, prey up to size seals and penguins, burrows. | 1 page; only large quadruped found on any remote island, tame, curious, decreasing around settlement. |
| Variation | Larger & darker in east; difference ½–⅓ that between west Falkland and S. American foxes. Similar variation found with domestication. | Different from S. American Foxes, Lowe says smaller and redder in west; but 4 specimens all the same. |
| Origin | Arrive on ice or driftwood via Falkland Current; adapted to climate. | No comment. |
| Specimens | British Museum. | Captain Fitzroy gave two to B.M. |
| Authorities | Simson, Byron, Burney, Bougainville, Forster. | Byron, Bougainville, Molina, Lowe. |

differed on East and West Falkland by the Scottish sealer William Low(e). Low(e) surely deserves at least as much of the credit for the discovery of evolution as Nicholas O. Lawson of Charles Island who subsequently reported the occurrence of similar variation in the animals of the Galapagos (Lack, 1963). On this occasion, which must have been the first time that Darwin encountered variation in the animals of islands, it appears to have been FitzRoy who speculated how the fox could have arrived there and become modified, comparing it to the variation of domestic animals, whereas Darwin was mainly interested in whether it was confined to the islands and questioned whether it showed any variation.

Thus, while Darwin (1839) had already speculated how animals might have reached St Paul's Rocks in the central tropical Atlantic, the fact that he had not yet begun to consider the possibility of local evolution on islands is shown by his unpublished manuscript zoological notes,[1] which include on sheet 228 an account of the fox dated April 1834, when they finally left the Falklands for the last time, expressed in much the same terms as his published accounts (Darwin 1838-43, 1839), but terminating: "—very few of the foxes are found in the NE peninsula of East Island (between St Salvador Bay & Berkeley Sound)—very soon these confident animals must all be killed: How little evidence will then remain of what appears to me to be a centre of creation", a statement which implies that at that time he must still have thought the fox originated locally. Indeed in the Galapagos in September 1835 he was still writing on sheet 337 of these notes: "It will be very interesting to find for future comparison to what district or "centre of creation" the organised beings of this archipelago must be allocated", whereas by the time he published his formal account of the fox in 1838[1] he had modified the end of the prophetic first passage to say no more than that the animal seemed unlikely to last as long as its portrait.

The fact that Darwin initially appears to have been slow to accept the significance of the distribution and variation of the Falkland Fox, which at first received more attention from FitzRoy, when it conflicted with the idea that animals originated in "centres of creation", may help explain, firstly, why he subsequently failed to give it much prominence in the history of the development of his views on the origin of species, and secondly the statement in his autobiography (Barlow 1958: 123):

> I had, also, during many years, followed a golden rule, namely, that whenever a published fact, a new observation or thought came across me,[2] which was opposed to my general results, to make a memorandum of it without fail and at once; for I had found by experience that such facts and thoughts were far more apt to escape from the memory than favourable ones.

## DARWIN'S FINCHES

Personally I originally assumed that since Darwin was popularly believed to have derived many of his ideas about evolution from the study of the birds of the Galapagos (Lack, 1947) he must have been interested in birds, and that his notes on them should be interesting. Actually as pointed out by Sulloway (1982a) the extent of his ornithological expertise appears to have been exaggerated, and his true attitude to birds is revealed in a letter to his cousin, contemporary at Cambridge and lifelong confidant W.D. Fox from Maldonado on 23 May 1833 (Burkhardt and Smith, 1985: 316):—

> —You ask me about Ornithology; my labours in it are very simple.—I have taught my servant to shoot & skin birds, & I give him money.—I have only taken one bird which has much interested me: I dare

say it is as common as a cock sparrow, but it appears to me as if all the Orders had said "let us go snacks in making a specimen".

Apparently this was a Least Seed-snipe *Thinocorus rumicivorus* (Barlow, 1963, addendum 1), in which case its importance seems exaggerated.

In fact, the official *Beagle* natural history collection made by a number of people appears to be the one passed by FitzRoy to William Burnett of Haslar Hospital for presentation to the British Museum. Unfortunately, the British Museum was then going through one of its intermittent periods of neglect of ornithology, during which apparently it often only retained a male, female and juvenile of each species, so that it had no further use for Darwin's largely duplicate collection. This may explain why he later presented it to the Zoological Society of London instead, though some of his specimens eventually also reached the British Museum after it had adopted a broader policy when the Zoological Society sold its collection in 1855.

It has also been shown by Sulloway (1982a, b) that Darwin was in the habit of numbering his specimens instead of labelling them, and recording the details elsewhere, apparently often some time after they were collected.[1] Confirmation is provided in a letter to J.S. Henslow from the Falklands in March 1834 (summarised by Burkhardt and Smith, 1985: 368–371):

> —I have been alarmed by the expression cleaning all the bones, as I am afraid the printed numbers will be lost . . . my entire ignorance of comparative Anatomy makes me quite dependent on the numbers: so that you will see my geological notes will be useless without I am certain to what specimens I refer . . . What I have said about the fossils, applies to every part of my collections—Vidilicet. Colours of all the fish: habits of birds &c &c.

On checking some critical Darwin skins said to come from the Falklands (Woods, 1988) in the Natural History Museum, in the light of this, an Austral Canastero, *Asthenes anthoides*, No. 20211 from the Seebohm Collection appears to be No. 2021 from Chile (Barlow, 1963), and it also seems doubtful whether we can place much more faith in the only supposed Falkland Andean Tapaculo, *Scytalopus magellanicus*, donated by Darwin to the Zoological Society, which like most Darwin specimens (Sulloway, 1982a) has now lost its number entirely.

On examining the route of the *Beagle* and Darwin's manuscript notes[1] in the light of these considerations, it appears that he first met bird sibling species when he discovered the Spanish and Iago Sparrows *Passer hispaniolensis* and *P. iagoensis* living together in the Cape Verde Islands, where their fluctuating relationship is now attracting growing attention (Bourne, 1986–87), but unfortunately made no useful comments on them.

He apparently also found two more sibling species, the Black-chinned and Yellow-bridled Finches, *Melanodera melanodera*, and *M. xanthogramma*, occurring together in the Falklands, although following ecological changes (Bourne, 1988) the second has now disappeared from the islands, so that his notes could again be of considerable interest. Unfortunately, a similar confusion to that found by Sulloway (1982b) with the Geospizinae also occurs with a number of inadequately labelled specimens of these rather variable birds collected on both the islands and the mainland by several different people in the *Beagle*, and his notes are once more unhelpful, although he eventually concluded that while they often mingled in the same flock the Yellow-bridled Finch tended to occur higher in the hills (Darwin, 1838–43). Apparently little more is known about them even now, and since both their ranges are still shrinking they may also merit further study.

Like most country gentlemen of his day Darwin was a keen shooter, and Keith (1955: 223) records how he used to keep a meticulous record of every bird killed by tying a knot in a piece of string attached to a button-hole. None the less it is well known that when the *Beagle* reached the Galapagos in the dry season of 1835 Darwin initially failed to record on which islands the birds were collected (Sulloway, 1982b). Although he comments in his original zoological notes[1] that the thenca or mockingbird *Nesominus* differed from that on the mainland and varied on the different islands (Barlow, 1935), he merely says that all the geospizine finches were feeding together in the dry grass on dormant seeds except for *Geospiza scandens* which frequented the cacti, and that the main variation occurred in the amount of black in the plumage;[1] when John Gould asked if they differed in any other ways he was unable to say more (Darwin, 1837). It was therefore once more FitzRoy (1839: 503) who started to speculate why "Infinite Wisdom" had seen fit to provide them with the varied bills now known to play such an important role in both their ecology and ethology (Barlow, 1935; Lack, 1947; Bowman, 1961).

CONCLUSION

Initially Charles Darwin seems to have been primarily interested in geology, which takes pride of place in his field pocketbooks.[1] An examination of his first report of an example of animal evolution on islands, involving the Falkland Foxes quoted in the introduction (Barlow, 1935; 1963; 282), indicates that, as shown by Sulloway (1982a, b, c) for the Galapagos finches, in other branches of science he collected specimens but at best only numbered them and made separate notes of his observations on them after returning to the *Beagle*, by which time he had sometimes forgotten such details as the collecting localities. He then usually relied upon experts to interpret his discoveries after his return home (Darwin, 1838–43) until he was unable to find one for barnacles, at which point he started to show his true capacity in a work comparable to a Ph.D. thesis (Darwin, 1851).

Little attention has been paid to the role of FitzRoy in his activities, despite the fact that after some prompting (Stanbury, 1977: 19–20) he eventually acknowledged his "steady assistance" in both the preface to his journal (Darwin, 1839), and long afterwards his autobiography (Barlow, 1958). FitzRoy has sometimes been represented as an eccentric religious fundamentalist who provoked Darwin to seek a more rational explanation for the origin of species (Barlow, 1932). Actually when they met FitzRoy was already an experienced hydrographer with wide general interests engaged upon a monumental survey of the coast of South America, who apparently on finding his official surgeon-naturalist Robert McCormick inadequate[3] invited Darwin to join him "more as companion than mere collector" (Barlow, 1932; Gruber, 1969; Burstyn, 1975). In fact FitzRoy was not yet unusually religious, so that Darwin, who was himself training for the Church, and was fairly outspoken in his private autobiography written for his family (Barlow, 1932; 1958), only complains about his temperament, including both a tendency to rages and an "austere silence produced from excessive thinking", which implies that he must have been a manic-depressive, and helps explain his eventual suicide.

Thus it might be better to regard FitzRoy as the senior scientist who set up the investigation, recruited Darwin, and bore much the same relation to him that a modern university supervisor does to his research students. Darwin was, however, never an ideal student. Before the *Beagle* reached the Falklands it had only visited islands briefly. It now made two visits a year apart each lasting several weeks to this bleak archipelago abounding with dramatic and easily-accessible wildlife, the longest period it spent at any remote island, which must have provided ample opportunity for those members of the crew including Darwin who remained with the ship while the boats were away surveying to explore the vicinity and discuss their observations with its inhabitants. Yet although he claimed to have made a full bird collection Darwin as usual missed most of the seabirds which still occur nearby and does not supply much more information about the landbirds either. Instead remarking at an early stage "we have never before stayed so long at a place & with so little for the journal" (Barlow, 1933), he appears to have devoted much time to his correspondence (Burkhardt and Smith, 1985).

The person who therefore appears to have collected most information about the islands and their fox, including speculations about its origin dating back to the seventeenth century, was apparently FitzRoy, whereas Darwin still had to ask FitzRoy the origin of both his fox and Galapagos finch specimens after they returned home (Sulloway, 1982a, b, both Fig. 5). In fact, although FitzRoy must have been rather busy running the *Beagle* and its surveying programme, he also appears to have kept in regular touch with its other investigations throughout the voyage as well. As suggested by Barlow (1932), five years of this association with meticulous hydrographers (Ritchie, 1967) must have been highly educational for young Darwin, causing him to adopt a much more serious approach, so that his formidable father, who was evidently becoming impatient with him when he left, could hardly recognise him, and was filled with approval, on his return (Keith, 1955; Barlow, 1958). He then developed such an obsession with his studies that he periodically seems to have collapsed from the strain (Colp, 1977), possibly as the result of the exacerbation of a multiple allergy (Smith, 1990).

It also appears that by the time he reached the Falklands Darwin was already becoming increasingly interested in the relationship between different animals, as set out in his autobiography (Barlow, 1958: 118–119):—

> During the voyage of the *Beagle* I was deeply impressed by discovering in the Pampean formation great fossil animals covered with armour like that on the existing armadilloes; secondly by the manner in which closely allied animals replace one another in proceeding southwards over the Continent; and thirdly by the South American character of most of the productions of the Galapagos Archipelago, and more especially by the manner in which they differ slightly on each island of the group, none of these islands appearing to be very ancient in a geological sense.
>
> It was evident that such facts as these, as well as many others, could be explained on the supposition that species gradually become modified; and the subject haunted me.

Thus the Falklands may represent an important stage in the sequence of observations along the route of the *Beagle* from Britain to the Galapagos, where Darwin reviewed his progress for his correspondents, moved on from the study of fossils to surviving species, and first encountered reports of the variation of animals on islands. Apparently at this point it was Robert FitzRoy who first questioned how foxes reached the Falklands and why they had become modified there, comparing it to the variation found under domestication subsequently discussed at length in the first chapter of the *Origin of Species* 25 years later (Darwin, 1859); whereas Darwin, who was still speculating about "centres of creation", was initially sceptical, although it would also appear that his earlier experiences on islands must at least have had a sufficient impact on him to cause him to approach the Galapagos with considerable curiosity.

Darwin himself remarked at the end of his autobiography (Barlow, 1958: 140) "I have no great quickness of apprehension or wit which is so remarkable in some clever men, for instance Huxley. I am therefore a poor critic: a paper or book, when first read, generally excites my admiration, and it is only after considerable reflection that I perceive the weak points . . .". Similarly, it would appear that Darwin, who doubt-less had a clearer appreciation of the consequences, did not respond nearly as rapidly as FitzRoy to Lowe's report of the occurrence of variation in the Falkland Fox, until one of the first things that they encountered in the Galapagos was a similar report of variation in the tortoises and birds there from Lawson which he was able to confirm for himself. By the time he started to take the issue seriously FitzRoy apparently had already lost interest in insular variation and gone on to speculate why the finches had such strange beaks, though scientists failed to follow that up for another century.

If Darwin originally questioned FitzRoy's unorthodox conclusions about the varia-tion of the Falkland Fox until he encountered similar phenomena in the Galapagos, it is hardly surprising that insular variation then started to "haunt" him, so that he made much more careful notes about discrepancies with his theories ever after, and still treated the subject cautiously in Chapter 12 (later 13) of the *Origin*; or that FitzRoy with his religious inclinations and difficult temperament became sensitive when he eventually realised the ultimate consequences of an expedition which he originally organised and invited Darwin to accompany as his guest (Mellersh, 1968).

In general, it appears that FitzRoy must have been a brilliant manic-depressive who thought rapidly on his good days, set up the voyage of the *Beagle*, and instantly spotted the most remarkable features of both the Falkland Fox and Galapagos finches. However, he paid so little attention to the implications of his actions that he failed to recognise what he had seen, continually got into trouble, and although he rose to be a vice admiral and pioneer the study of meterology, died by his own hand in debt. Darwin, on the other hand, seems to have been such a cautious obsessional who scrutinised all his options so carefully that he failed to respond to the bell calling him to immortality until it had rung twice, although he then worked himself sick for twenty years to establish his case, and as a result of his general prudence (Keith, 1955) became a rich man and was buried in Westminster Abbey.

ACKNOWLEDGEMENTS

I am indebted to Mark FitzRoy who lent me material relating to his great-great-great-uncle, Dr Juliet Clutton-Brock and staff of the Natural History Museum who showed me his foxes, Ann Datta and Richard Wilding who provided useful refer-ences, and David Stanbury for helpful comments as a referee.

## NOTES

[1] Darwin's ornithological manuscripts in Cambridge University Library are described by Barlow (1963). His field pocketbooks contain far more notes about geology than anything else, and it would appear that he usually brought his biological specimens back to the ship and may not have listed them and made notes on them on separate sheets of paper until they went back to sea. He then copied the information out again on the voyage home in a fairly similar form in the notes on each group compiled for the use of the experts who would be asked to report on them. In consequence some information may have become lost or garbled owing to initial delays in recording the observations or in subsequently copying it out, both into his notes and on to fuller labels for his specimens (Sulloway, 1982a, b). The five volumes of his final report (Darwin, 1838-43) eventually appeared in parts, so that it is also necessary to exercise caution in dating their contents; thus the account of the Falkland Fox in which Darwin had already suppressed his terminal comment of 1834 about it providing evidence for a "Centre of Creation" actually appeared among the first in 1838 (Freeman, 1977).

[2] Smith (1990) remarks that not only do Darwin's symptoms suggest that his ill-health may have been due to a multiple allergy, but that his fairly frequent grammatical lapses, such as where he says "whenever a published fact, a new observation or thought came across me" when he presumably meant he came across it, may imply that he also had mild dyslexia, which is frequently associated with allergies, and would also have escaped recognition in the past. This has a number of possible implications; thus it may help explain his reluctance to take up a regular profession or indulge in public controversy, his slow academic development, his preference for having people read aloud to him rather than read himself when he was relaxing, his extraordinarily laborious working routine, with frequent rests and walks in the garden, leading periodically to a collapse, the length of time that he says he took to appraise a publication, and quite possibly the quality of his thought, since in order to comprehend a statement he may have had to study it much more carefully than ordinary people.

[3] Gruber (1969) may be a little unfair to Robert McCormick, the original surgeon-naturalist on the *Beagle*, who was the subject in his old age of a much more sympathetic portrait by Sharpe (1906). Although he appears to have been a lightweight chatterbox on board and marsh cowboy ashore, he also put in a lifetime of devoted service as a naval surgeon who appears to have got on well with his shipmates on several other heroic expeditions. He must have been placed in an intolerable situation when FitzRoy imported a better-qualified amateur who lived in the same cabin and was allowed to wander freely ashore, while McCormick doubtless had to eat in the wardroom and found that he was obliged by his medical duties to stay close to the ship. Sir James Ross (who wished to write up his results himself) later decided that he preferred McCormick to another "philosopher" (Ross, 1982).

## REFERENCES

BARLOW, N., 1932 Robert FitzRoy and Charles Darwin. *Cornhill Magazine* **72**: 493-510.

BARLOW, N. (Ed.) 1933 *Charles Darwin's Diary of the Voyage of H.M.S. Beagle.* Cambridge.

BARLOW, N., 1935 Charles Darwin and the Galapagos Islands. *Nature* **136**: 391.

BARLOW, N., (Ed.) 1958 *The Autobiography of Charles Darwin 1809-1882.* London.

BARLOW, N., (Ed.) 1963 Darwin's ornithological notes. *Bulletin of the British Museum (Natural History) Historical Series* **2** (7): 201-278.

BOURNE, W.R.P., 1986-7 Recent work on the origin and suppression of species in the Cape Verde Islands, especially the shearwaters, the herons, the kites and the sparrows. *Bulletin of the British Ornithologists' Club* **106**: 163-170, **107**: 4.

BOURNE, W.R.P., 1988 The effect of burning and grazing on the grassland birds of northwest Britain, the Falklands, and other oceanic islands. *International Council for Bird Preservation Technical Publication* **7**: 97-104.

BOWMAN, R.I., 1961 Morphological differentiation and adaptation in the Galapagos finches. *University of California Publications in Zoology* **58**: 1-302.

BROAD, W.J., Survival of the Fittest in the Falklands. *Science* **216**: 1389-1392.

BURKHARDT, F. and SMITH, S. (Eds.), 1985 *The Correspondence of Charles Darwin*, Vol. 1 1821-1836, Cambridge.

BURSTYN, H.L., 1975 If Darwin wasn't the Beagle's naturalist, why was he on board? *British Journal for the History of Science* **8**: 62–69.

CLUTTON-BROCK, J., 1977 Man-Made Dogs. *Science* **197**: 1340–1342.

COLP, R., 1977 *To be an invalid: the illness of Charles Darwin*. Chicago.

DARWIN, C., 1837 (Remarks on the Habits of the Genera *Geospiza, Camarhynchus, Cactornis* and *Certhidea* of Gould). *Proceedings of the Zoological Society of London* **5**: 49.

DARWIN, C. (Ed.) 1838–43 *The Zoology of the Voyage of H.M.S. Beagle, under the command of Captain FitzRoy R.N. during the years 1832 to 1836*. London.

DARWIN, C., 1839 *Journal of Researches into the Geology and Natural History of the Various Countries visited by H.M.S. Beagle under the Command of Captain FitzRoy R.N. from 1832 to 1836*. London.

DARWIN, C., 1851 *A Monograph of the sub-class Cirripedia, with figures of all the species*, 2 vols. London.

DARWIN, C. 1859 *Of the Origin of Species by means of Natural Selection, or the Preservation of Favoured Races in the Struggle for Life*. London.

FITZROY, R., 1839 *Narrative of the Surveying Voyages of His Majesty's Ships Adventure and Beagle, between the years 1826 and 1836. Describing their examination of the Southern Shores of South America, and the Beagle's Circumnavigation of the Globe*. Vol. 2. *Proceedings of the Second Expedition, 1831–1836, under the Command of Captain Robert Fitz-Roy R.N.*. London.

FREEMAN, R.B., 1977 *The Works of Charles Darwin. An Annotated Bibliographical Handlist*. 2nd. ed., Folkestone.

GORHAM, S.W., 1972 History of the so-called Falkland Island "Wolf" *Dusicyon australis* (Kerr) 1792 (formerly *Dusicyon australis* Bechstein 1799). *Falkland Island Journal* 26–39.

GRUBER, J.W., 1969 Who was the *Beagle's* Naturalist? *British Journal for the History of Science* **4**: 266–282.

HAMILTON SMITH, C., 1843 *Mammalia; dogs, part 1st*. In Jardine, W. (Ed.) *The Naturalist's Library*. Edinburgh.

KEITH, A., 1955 *Darwin revalued*. London.

LACK, D., 1947 *Darwin's Finches: An Essay on the General Biological Theory of Evolution*. Cambridge.

LACK, D., 1963 Mr Lawson of Charles. *American Scientist* **51**: 12–13.

MELLERSH, H.E.L., 1968. *Fitzroy of the Beagle*. London.

RITCHIE, G.S., 1967 *The Admiralty chart. British naval hydrography in the nineteenth century*. New York.

ROSS, M.C., 1982 *Ross in the Antarctic — the voyage of James Clark Ross in Her Majesty's Ships Erebus and Terror*. Whitby.

SHARPE, R. BOWDLER, 1906 Birds. In Lankester, E.R. (Ed.) *The History of the Collections Contained in the Natural History Departments of the British Museum*. London.

SMITH, F., 1990 Charles Darwin's Ill Health. *Journal of the History of Biology* **23**: 443–459.

STANBURY, D. (Ed.), 1977 *A Narrative of the Voyage of H.M.S. Beagle*. London.

SULLOWAY, F.J., 1982a Darwin and his finches: the evolution of a legend. *Journal of the History of Biology* **15**: 1–53.

SULLOWAY, F.J., 1982b, The *Beagle* collections of Darwin's finches (Geospizinae). *Bulletin of the British Museum (Natural History) (Zoology)* **42**(2): 49–94.

SULLOWAY, F.J., 1982c Darwin's conversion: the *Beagle* voyage and its aftermath. *Journal of the History of Biology* **15** 325–388.

WOODS, R.W., 1988 *Guide to the Birds of the Falkland Islands*. Oswestry.

(Accepted 27 June 1991.)

# INTRODUCTION

## By Sir Gavin de Beer, F.R.S.

In his invaluable Darwin Bibliography,[1] R. B. Freeman describes the pamphlet *Questions about the Breeding of Animals* as the only one of Darwin's printed works which "in any form of its text, can surely be described as rare". He has not seen it reprinted, described, or mentioned in any work relating to Darwin, and the present occasion is therefore a fitting one for rescuing it from oblivion, especially as it is of great interest in itself.[2]

The pamphlet raises a number of problems. First, there is the date of its production. Next, it introduces a method of obtaining information of which Darwin made great use. Thirdly, it is an eloquent authentic document reflecting the state of ignorance of genetics at that date. Lastly, it serves as a round-up of the myriad questions which Darwin asked himself in his Notebooks on Transmutation of Species,[3] questions the answers to which should have been reflected in his Essay[4] of 1844 if he received any.

### The Date of the Pamphlet

The date of the production of the pamphlet falls within a known time-bracket from the fact that it was issued from the address 12 Upper Gower Street. This was Darwin's

---

[1] R. B. Freeman, *The Works of Charles Darwin. An annotated bibliographical Handlist*, London: Dawsons of Pall Mall, 1965.

[2] The pamphlet consists of eight pages containing twenty-one numbered paragraphs in which there are forty-four queries. The type is set in one column on the inner side of each page, so that the inner margin is narrow and the outer margin wide, presumably to take the answers to the questions. There is no indication of printer, date, or place. The copy in the Zoological Library of the British Museum (Natural History) is the only one known, and is reproduced here in exact facsimile by kind permission of the Trustees.

[3] "Darwin's Notebooks on Transmutation of Species. Parts I to IV", edited by Sir Gavin de Beer, *Bull. Brit. Mus. (Nat. Hist.)* Historical Series, **2**, 1960, pp. 23–183; "Addenda and Corrigenda", edited by Sir Gavin de Beer and M. J. Rowlands, *op. cit.*, **2**, 1961, pp. 185–200; "Part VI. Pages excised by Darwin", edited by Sir Gavin de Beer, M. J. Rowlands, and B. M. Skramovsky, *op. cit.*, **3**, 1967, pp. 129–176. Hereafter referred to as "Notebooks".

[4] Darwin's Essay of 1844, reprinted in *Evolution by Natural Selection*, with a foreword by Sir Gavin de Beer, Cambridge at the University Press, 1958. Hereafter referred to as "Essay".

address from 1 January 1839 until 17 September 1842 when he went to live in the country at Down House in the village of Downe [*sic*], Kent. In the printed *Catalogue of the Books, Manuscripts, Maps and Drawings in the British Museum (Natural History)*, (Vol. VI. Supplement A–I. 1922) the pamphlet is dated "1840?". A search through Darwin's letters during this period tends to confirm this date.

In a letter[5] to his cousin William Darwin Fox, who was well informed on general matters relating to natural history, quoted by Francis Darwin as "written in June", without specifying the year, but printed on a page bearing 1838 in its heading, Darwin wrote: "I am delighted to hear you are such a good man as not to have forgotten my questions about the crossing of animals. It is my prime hobby, and I really think some day I shall be able to do something in that most intricate subject, species and varieties."

As it was only on 28 September 1838 that Darwin, on reading Malthus's *Essay on the Principle of Population*, saw how natural selection works,[6] before which time he refused to hazard his arm on this subject or speak to anyone about it at all, this letter to Fox cannot have been written in June 1838. While he started his Notebooks on Transmutation of Species in July 1837, his Journal shows that during that year, 1838, and during 1839 he was mostly engaged in geological work, preparing his books for publication, visiting and meditating on the Parallel Roads of Glen Roy, and editing the *Zoology of the Voyage of H.M.S. Beagle....* There was also his marriage and the establishment of his home in Upper Gower Street. The letter must probably be dated June 1840, which agrees with the entry in the Journal[7]: "During the summer when well enough did a good deal of Species work."

In January 1841 Darwin[8] wrote again to Fox, saying, "I continue to collect all kinds of facts about 'Varieties and Species', for my some-day work to be so entitled; the smallest contributions thankfully accepted; descriptions of offspring between all domestic birds and animals, dogs, cats, &c., &c., very valuable." This is exactly what the pamphlet is about, and this letter to Fox was clearly a reminder—"I send you this P.S. as a memento"—of his former letter.

The Notebooks on Transmutation of Species, scribbled between July 1837 and July 1839, are riddled with queries about problems of breeding and crossing in plants and animals, hybridism, and inheritance. The pamphlet represents a consolidated questionnaire as regards the animal kingdom.

[5] *Life and Letters of Charles Darwin*, edited by Francis Darwin [hereafter referred to as *Life & Letters*], London: John Murray, 1887, **1**, p. 298.

[6] "Notebook III", MS p. 135; *op. cit.*, **3**, 1967, p. 162.

[7] "Darwin's Journal", edited by Sir Gavin de Beer, *Bull. Brit. Mus. (Nat. Hist.)* Historical Series, **2**, 1959, pp. 1–21; reference on p. 9.

[8] *Life & Letters*, **1**, p. 301.

vi

*The Questionnaire Method*

In the Introduction to *Variation in Animals and Plants under Domestication*,[9] Darwin described his method: "In treating the several subjects included in the present and my other works, I have continually been led to ask for information from many zoologists, botanists, geologists, breeders of animals, and horticulturists, and I have invariably received from them the most generous assistance. Without such aid I could have effected little. . . . I cannot express too strongly my obligations to the many persons who have assisted me, and who, I am convinced, would be equally willing to assist others in any scientific investigation."

So it was a deliberate method, and Darwin might well be grateful to his correspondents for the information which they gave him, because he pursued them unmercifully with cataracts of questions. One of his most profitable quarries was William Bernhard Tegetmeier[10] who, for over twenty years, was continually bombarded with questions about breeds of fowls, Turkish fowls, Indian jungle fowls, rumpless fowls, eggs with chicks just hatching, down in young birds, owls' eggs, laughing pigeons, runts, carriers, skanderoons, rabbits, length of cats' teeth, bees, sex-ratios at birth, race-horse records, and what not besides.

The pressure which Darwin maintained was unrelenting, and must be regarded as a measure of that with which he attacked his subject, which occupied the entire focus of his mind to the exclusion of everything else. 21 November 1857: "Will you keep in mind Malay eggs?" 5 February 1860: "Have you quite thrown me over as too troublesome?" Next, he gets impatient, and on 27 February 1864 he asks for the return of a manuscript which Tegetmeier "has kept so long for him". 7 April 1865, Tegetmeier is asked to get Mr L. Wells to draw some fowls and to submit the drawings to Tegetmeier for his approval: this for the book on *Variation*. 16 January 1866, will Tegetmeier please ask Wells to hurry up with his drawings. 21 February 1868: "I suppose you are too busy a man to try whether a magenta-coloured pigeon would please or disgust his associates." The pigeon was to be stained. 18 April 1869, Darwin wants more information "if it would not cause you too much trouble"; he is ready to receive the eggs which Tegetmeier is to send him, "and the sooner the better". 14 May 1872: "Pray add to your kindness by hereafter telling me the sex of the single bird." 8 August 1875: "You have helped me, and can you do so again?"

Tegetmeier was far from the only quarry. Lawrence Edmondstone,[11] in the Shetland Islands, was asked, 11 September 1856, "Is the Rabbit wild in the Shetlands?. . . . A

---

[9] 1st ed., 1868, p. 14; popular ed., p. 16.

[10] Darwin's letters to W. B. Tegetmeier are preserved in the Library of the New York Botanic Gardens. A microfilm, obtained by Dr Sydney Smith, is in the Cambridge University Library, to the authorities of which I am grateful for communicating it to me.

[11] "Some unpublished letters of Charles Darwin", edited by Sir Gavin de Beer, *Notes & Records of the Royal Society*, **14**, 1959 [hereafter referred to as *Notes & Records*]; the reference is on p. 30.

vii

Shetland specimen put in a jar with lots of salt would be a treasure to me. . . . I fear that you will think that you have fallen on a most troublesome petitioner." He got his rabbit. On 2 August 1857, the barrage continues,[12] introduced in a disarming way: "I thought I had already trespassed to a *quite* unreasonable extent on your kindness: but as you offer with so much good nature to assist me further, I will ask you my question, as I do not think that it can cost very much trouble." It was whether dun-coloured Shetland ponies have a black stripe along the spine.

Henry Tibbats Stainton[13] was at the receiving end of a barrage of questions about insects to which Darwin wanted answers for his work on sexual selection. 18 February 1858: "I am going to be very unreasonable and beg from you any little information which you can give me on some points, which can hardly fail to be very doubtful. I must trust to your kindness to excuse me." There follow seven questions. "Now you will think me, I fear, the most unreasonable and troublesome man in Great Britain; and I can hardly expect you to go seriatim through my queries. But I should be truly obliged for any hints, with permission to quote you."

So it went on, in one subject after another, with Darwin clinging like a leech to the problem on which he was at work. Nothing was allowed to stand in his way, and he realised this himself, as when he wrote[14] to Hooker: "It is an accursed evil to a man to become so absorbed in any subject as I am in mine." For a man as kind as he was, it was this absorption in his subject which blinded him to the probability that some of his correspondents must have shuddered when they received a letter bearing the Downe postmark.

Occasionally in his letters there is evidence of the insertion of questions in periodical journals. On 11 June and 20 June, he sends questions to the long-suffering Stainton[15] with the request that they be inserted and printed in *The Intelligencer*. On 20 December 1862, he tells Tegetmeier that he has published a query on the running powers of the penguin duck. While reaching a wider public, this method of publishing printed questions reduced the pressure on the recipient of a letter.

Apart from these insertions of isolated questions (a collection of which would be of great interest), there is evidence of two questionnaires, additional to the present pamphlet, printed in leaflet or pamphlet form. One of these relates to the expression of the emotions. On page fifteen of the book bearing this title,[16] Darwin reprinted sixteen questions which he said that he had circulated in 1867 to persons who had been in touch

[12] *Notes & Records*, p. 32.

[13] *ibid.*, p. 56.

[14] *Life & Letters*, **2**, p. 139.

[15] "Further unpublished letters of Charles Darwin", edited by Sir Gavin de Beer, *Annals of Science*, **14**, 1958, p. 107.

[16] *The expression of the Emotions in Man and Animals . . .*, London: John Murray, 1872.

viii

with non-European peoples. The questions, with differences and an additional question, were reprinted by the Smithsonian Institution of Washington.[17]

The second questionnaire related to sexual selection in man. To David Forbes, Darwin wrote[18] in March 1868; "I forgot to remind you that any notes on the idea of human beauty by natives who have associated little with Europeans would be very interesting to me. Also if by any strange chance you should have observed any facts leading you to believe that the women of savage tribes have influence in determining which man shall steal them or buy them or run away with them I should much like to hear such facts. I have lately been sending the enclosed queries to all parts of the world and I send a copy to you." It would be pleasant to think that it was from this questionnaire that Darwin's attention was drawn to Captain Burton's observation, which Darwin included in *The Descent of Man*[19] with the words "The Somal men 'are said to choose their wives by ranging them in a line, and by picking her out who projects farthest *a tergo*.'" One wonders if this was the passage which, in a letter[20] to John Murray of 29 September 1870, Darwin agreed to a change in the text to make it less coarse. It does not appear that this questionnaire, or that on the expression of the emotions, has been found.

### *The Questions about the Breeding of Animals and their Answers*

In the first edition of the *Origin of Species*, published in 1859, Darwin was still obliged to say[21] that "The laws governing inheritance are quite unknown", as, indeed, they were to remain, not only until Mendel's discoveries were rediscovered, but until their significance was appreciated,[22] in 1930. Darwin's statement means, in effect, that whatever replies he may have received to his questions about the breeding of animals (which will never be known, except, perhaps, for some references in his books, because all letters received by him before 1862 were destroyed by him), they failed to provide him with any valid theory. Here, all that can be attempted is to point out some passages in Darwin's Essay of 1844, his first round-up of information after broadcasting his questions about the breeding of animals, which reflect some of the questions asked in that pamphlet.

[17] *Annual Report of the Smithsonian Institution for 1867*, 1868, p. 324.

[18] *Notes & Records*, p. 33.

[19] 2nd ed., 1874 (and reprinted) p. 882.

[20] John Murray Archives, Darwin Letters, 212–213.

[21] p. 13. In the 6th ed. (1872, reprinted in World's Classics, p. 13) the sentence runs: "The laws governing inheritance are for the most part unknown." It is not clear what had been learnt during the interval.

[22] Appreciation of the significance of Mendel's discoveries and of the importance of the particulate nature of inheritance dates from Sir Ronald Fisher's *The Genetical Theory of Natural Selection*, Oxford at the Clarendon Press, 1930.

ix

Question 1 is echoed by the section on "Crossing Breeds" in the Essay, but the answer eludes his correspondents (if any) and him: "When two well marked races are crossed the offspring in the first generation take more or less after either parent or are quite intermediate between them, or rarely assume characters in some degree new."[23] In other words, no answer.

Question 2 fares better in the Essay, though not in modern terms. "Although intermediate and new races may be formed by the mingling of others, yet if the two races are allowed to mingle quite freely, so that none of either parent race remains pure, then, especially if the parent races are not widely different, they will slowly blend together, and the two races will be destroyed, and one mongrel race left in its place. This will of course happen in a shorter time, if one of the parent races exists in greater number than the other."[24] This passage serves to show the impenetrability of the fog which the theory of blending inheritance imposed on biologists before Mendel's demonstration of particulate inheritance and Sir Ronald Fisher's equally important demonstration that variance in populations is conserved, subject to natural selection.

Question 3 relates to the efficacy in practice of artificial selection of variants. It is reflected in the "Summary of First Chapter" of the Essay[25]: "Races are made under domestication . . . by man's selecting and separately breeding certain individuals, or introducing into his stock selected males, or often preserving with care the life of the individuals best adapted to his purpose." This was the basis from which Darwin's whole system stemmed.

Question 4 brings up the subject of what Darwin called "Yarrell's Law". In his First Notebook on Transmutation of Species, Darwin wrote[26]: "Mr Yarrell[27] says that old races when mingled with newer, hybrid variety partakes chiefly of former."

"Yarrell's Law" underwent vicissitudes in Darwin's mind, even in the years 1838 and 1839. For instance, in the Second Notebook: "I am sorry to find Mr Yarrell's evidence about old varieties is reduced to scarcely anything—almost all imagination."[28] But in the Fourth Notebook: "Yarrell's Law must be partly true, as enunciated by him to me, for otherwise breeders who care only for first generations, as in horses, would not care so much about breed."[29]

It would be tedious to go through all the questions, which started from a false basis; but numbers 13 and 14 are worth a mention. They refer to telegony, the supposed

[23] Essay, p. 100.
[24] *ibid.*, p. 101.
[25] *ibid.*, p. 110.
[26] Notebook I, MS p. 138; *op. cit.*, **2**, 1960, p. 57.
[27] William Yarrell (1784–1856), zoologist, one of Darwin's oldest friends; he helped him to buy his kit for the voyage of the *Beagle*.
[28] Notebook II, MS p. 121; *op. cit.*, **2**, 1960, p. 94.
[29] Notebook IV, MS p. 112; *op. cit.*, **2**, 1960, p. 173.

x

appearance in offspring from a second mating of the characters of a first sire. The classical case of this error was that of "Lord Morton's mare", which, when crossed first with a quagga and later with a horse, produced by the horse a foal with striped legs. It was really bad luck on Lord Morton that in a certain percentage of cases, foals which have never had any quagga history in the dams out of which they were foaled by horses, have striped legs. Lord Morton's "Law" figures frequently in the Notebooks on Transmutation of Species.[30] It also appears in the Essay[31] of 1844, in all six editions[32] of the *Origin of Species*, and in *Variation in Animals and Plants*.[33] Some superstitions die very hard.[34]

### The Recipients of the Pamphlet

There remains only one problem to consider: to whom did Darwin distribute this questionnaire, in 1840, when his circle of correspondents was still not very large, and most of his friends, not being breeders or farmers, would have been unable to answer them? What can they have thought of the questions, some of which are decidedly imperious, as, for instance, number 4 which ends like an examination paper with the words: "Please to mention in detail any instances you may be acquainted with." How did he distribute the pamphlet; was it through the post to selected addressees, or with the publications of the Zoological Society?

Darwin was not the only man to issue questionnaires. Francis Galton sent one[35] to men of science in 1873, asking for personal details of their lives, physical characters, and mental traits. There have been recent revivals of the practice, trying to arrive at the truth by the method of majority opinion.[36] Darwin's questions had at least the merit of objectivity.

---

[30] Notebook II, MS p. 98; Notebook III, MS p. 8, 152, 165, 168; Notebook IV, MS p. 79.

[31] Essay, p. 133.

[32] 1st ed., p. 165; 2nd ed., p. 165; 3rd ed., p. 183; 4th ed., p. 193; 5th ed., p. 201; 6th ed., p. 129; World's Classics ed., p. 168.

[33] 2nd ed., **1**, p. 435; popular ed., p. 518.

[34] Professor F. A. E. Crew, F.R.S., told me that when he was Head of the Animal Breeding Research Department at Edinburgh, he was consulted by the owner of a bull which had served a pedigree heifer whose owner sued him for damages on the grounds that the heifer was ruined for pedigree breeding. Professor Crew advised the owner of the bull to enter a counter-claim for contamination of the male.

[35] *Life & Letters*, **3**, p. 177.

[36] The questionnaire method has been used for grave abuses. Henri Corbières, of Paris, circularised men of science and asked them to contribute a letter, for him to print and publish, on whether science was compatible with morality. I replied declining to supply him with any·such letter, because I rejected the method of aiming at truth by referendum. This did not prevent him from printing and publishing the letter in which I said this. A Belgian biochemist has sought to advance science by asking zoologists all over the world (on a basis of selection which is not clear) who, in their opinion, was the greatest living zoologist. I was fascinated to find myself listed in forty-first place in this non-competition.

1

2

QUESTIONS

ABOUT THE

# BREEDING OF ANIMALS.

1. IF the cross offspring of any two races of birds or animals, be interbred, will the progeny keep as constant, as that of any established breed ; or will it tend to return in appearance to either parent ? Thus if a cross from the Chinese and common pig be interbred, will the offspring have a uniform character during successive generations, that is, as uniform a character, as the pure-bred English or Chinese ordinarily retains? Thus, again, if two mongrels, (for instance of shepherd dog and pointer) which are like each other, be crossed, will the progeny, during the succeeding generations retain the same degree of constancy and similarity, which might have been expected from pure-bred animals ? Is it known by experience, that when an attempt has been made to improve any breed by a cross with another, that the offspring are apt to be uncertain in character, and that *unusual* care is required in matching the descendants of the half-bred among themselves, in order to keep the character of the first cross ?—Always please to give as many examples as possible, to illustrate these *and the following* questions.

2. **If by care,** the character of half-bred animals (mongrels or hybrids) be pre-

served through some two, three, or more generations, is it then generally found, that the character becomes more permanent, and less care is required in matching the offspring? If this be so, how many generations do you suppose is requisite to form a mixed race, into what is ordinarily termed a permanent variety or well-bred race ?

3. Supposing some new character to appear in a male and female animal, not present in the breed before, will it become more permanent, and less likely to disappear, after it shall have been made to pass through some successive generations, by. picking out and crossing those of the offspring, which happened to possess the character in question ?

4. In crossing between an old-established breed, or local variety, which from time immemorial has been characterized by certain peculiarities, or the animal in its aboriginal state, with some new breed, does the progeny in the first generation take more after one than the other? or if not so, is the character of one more indelibly impressed on the successive generations, than that of the other? Or, which is the same question, is the *breed* of the parents of more consequence, when a *breeding* animal is wanted, than when merely a fine animal is wanted in the *first* generation ? The effect should be observed both in a female of the old race crossed by the new, and a female of the new crossed by a male of the old ; for otherwise the greater or less preponderance of the peculiarities in the progeny might be attributed to the power of the sex, thus characterized in transmitting them ; and not to the length of

3

time the breed had been so characterized. Thus to take an extreme example, we may *presume* that an Australian Dingo is an older breed than a pug-dog : if both were crossed with Spaniel bitches, would the litter in the one case more resemble the Australian, than in the other case the pug : and however this may be, would the pug, or Australian character be most persistent under similar circumstances in successive generations ? How would this be in the various breeds of cattle? Thus if a Bull (or cow) of a breed which had long been known to have been white with short horns, were crossed with a black cow with long horns, (or Bull, if the first were a cow) which had accidentally sprung from some breed, not thus characterized, would there be any marked leaning in the character of the calves to either side ; or would *successive* generations have a stronger tendency to revert to one than the other side ? Please to mention in detail any instances you may be acquainted with.

5. What would the result be, in the foregoing respects, in crossing a wild animal with a highly domesticated one of another species, supposing the half cross to be fertile ? Thus if a fox and hound were crossed with pointer-bitches, what would the effect be both in the first litter and in the successive ones of the half-bred animals ? To form a judgment on this latter point, the subsequent crosses in each case should be relatively the same ; thus the half-bred fox and half-bred hound should be recrossed with the pointer, or with some other, but the same breed.

6. Where *very* different breeds of the

4

same species are crossed, does the progeny generally take after the father or mother ?

7. When two breeds of dogs are crossed, the puppies of the same litter occasionally differ very much from each other, some resembling the bitch and some the dog. In the mule between the ass and horse, this great variation does not appear commonly to occur. Do you know any cases, where two *varieties* have been often crossed, and *mongrels* have been uniformly produced similar to each other within small limits, and intermediate between their parents? And on the other hand, do you know of *hybrids*, between such animals as are generally considered distinct *species*, varying in this manner?

8. When breeds extremely different (as the grey-hound and bull-dog, the pouter and fantail-pigeon,) are crossed, are *their offspring* equally prolific, as those from between nearer varieties (such as from the grey-hound and shepherd-dog). Is the half-bred Chinese pig as prolific as the full-bred animal ? Does a slight cross increase the prolifickness of animals?

9. Do you know of instances of any character in the external appearance, constitution, temper, or instinct, appearing in half-bred animals, whether mongrels or hybrids, which would not be expected, from what is observable in the parents?

10. In those rare cases, where hybrids *inter se* have been productive; have the parent hybrids resembled each other; or have they been somewhat dissimilar, partaking unequally of the appearance of their

5

pure-bred parents. Also, what has been the character of the progeny of such hybrids?

11. When wild animals in captivity, cross with domesticated ones, is it most frequently effected by means of the male or female of the wild one?

12. Amongst animals (especially if in a free, or nearly free condition,) do the males show any preference, to the young, healthy, or handsome females? or is their desire quite blind?

13. Where a female has borne young to two different breeds or kinds of animals, do you know of any instances, of the last born partaking of any part of the character of the first born, and to what extent?

14. When a female of one breed has been crossed by a male of another breed *several times*, do the last-born offspring resemble the breed of the father, more than the first-born, and therefore are they more valuable in those cases, where the peculiarity of the father is desired?

15. Do you know instances of any peculiarities in structure, present for the first time in an animal of any breed, being inherited by the grand-children, and *not by the children?* It cannot be said to be *inherited* without it appear in more than one of the grand-children, or without it be of an extremely singular nature; for otherwise it ought to be considered as the effect of the same circumstances, which caused it to appear in the first case.

6

16. What are the effects of breeding in-and-in, very closely, on the males of either quadrupeds or birds? Does it weaken their passion, or virility? Does it injure the secondary male characters,—the masculine form and defensive weapons in quadrupeds, or the plumage of birds? In the female does it lessen her fertility? does it weaken her passion? By carefully picking out the individuals most different from each other, without regard to their beauty or utility, in every generation from the first, and crossing them, could the ill effects of inter-breeding be prevented or lessened?

17. Where any animal whatever (even man) has been trained to some particular way of life, which has given peculiarity of form to its body by stunting some parts and developing others, can you give any instances of the offspring inheriting it? Do you know any such case in the instincts or dispositions of animals? If an animal's temper is spoilt by constant ill usage, or its courage cowed, do you believe the effect is transmitted to its offspring? Have any cases fallen under your observation, of quadrupeds (as cats or pigs, &c.) or birds (fowls, pigeons, &c.) born in this country, from a foreign stock, which *inherited* habits or disposition, somewhat different from those of the same variety in this country? If removed early from their parents, there are many habits, which we should be almost compelled to believe were inherited, and not learnt from them; and if transmitted to any half-breed we should feel sure of this.

18. Can you give any detailed account of the effects on the mind, instincts or dis-

163

7

position of the progeny, either in the first or in the succeeding generations from crossing different breeds, (for instance carrier and tumbler pigeons, grey-hounds and spaniels) or different species, (as fox and dog.) Do they show an aptness to acquire the habits of both parents? Or do they partake strongly of the habits of one side, (if so, which side?) with some peculiarity showing their hybrid origin? Or do they entirely follow one side?

19. Can you give the history of the production in any country of any new but now permanent variety, in quadrupeds or birds, which was not simply intermediate between two established kinds?

20. Do you know any cases of different breeds of the same species, (as of dogs &c.) being differently affected by contagious or epidemic diseases, and which difference cannot be attributed merely to a greater vigour in the one breed than in the other? In countries inhabited by two races of men, facts of this kind have been observed.

21. All information is valuable, regarding any crosses whatever, between different wild animals, either free or in confinement, or between them and the domesticated kinds; *equally so* between any different *breeds* of the same species, especially the less known kinds, as Indian with common cattle, different races of Camels, &c. Please to state all or any particulars, for what object the cross was made and whether it is habitually made; whether the female had offspring before; whether she produced as many of the half-breed at one birth, (if more than one be produced) as she probably would have done of the pure

8

breed; whether the progeny were fertile *inter se,* or with their parents whether they resembled one stock more than the other and in what respects, and which; and whether the favoured side was the male or female. State, if known, whether the progeny differ when stock (A) is the father and (B) the mother, and from what it does where (A) is the father and (B) mother. If the half-bred are fertile, *inter se* or with the parent stock, describe the offspring whether like their parents and all like each other, or whether they revert to either original stock, or whether they assume any new character?

C. DARWIN.

12, *Upper Gower Street, London.*

*J. Soc. Biblphy nat. Hist.* (1969) **5** (3): 220–225.

# Darwin's *Questions about the Breeding of Animals,* with a Note on *Queries about Expression*

By R. B. FREEMAN and P. J. GAUTREY

Department of Zoology,
University College London, and
The University Library,
Cambridge, England.

The Society, through its Sherborn Fund, has recently published an offset facsimile of Charles Darwin's *Questions about the Breeding of Animals* (Darwin, 1968), with an introduction by Sir Gavin de Beer. This facsimile was based on the only copy known at the time, that in the Zoological Library of the British Museum (Natural History). The original is number 82 in the Darwin *Handlist* (Freeman, 1965). It has been bound so that the sewing cannot be seen, and has been slightly cut by the binder, giving a page height of 280 mm to the nearest 2·5 mm.

Another copy was found in September 1968 in the Darwin archive in the University Library, Cambridge. This one is in its original state, uncut and with the sewing intact. Of the twenty-one numbered items nineteen have been answered in manuscript in the wide outer margins, and the copy is dated 10 May 1839 in Darwin's hand. Also in the Darwin archive is a second set of answers to five of the items. These are not on a copy of the questionnaire itself, but on a half sheet of paper folded once, making four pages, the last of which is dated 6 May 1839. Between them, these two documents make it possible to give a more accurate bibliographical description of the *Questions* than was previously available. This new material was discovered just too late to be incorporated in de Beer's introduction to the facsimile, and does not entirely support his conclusions.

Because the facsimile exists, there is no need to give a detailed description of the original. It will be sufficient to add that the uncut Cambridge copy is 290 mm high by 230 mm wide, and consists of a single wove medium sheet. It is folded in quarto, forming four leaves of eight pages, and sewn through the fold. Neither surviving copy has any wrappers and it is probable that it was issued without. Its date of earliest circulation can now be narrowed down to between 1 January 1839, the day that Darwin moved into 12 Upper Gower Street, and 6 May 1839, the earlier date on the answers. Other copies may of course have been circulated later. The date of composition and printing could have been during the last quarter of 1838, because Darwin then knew that he would be moving to Upper Gower Street, the address given on the questionnaire, on 1 January. Early in 1839 is, however, more probable.

A date "?1840" is given in the printed *Catalogue of the Books, Manuscripts, Maps and Drawings in the British Museum (Natural History)*, Vol. VI, Supplement A–I, 1922. The date [1840] is printed on the title page of the facsimile, but de Beer is cautious in his text. He concludes that: "A search through Darwin's letters during this period tends to confirm this date." He refers to a letter from Darwin to his second cousin William Darwin Fox, dated June, which is printed in *Life and Letters* (Darwin, F., 1887) under a date heading of 1838. This letter mentions "my questions about the crossing of animals", probably in

220

reference to the printed pamphlet. He shows that a date of 1838 must be wrong and implies one of 1840; the present evidence would suggest 1839 as more probable.

On page ix de Beer states: " . . . whatever replies he may have received to his questions about the breeding of animals (which will never be known, except, perhaps, for some references in his books, because all letters received by him before 1862 were destroyed by him), they failed to provide him with any valid theory". The evidence for this destruction is Francis Darwin's statement in the preface to *Life and Letters*: "This process [of burning letters], carried on for years, destroyed nearly all letters received before 1862." Although de Beer was wrong to include documents with letters, the two sets of answers do exactly illustrate the point that he makes on page x in quoting a passage from Darwin's *Essay of 1844* (in Darwin, F., 1909): "This passage serves to show the impenetrability of the fog which the theory of blending inheritance imposed on biologists before Mendel's demonstration of particulate inheritance." The fog was made deeper because such characters as configuration, hardiness, fat content, speed or even coat colour are not suitable for the demonstration of particulate inheritance. All depend on the interaction of numerous genes, and the breeders do not seem to have recognized any of the few characters, such as hornlessness in cattle, which may be due to single factors. These breeders were highly successful, and the improvement of British livestock during this period was great, but the methods and results could not contribute towards what Darwin needed.

The first, and more complete, set of answers is headed in pencil in Darwin's hand: "Mr George Tollet, Betley Hall, 10th of May 1839". George Tollet, or Tollett as it is sometimes spelled, was born George Embury on 3 August 1767, and assumed the name of Tollet on inheriting estates at Betley from Charles Tollet, a cousin. He was a Justice of the Peace, Deputy Lieutenant for Staffordshire, and Recorder of Newcastle-under-Lyme 1792–1800 (Simms, 1894). "Mr Tollet was long before the public as an active magistrate and devoted agriculturist; one of the promoters of that agricultural movement which has produced such great and beneficial results to the Kingdom at large" (Hinchliffe, 1856). He was a member of the group of agricultural reformers whose meetings were originated by Thomas William Coke, later Earl of Leicester, the first being held at the latter's house Holkham Hall, in Norfolk. Another member was Sir John Sebright, who is mentioned in the answers. George Tollet died in 1855. He is not to be confused with George Tollet the Shakespearean scholar, and elder brother of Charles. He also lived at Betley Hall, but belonged to the previous generation.

Betley is a small village in Staffordshire, only a few miles from Maer, the home of Josiah Wedgwood (1769–1843), Darwin's uncle and father-in-law. It is known (Darwin, F., 1887) that Charles and Emma his wife were at Maer and at Shrewsbury, Charles's family home, between 26 April and 13 May 1839, and it is possible that the questionnaire may have been given or sent to Tollet at that time. This may answer in part de Beer's query as to how Darwin distributed it.

George Tollet's answers are as follows:

> 1. It will not I think keep constant; it will tend in appearance to one parent or the other but will more commonly shew in individuals an evident admixture of both—In pigs some would be more like the Chinese others more like the common breed but all would shew the cross in some degree. The Chinese character being probably the older race would prevail. As before stated the progeny will shew the admixture for some generations even under the greatest care taken in matching them.

2. There is little doubt but by great care in selecting for several generations the character would become more permanent, but for a long time the characters of the different varieties of the first cross would frequently occur—but it must have been by a long perseverance that the different varieties or races have been formed—Thirty years ago in order to get a large breed of fowls I crossed *once* with the large long-leg'd *Malay-breed*. Having endeavoured to breed fowls with compact bodies and short legs from the time of the first cross yet the long-leg'd character of the Malay fowl together with a tenderness of the feet *from cold* is frequently breaking out. This year a hen of my sort was crossed with a particularly handsome short leg'd Dutch fowl belonging to my son. But the Malay blood shewd itself most pointedly in two young cocks the produce of this cross: So that it is difficult to say how many generations would be requisite [to] form a mixed race into a permanent variety.

3. I think a new character appearing in a male & female of an established race would probably be made permanent after successive generations by carefully selecting and breeding from those which happen to possess the character in question. To affect this after a certain time the choice out of a considerable number would be desirable.

4. If the Australian Dingo is an older or aboriginal breed the cross from the spaniel bitches would I think more resemble the Australian than the pug: and would be most persistent under similar circumstances in successive generations.
What has been said of the Malay fowl—I think shews it to be much nearer the aboriginal breed—probably from proximity to the jungles of India.

5. If you can suppose such a cross to be fertile I think the character of the fox would prevail for a great length of time.

6. I should be inclined to think they would be more apt to take after the father than the mother but it would be by no means certain.

8. I think every cross has a tendency to make the offspring prolific. But if the Chinese pig were crossed with a race less prolific the offspring would most likely not be so prolific as the original Chinese tho more prolific than the race with which it was crossed.

  * *Breeding in and in as it is called* has a manifest tendency to decrease the prolifickness of animals.

  * Sir John Sebright many years ago published an interesting little book on this subject [see Seebright, 1809]. I have mislaid it or I would have referred to it.

9. No experience.

10. No experience.

11. I should suppose by means of the male. This I understand is generally the case in breeding hybrid Canary birds.

13. No experience.

14. No experience but I should think it would be a mere chance.

15. No experience.

16. In bulls it has a tendency to weaken the masculine form and I suspect also the virility. I have seen very high bred bulls that had lost many of the characteristics of the males. It was thought an improvement but it was found to be an imperfection, and the more masculine appearance was again preferred—The same circumstances happened in the breed of sheep— *Breeding in and in* or *closely* in an improved stock greatly lessens the fertility of cattle. Mr Mason of Chilton sold a high bred Durham heifer for 1000 guineas upon a warranty that she was to produce 3 calves or the bargain was to be off—She was too high bred to do this—Some of the defects of interbreeding might be lessened in a large stock by judicious selection.

17. I think timidity—ill temper etc are hereditary—and I should not choose to breed from animals that had those or similar defects.

18. No experience.

19. No exp[erience].

20. No experience.

21. In crossing cattle which may be very advantageous under certain circumstances, the agriculturist would never by choice go on breeding from mongrels. In dairy cattle crossing may be generally advantageous and the choice would be of breeds yielding either in quality or quantity or both the best return of milk butter & cheese. The *Holstein, Dutch* or *Holderness* cow is large & gives a large quantity of milk but not of rich quality. The Ayrshire cow is smaller and gives a good return of milk of better quality. If from circumstances a smaller animal with better milk were desirable the dairy farmer would put *an Ayrshire\** bull to his Yorkshire short horned cows. But to carry on the breed he would not use a bull of the mixed breed but would go on with an Ayrshire bull till his object was attained. If he afterwards wished for larger size he would use a Yorkshire bull & so keep changing to suit his purpose. The Alderney cow is still smaller than the Ayrshire & the milk is still richer. From one or other of these crosses, which may all be reckoned *short horned varieties* the best sort of milking cow may be obtained. N.B. The first cross generally gives so much vigour that the produce is apt to be superior to either of the parent breeds.

\* N.B. Where the choice can be had it is always better to have the cows of the larger and the bull of the smaller breed.

The second set of answers is signed at the end "R. S. Ford, Swynnerton May 6th 1839." Richard Sutton Ford was baptized in February 1785 and died about 1850. At the time of his marriage in 1807 he was described as a farmer of Newstead near Trentham, and he was for many years Agent to the Fitzherbert Estate at Swynnerton. Swynnerton and Trentham are both Staffordshire villages close to Maer.

R. S. Ford's answers are as follows:

1. In crossing varieties of cattle & sheep, I have observed generally, that although the first cross has usually produced a satisfactory result—such in fact as might have been expected—a remarkable inconstancy has often attended subsequent crossings between this progeny and either of the parent stocks, as well as the breeding from the produce of the cross exclusively.

4. In crossing between an old established variety, and a new, or mixed breed, the progeny will usually take more after the former than the latter. I may instance the effect produced by crossing our variously bred mares with an Arabian stallion, in which the peculiarities of the sire have been remarkably impressed on the offspring, and continued through many generations, though of course becoming gradually weaker in each. Nor am I inclined to attribute this to the power of the sex, although not aware that we have any examples of Arabian mares having been put to our native stallions. But in breeding sheep, which I have crossed *both ways*, between several of our old varieties and the new Leicestershire breed, I have not been able to determine, on the whole, that there was any preponderance of character in favour of either sire or dam, though in numerous cases the offspring has taken much more after the one parent than the other. [Marginal note in C.D.'s hand—C. Did they take more after old breed than new??] The colours of the progeny of domestic animals are extremely capricious. In crossing a black-sided long-horned cow, with a red and white mottled bull, of the short-horned Durham breed, the produce in one year was a calf almost entirely red, and in the following year a white one. And I have now in my possession a brown horse, the sire and dam of which were both grey: nor do I mention these as being at all rare instances of the kind.

12. The sexual passion of the males, in cattle and sheep at least, appears to be wholly indiscriminate, without regard to age, symmetry, or colour—the females, if not in almost perfect health, are never in season for the male—It may be remarked however, that a bull or ram, having served one female, will commonly prefer a fresh one to the same again; and sometimes

when he has such choice, after having served the whole, he will continue to follow a particular one, though I think this is wholly incidental, and not the effect of partiality or design. Whether or not any preference of a more marked kind is shewn by these male animals in their wild state, I am unable to say.

16. Most of our fine breeds of cattle and sheep (or what is termed by Breeders, "High Blood") have been raised by breeding "in and in"; and when this has been pursued to a great extent— that is, through many generations—the result has been, feeble virility with effeminate appearance in the males; weak passion in the females, which even in sheep rarely produce twins; and diminutive size and great delicacy in the offspring during the first month; to which may be added, that a deficient supply of milk is usually a consequence of this high breeding. But whether these habits arise altogether from breeding in and in, or are more or less occasioned by the finest-boned and smallest-headed males being selected to breed from—such animals being more disposed to take on fat—I do not undertake to determine. [Marginal note in C.D's hand—No because of dogs and pidgeons]. Certain it is however, that in these extreme cases, the procreative powers of the stock have been in a great measure sacrificed in order to attain the greatest possible tendency to fatten. I will here notice one curious and well established fact, which so far as I am aware is peculiar to cattle: if a cow produce at one birth, two male calves, or two females, in either case both animals will be fertile; but if she produce a male and a female calf at the same birth, though the male will possess the power of propagating his species, the female is invariably barren.

18. That the dispositions of animals are not altogether the effect of training, but in some degree at least hereditary, may be gathered from the circumstance, that if the eggs of wild fowl—as of the wild-duck for example—be hatched under a tame duck, or dunghill hen, the produce will shew in a striking manner their wild habits: on the approach of intruders which would scarcely attract the notice of ducklings regularly descended from any of our tame breeds, they will endeavour to escape, or hide themselves; and on arriving at maturity, unless previously pinioned, and well guarded, they will desert their foster friends, and rejoin their original kindred, if such there be in the neighbourhood.

The maxim, "Like produces like", is generally true; and I think this applies equally to temper, disposition, constitution and habits, as to form and size, though all these may be varied by incidental or artificial means. With respect to colour—we find that most animals in their wild, or native state, are true to that of their respective breeds; or if their colours vary, the variations are, with rare exceptions, very slight; and there may be a few species which even domestication does not appear to have altered in this respect—the Guinea fowl for example. But in general, tame animals are of an almost infinite variety of colours: whether this is to be accounted for on the principle laid down in Genesis Ch. XXX v. 37 and following*, or by the admixture of breeds which is continually taking place amongst them, I do not presume to decide: though I am of opinion that the latter cause must operate very powerfully; whilst I do not mean to deny the influence of the former.

On the subjects of most of these questions, many opinions are current which I believe to be

---

* Footnote added by us—R.B.F. & P.J.G. 37 And Jacob took him rods of green poplar, and of the hasel and chesnut tree, and pilled white strakes in them, and made the white appear which *was* in the rods.     38 And he set the rods which he had pilled before the flocks in the gutters in the watering-troughs, when the flocks came to drink; that they should conceive when they came to drink.     39 And the flocks conceived before the rods, and brought forth cattle ring-straked, speckled, and spotted.     40 And Jacob did separate the lambs, and set the faces of the flocks toward the ring-straked, and all the brown in the flock of Laban: and he put his own flocks by themselves; and put them not unto Labans cattle.     41 And it came to pass, whensoever the stronger cattle did conceive, that Jacob laid the rods before the eyes of the cattle in the gutters, that they might conceive among the rods.     42 But when the cattle were feeble, he put *them* not in: so the feebler were Labans and the stronger Jacobs.—*Authorized Version*, 18c Edition.

vulgar errors. I have therefore not asserted above any thing as fact but what I think I have proved by my own experience.

de Beer, pages viii–ix, refers to two other sets of questions by Darwin. The first of these, on how human males select their females, is conjectural, and, from the date of the letter which he prints in support of its existence, we would suggest that this letter refers to the second set. The second is *Queries about Expression*. de Beer says: "It does not appear that this questionnaire, or that on the expression of the emotions, has been found." This is number 231 in the *Handlist*, and in 1965, when that list was published, no copy of the original printing was known. One copy of it was discovered in the Darwin archive in the University Library, Cambridge, in October 1967, and another and a corrected proof have been found there since. The description of it given in the *Handlist*, which was conjectural, is correct except that the title is as given above and not as in the Smithsonian Institution printing of 1868 (Darwin, 1868). The text is not that of the Smithsonian printing, nor the same as that which appears in all editions of *Expression of the Emotions* (Darwin, 1872). There is also at Cambridge a consolidated list of the replies received, with the correspondents' names. We intend to publish some notes on this material, with a facsimile of the original printing, at a later date.

ACKNOWLEDGEMENTS

We are grateful to Sir Robin Darwin and to the University Librarian, Cambridge, for permission to make use of the material in the Darwin archive. We are grateful also to the Librarian of the University of Keele and to the Librarian of the William Salt Library, Stafford, for their help with local biography.

REFERENCES

DARWIN, C., 1868. Queries about Expression for Anthropological Enquiry. *Rep. Smithson. Instn* (1867): [324].

——, 1872. *The Expression of the Emotions in Man and Animals*. London.

——, 1968. *Questions about the Breeding of Animals*. [1840]. *With an Introduction by Sir Gavin de Beer*. Society for the Bibliography of Natural History, London.

DARWIN, F. (Edit.), 1887. *The Life and Letters of Charles Darwin, including an Autobiographical Chapter*. 3 vols, London.

——, (Edit.), 1909. *The Foundations of the Origin of Species. Two essays written in 1842 and 1844*. Cambridge. Also published in *Evolution by Natural Selection*. Cambridge, 1958.

FREEMAN, R. B., 1965. *The Works of Charles Darwin. An Annotated Bibliographical Handlist*. London.

HINCHLIFFE, E., 1856. *Barthomley: in Letters from a Former Rector to his Eldest Son*. London.

SEBRIGHT, Sir John S., 1809. *The Art of Improving the Breeds of Domestic Animals. In a Letter addressed to Sir Joseph Banks, K.B.* London.

SIMMS, R., 1894. *Bibliotheca Staffordiensis*. Lichfield.

5—B.N.H.

170

*Journal of the Society for the Bibliography of Natural History* **8** (2): 184 (1977).

**Charles Darwin's** *Questions about the breeding of animals,* **[1839].**

There is no indication on this rare Darwin pamphlet of either the printer or the place of printing. In our article about it (*J. Soc. Biblphy nat. Hist.* Vol. 5, pp. 220–225, 1969) both these are left unknown, and they remain unrecorded in the second edition of the *Darwin handlist,* 1977. An examination of Darwin's manuscript personal accounts, which are kept at Down House, shows an entry for 4th June, 1839, which reads "Messrs Stewart & Murray (for printing Questions [£] 2 5 6". The *Post Office, London directory* for 1839 shows a printing firm of this name at Green-arbour Court, Old Bailey. This solves two missing points and the entry in the *Handlist* should read "[London, Stewart & Murray printed]". The entry also helps towards a more accurate dating of the pamphlet. Normal commercial practice, at least to a previously unknown customer, would demand a monthly account, and Darwin was meticulously prompt in financial matters. We know that he received a set of answers from R. S. Ford, of Swynnerton near Maer, on 6th May. He probably received his copies in London shortly before he went to Maer with his bride on 26th April. They returned to London on 20th May.

R. B. Freeman
P. J. Gautrey

*Archives of natural history* **13** (1): 98 (1986).

**14 Darwin again.** Do members have any anecdotes about odd ways in which natural history equipment, such as geological hammers, can be used? In a letter to Charles Darwin, dated 10 May 1843, Captain Bartholomew James Sulivan retold the following story with some gusto:

> We only met with the wild pigs on one island of the Falklands—they were all black, and had little resemblance to a tame pig, being completely wild Boar, the ridge along the back and head bristling up and the head deep and narrow with large tusks . . . I had rather a formidable encounter with one large boar, & had to thank my being a little bit of a geologist for my victory for holding him cheap and getting in his path. After putting my two dogs to flight he made at me and tho I put a ball through him and a charge of small shot in his face he still came at me till just as he got within two feet and was jumping over a bunch of tussack I recollected my geological hammer in my belt and got it out in time to strike him so fairly on the Fore Head that he fell dead.

*Janet Browne, Darwin Letters Project, Cambridge University Library, Cambridge*

*Archives of Natural History* (1986) **13** (2): 165–168

# Charles Darwin and 'Ancient Seeds'

By DUNCAN M. PORTER

Department of Biology,
Virginia Polytechnic Institute & State University,
Blacksburg,
Virginia 24061,
U.S.A.

One of the many phenomena that caught Charles Darwin's interest was that of seed viability. This interest was tied in closely with study of dispersal mechanisms and was discussed in detail in chapter XI of *The Origin of Species* (Darwin, 1859). Earlier, in 1855 and 1856, Darwin published seven mostly short papers on viability of seeds. Most were concerned with the action of salt water on seeds, but two dealt with germination upon exposure through disturbance of seeds long buried in the soil.

The first (Darwin, 1855a) reported that seeds of charlock (*Sinapsis arvensis* L., Brassicaceae) buried eight or nine years in a field planted to trees germinated following disturbance through removal of some shrubs. When Darwin observed what had happened, he had three two-foot-square plots in other parts of the wood cleared of plants and excavated to the depth of a spade. In each of these charlock seedlings appeared, presumably from dormant seed germinated *in situ*.

In a footnote to a July 1855 letter from Charles Darwin to Joseph Dalton Hooker (1817–1911), Assistant Director of the Royal Botanic Gardens, Kew concerning longevity of charlock seeds, Darwin's son Francis (1848–1925) added the following information. A pasture was planted to grass in 1840, and trees were planted in 1846 to form the famous 'sand walk' at Down House. In 1855 charlock seedlings appeared here on soil exposed following digging. 'The subject continued to interest him, and I find a note dated July 2nd, 1874, in which my father recorded that forty-six plants of Charlock sprang up in that year over a space (14 × 7 feet) which had been dug to a considerable depth.' (F. Darwin, 1887: 424). It is now known that charlock seeds will remain viable in the soil for at least 11 years, and there is strong evidence for their doing so for 50 years or longer (Fogg, 1950).

The second paper (Darwin, 1855b) was not based on personal observation. It indicated that the German physician and botanist Carl Friedrich von Gaertner (1772–1850) reported (Gaertner, 1849) that viable seeds of heliotrope (*Heliotropium europaeum* L., Boraginaceae, reported as *H. vulgare* Rota), cornflower (*Centaurea cyanus* L., Asteraceae) and hop trefoil (*Trifolium campestre* Schreber, Fabaceae, reported as *T. minimum* Barton) had been found in European graves over a thousand years old. The validity of this report is questionable, although seeds of the sacred lotus (*Nelumbo nucifera* Gaertner, Nelumbonaceae) from China, accurately dated at about 700 years old, recently were successfully germinated (Shen-Miller *et al.*, 1983).

While pursuing research at the Cambridge University Herbarium in 1980–81 on Darwin's plant collections from the voyage of HMS *Beagle*, I discovered several specimens that indicated an even earlier interest in seed viability on his part. The sheets all bear the annotation label 'Cambridge Botan. Museum. Herb. J. Lindley, Ph.D. Purchased in 1866'. Dr John Lindley (1799–1865) was Professor of Botany at University College, London, having previously been Secretary to the Horticultural Society. The specimens were grown in the Horticultural Society's garden from seed supplied to Lindley by Darwin. Presumably, originally they were deposited in the herbarium of the Society, which was dispersed in 1856.

Two of the sheets bear specimens of common orache (*Atriplex patula* L., Chenopodiaceae). The third has sheep's sorrel (*Rumex acetosella* L., Polygonaceae). The first sheet is labelled 'HHS. [Herbarium of the Horticultural Society] 1843 ancient seed from Mr C. Darwin'. It bears the annotation label 'Atriplex patula', in Lindley's handwriting, which has been cut from another sheet and pasted on this one. The second is labelled 'HHS. 1844 Darwin *'seeds sent in 1843'* J. L. Atriplex patula', and is annotated 'raised most certainly from seed of *'*. That is, raised from the 1843 *A. patula* seed. The collection of *Rumex* is unidentified and is labelled '1844 Ancient seeds, with Atriplex patula? C. Darwin Esq. Garden of the Horticultural Society'.

Darwin sent these seeds to Lindley in April 1843, stating in a covering letter that

They have been sent to me by Mr. W. Kemp of Galashiels [Borders, Scotland], a (partially educated) man, of whose acuteness and accuracy of observation, from several communications on geological subjects, I have a *very high* opinion. He found them in a layer under twenty-five feet thickness of white sand, which seems to have been deposited on the margins of an anciently existing lake. These seeds are not known to the provincial botanists of the district. [Both species, however, are native to Scotland.] He states that some of them germinated in eight days after being planted, and are now alive. Knowing the interest you took in some raspberry seeds, mentioned, I remember, in one of your works, I hope you will not think me troublesome in asking you to have these seeds carefully planted, and in begging you so far to oblige me as to take the trouble to inform me of the result. (Darwin & Seward, 1903: 243).

In September he reported to his friend the geologist Sir Charles Lyell (1797–1875) that

These seeds germinated freely, and I sent some to the Horticultural Society, and Lindley writes to me that they turn out to be a common *Rumex* and a species of *Atriplex*, which neither he nor Henslow (as I have since heard) have ever seen, and certainly not a British plant! Does this not look like a vivification of a fossil seed? It is not surprising, I think, that seeds should last ten or twenty thousand [years], as they have lasted two or three [thousand years] in the Druidical mounds, and have germinated. (Darwin & Seward, 1903: 224; brackets in the original.)

This same month, Darwin wrote to his old Cambridge mentor the Rev. John Stevens Henslow (1796–1861), Professor of Botany, regarding the seeds. Two more letters mentioning them followed in October and November 1843. These three letters were printed by Barlow (1967).

Lindley was unable to identify the *Atriplex* and sent it to Henslow. There is no evidence indicating that Henslow identified it for either Lindley or Darwin. In October, Darwin sent seeds of it to Charles Cardale Babington (1808–1895), later to become Henslow's successor as Professor of Botany at Cambridge. Babington was well known as a taxonomic splitter and identified the plant as *Atriplex angustifolia* J. Sm., now considered to be a synonym of *A. patula*.

In his last letter to Henslow regarding these seeds, written on 5 November 1843, Darwin closed the story with the following remarks:

Until your last note I had not heard that Mr. Kemp's seeds had produced two Polygonums. He informs me he saw each plant bring up the husk of the individual seed which he planted. I believe myself in his accuracy, but I have written to advise him not to publish, for as he collected only two kinds of seeds— and from them two Polygonums, two species or varieties of *Atriplex* and a *Rumex* have come up, any one would say (as you suggested) that more probably all the seeds were in the soil, than that seeds, which must have been buried for tens of thousands of years, should retain their vitality. If the *Atriplex* had turned out new, the evidence would indeed have been good. I regret this result of poor Mr. Kemp's seeds, especially as I believed, from his statements and the appearance of the seeds, that they did germinate, and I further have no doubt that their antiquity must be immense. (Darwin & Seward, 1903: 246).

I found no *Polygonum* (knotgrass, Polygonaceae) specimens at Cambridge.

In spite of Darwin's lack of doubt as to the great age of these seeds, there is little evidence that seeds remain viable for as long as he would have liked. However, sheep's sorrel seeds have survived burial for 26 years and common orache in a meadow for about 58 years (Harrington, 1972). So the possibility remains that 'poor Mr. Kemp's seeds' may have remained dormant in the soil, although not for so long as hypothesized. On the other hand, their seeds, and those of charlock, pass unharmed through the alimentary tracts of birds (Salisbury, 1961). Thus the possibility also exists that both Darwin's and Kemp's seedlings originated by contamination from bird droppings, and not from seeds retained in the soil.

Let me, however, allow Darwin the last word on the subject. He ended the July 1855 letter to J. D. Hooker mentioned above:

A man told me the other day of, as I thought, a splendid instance,—and *splendid* it was, for according to his evidence the seed came up alive out of the *lower part* of the *London Clay*!!! I disgusted him by telling him that Palms ought to have come up. (F. Darwin, 1887: 425).

The tropical fossil flora of the London Clay is about 60 million years old. Obviously, Darwin was aware of the limits of seed longevity.

## ACKNOWLEDGEMENTS

Research at the Cambridge University Herbarium was funded by a grant from the National Geographic Society and was undertaken while I was a Visiting Fellow at Clare Hall.

## REFERENCES

BARLOW, N. (Ed.), 1967 *Darwin and Henslow, the growth of an idea. Letters 1831–1860*. Bentham-Moxon Trust/John Murray, London.

DARWIN, C., 1855a Vitality of seeds. *Gardner's Chronicle* No. 46, 17 November: 758.

DARWIN C., 1855b Longevity of seeds. *Gardner's Chronicle* No. 52, 29 December: 854.

DARWIN, C., 1859 *On the origin of species by means of natural selection, or the preservation of favoured races in the struggle for life*. John Murray, London.

DARWIN, F. (Ed.), 1887 *The life and letters of Charles Darwin, including an autobiographical chapter*. Vol. 1. D. Appleton, New York.

DARWIN, F. & Seward, A. C. (Eds), 1903 *More letters of Charles Darwin*. Vol. 2. John Murray, London.

FOGG, G. E., 1950 Biological flora of the British Isles. No. 146. *Sinapsis arvensis* L. *Journal of Ecology* **38**: 415–429.

GAERTNER, C. F. von, 1849 *Versuche und Beobachtungen über die Bastarderzeugung im Pflanzenreich.* K. F. Hering, Stuttgart.

HARRINGTON, J. F., 1972 Seed storage and longevity. *In* T. T. Kozlowski (Ed.), *Seed Biology.* Vol. 3. *Insects, and Seed Collection, Storage, Testing, and Certification.* Academic Press, New York & London. Pp 145–245.

SALISBURY, E., 1961 *Weeds and aliens.* Collins, London.

SHEN-MILLER, J., SCHOPF, J. W., & BERGER R., 1983 Germination of a ca. 700 year-old lotus seed from China: evidence of exceptional longevity of seed viability. *American Journal of Botany* **70**(5/2): 78.

*J. Soc. Biblphy nat. Hist.* (1971) **5** (6): 474–476.

# The reading of the Darwin and Wallace papers: an historical "non-event"

By J. W. T. MOODY, F.L.S.

Greenville College, Greenville,
Illinois 62246, U.S.A.

It is a truth of historical research that great events reflected upon and recalled many years after the fact, by men who were there, often gain grandeur and magnificence in the process. This is, to some extent, the case with the commentary upon the Linnean Society meeting of 1 July 1858 when the Darwin and Wallace theories on natural selection were given their first public exposure. In Sir Francis Darwin's *The life and letters of Charles Darwin*[1] the recollections of Sir Joseph Dalton Hooker upon this event picture a scientific meeting stunned by the remarkable and revolutionary theories read before them. However, there are very few contemporary comments[2] from others present at that event. Hooker's participation at that time and the subsequent importance of Darwinian ideas no doubt embellished his reflection upon the meeting.

The *General Meeting Book* of the Linnean Society records the secretary's minutes of that meeting and those in attendance. The fellows at the meeting of 1 July 1858 included the President, Professor Thomas Bell, F.R.S., V.P.Z.S., P.L.S., Professor of Zoology, King's College, London; Sir Charles Lyell, F.R.S., F.G.S.; Dr Joseph Dalton Hooker; Mr Nathaniel B. Ward, F.R.S.; Mr John M. Camplin; Mr Robert Heward; Dr Fredric D. Dyster, M.D., F.R.S.; Mr Daniel Oliver jun; Mr Samuel P. Pratt, F.R.S., F.G.S.; Mr S. James A. Salter, M.B. Lond.; Mr William Archer; Mr John Ball, M.R.I.A.; Mr John Thomas Syme, lecturer on botany at Westminster and Charing Cross Hospitals; Mr Frederick Curry, M.A., F.R.S.; Dr William Baird, M.D. of the British Museum; Dr William Henry Fitton, M.D., F.R.S., F.G.S.; Mr Samuel Stevens, Treasurer of the Entomological Society; Dr William Benjamin Carpenter, M.D., F.R.S., F.G.S.; Dr Bethold Seemann, Ph.D.; Mr Arthur Henfrey, F.R.S., Professor of Botany, King's College, London; Mr Benjamin W. Hawkins, F.G.S.; Dr William John Burchell, D.C.L.; Mr George B. Buckton, F.R.S.; Mr William M. Buckton; and Mr Black, associate member. Dr Dyster brought Dr Baly as his guest and Mr Ward brought Dr Melville. The scientific interests of this group of fellows included botany, zoology, entomology, geology, conchology, paleontology, physiology, embryology, and anatomy to mention only a few.

The meeting of 1 July was especially summoned by the President, Thomas Bell, to elect a new member to the Council of the Society. This member would fill the vacancy left by the recent death of Dr Robert Brown, D.C.L., F.R.S., V.P.L.S. The meeting was called to order and the first item of business was the announcement of donations and gifts to the library and museum of the society. These included *Mémoires de la Société Royale des Sciences de Liège*; *Journal of the Royal Society*; *Abstracts of the Proceedings of the Geological Society*; *The Philosophical Magazine*, and *Annals of Natural History* presented by Richard Taylor; *The Literary Gazette* presented by Lovell Reeve; a collection of original drawings of Tasmanian Orchidiae by William Archer, F.L.S., presented by Mr Archer;

*The Handbook of British Flora* by George Bentham presented by Mr Bentham; *The Grasses of Great Britain* (Part 9) presented by J. E. Sowerby; *Examination of Pavon's collection of Peruvian Barks contained in the British Museum* presented by the author, Mr J. E. Howard, F.L.S.; *Flora Melitensis* presented by the author, Dr Giovanni Carlo Grech Delicata; and specimens of Highgate Resin, presented by J. W. Netherall.

The business then proceeded with the election of a new member of the Council. Dr Seemann, Mr Archer, and Mr Heward were appointed scrutineers and the balloting began. The tally of ballots showed Mr George Bentham had been elected. The President nominated George Bentham for Vice-President to fill out the term of the late Robert Brown. Sir Charles Lyell seconded this and Bentham received the unanimous vote of the fellows.

A lengthy statement of sincere appreciation for the late Robert Brown was read. His more than sixty years of association with the society was especially praised. This eulogy was made a part of the minutes of the meeting.

Although this was a special meeting a number of papers which would have been held over until the 4 November meeting were read. The *Register of Papers Read* shows the date when papers were received, their author, subject, by whom presented, date of reading and "how disposed of". There were three papers received before the date of this special meeting and six papers received on 1 July. The three on hand were J. E. Howard's "Nueva Quinologia" received 27 April, T. L. Ralph's "On the Arborescent Ferns of New Zealand" received 8 May, but not read until 2 December, and Dr Welwitsch's "Letters on the Vegetation of Angola" received on 2 June. On 1 July the secretary received six papers and only one was not read at that meeting. This paper was submitted by George Bentham under the title "Notes on British Botany" and was not read until 4 November.

The meeting proceeded with the presentation of papers. The secretary records the following presentations in the minutes of the *General Meeting Book*:

> "Read 1st—A letter from Sir Charles Lyell and Dr. J. D. Hooker addressed to the Secretary, as introductory to the following papers, on the laws which affect the production of varieties, races, and species, viz.
>
> An extract from MS work on species by Charles Darwin, Esq., F.R.S. & L.S., sketched in 1839 and copied in 1844.
>
> An abstract of a letter addressed by Mr. Darwin to Prof. Asa Gray of Boston, U.S. in Oct. 1857.
>
> An essay on the tendency of varieties to depart indefinitely from the original type, by A. R. Wallace Esq.
>
> 2ndly—"Notes on the organization of *Phoronis hippocrepis*" by F. D. Dyster, M.D., F.R.S.
>
> 3rdly—"Observations on *Ammocaetus*" by Samuel Highley, Esq.
>
> 4thly—"On *Hanburia*, a new genus of Cucurbitacea" by Bethold Seemann, M.D., F.R.S.
>
> 5thly—A MS memoir, by the late Prof. Pavon, entitled "Nueva Quinologia" with observations by J. E. Howard, F.L.S.
>
> 6thly—Two letters "on the Vegetation of Angola" by Dr. F. Welwitsch addressed to W. W. Saunders."

Sir Charles Lyell and Dr J. D. Hooker had presented introductory remarks to emphasize the importance of the Darwin and Wallace communications. It would appear, however, that the sheer volume of contributions practically buried the Darwin–Wallace papers. The fellows were not so much stunned by new ideas as they were overwhelmed by the amount of information loaded upon them at the meeting. Much of the Darwin–Wallace concept of natural selection went over their heads. This in large measure was the result of insufficient time to concentrate attention and discussion upon the Darwin–Wallace papers. Some of the silence was no doubt boredom, a constant danger of long meetings.

The *Register of Papers Read* indicates that the Darwin–Wallace papers, the Dyster paper, and the Welwitsch paper were accepted for publication. Seemann's paper on *Hanburia* was referred to Dr Hooker and Howard's "Nueva Quinologia" was referred to the council meeting of 4 November. The Darwin–Wallace paper was printed in the *Journal of the Proceedings of the Linnean Society, Zoology* (**3**: 45–62), which was published in August of 1858. The Dyster paper on *Phoronis* was printed in *Transactions of the Linnean Society* (**22**: 251–256). This volume was published sometime between 4 November and 24 December, 1858.[3] The Welwitsch paper, "Letters on the Vegetation of West Equinoctial Africa", was printed in the *Journal of the Proceedings of the Linnean Society, Botany* (**3**: 150–157). This volume was published in August of 1858.

The meeting of the Linnean Society on 1 July 1858 has been said to be the beginning of the "Darwinian Revolution", "the beginning of Modern Biology", "the beginning of a new era in scientific thinking", and many other such labels which indicate that from this date, dynamic new ideas, concepts, and forces were turned loose in biology, philosophy, and theology. This important day may well be the beginning of all these things. However, the actual event seems to be second only to the presentation of Mendel's discovery of the laws of genetics as an historical "non-event".

NOTES AND REFERENCES

[1] Darwin, Sir Francis, 1887. *The life and letters of Charles Darwin*. London, Vol. 2, p. 126.

[2] Charles Darwin's own recollections of the meeting and its aftermath were more prosaic. In his autobiography, written in 1876 for his children, he recalled, "Nevertheless, our joint productions excited very little attention, and the only published notice of them which I can remember was by Professor Haughton of Dublin, whose verdict was that all that was new in them was false, and what was true was old." See p. 122 of Nora Barlow's *The autobiography of Charles Darwin 1809–1882*. Collins, London, 253 pp. (1958).

[3] Raphael, S., 1970. The publication dates of the *Transactions of the Linnean Society of London*, Series 1, 1791–1875. *Biol. J. Linn. Soc.* **2**: 61–76.

*Archives of Natural History* (1985) **12** (1): 153–159

# The reception of Darwinism in Norway: the early years 1861–1900

By THORE LIE

Botanisk Institutt,
Pb. 1045, Blindern,
Universitetet i Oslo,
Norway.

INTRODUCTION

When Darwin's work *On the origin of species* was published in 1859, it immediately pro-voked an intense debate in England. It would be difficult to find a book which, at the same time, was received with such enthusiasm and such strong opposition. The debate about Darwinism spread quickly to other countries. In Germany the *Origin* became known early in 1860 through the translation of the palaeontologist Heinrich Bronn. The debate started immediately and involved not only biologists but also theologians, philosophers and the public in general. If one studies reference books like R. B. Freeman's *Charles Darwin a companion* (1978), one can follow in chronological order the publication of various foreign editions of the *Origin* in a number of countries, both European and others. One country not mentioned in that list however is Norway.

THE SITUATION IN NORWAY

Study of the scientific literature in Norway that was published in 1859 and the immediately following years, can easily give one the impression that the theories of Darwin and the international debate around Darwinism passed quite unnoticed in the scientific community.

One also receives an equally negative impression when reading the memoirs published in 1910 by the discoverer of the Leprosy bacillus, Armauer Hansen (1841–1912). There he describes, among other things, his first contact with Darwinism during a stay abroad in 1870 and its effect on his scientific ideas. He also claims in his memoirs to be the first to have introduced Darwinism in Norway.

However, this was not the case. More than 20 years earlier, in March 1861, an anonymous article with the title *Darwin's nye Skabningslære* (Darwin's new theory of creation) was pub-lished in the journal *Budstikken*. It later became known that the article was written by P. Chr. Asbjørnsen (1812–1885), at that time one of the editors of *Budstikken* together with, among others, the botanist F. C. Schübeler.

Asbjørnsen, best known today as collector and publisher of Norwegian folk tales, was also a prominent naturalist with significant investigations in several fields of science, in par-ticular marine zoology. He was also familiar with the scientific trends of the time and was aware that "the influence of the new theory on the human culture would be enormous". His interest in Darwin should be seen in relation to his studies on animal breeding problems, especially the connection between the Norwegian cattle breed and the aurochs (*Bos primigenius*). To some extent it was also Darwin's breeding theories to which he particularly wished to draw attention. He was aware, nevertheless, of the more general aspects of Darwinism, even if he did not understand the principle of natural selection with all its

179

implications, and wrote at the end of his article: "it is difficult to foresee the enormous consequences of what the natural sciences will undergo as a result of Darwin's theories". Darwin does not mention the origin and development of man in *Origin*, but Asbjørnsen realised that Darwin's theories would be used in such connection. As a basis for his article Asbjørnsen used Bronn's German translation of *Origin* (1860).

### Introduction of Darwin to the scientific community

In this way Asbjørnsen introduced Darwin to Norway, but it was the theologian and zoologist Michael Sars (1804–1869) who made him known in the scientific community. This happened during a lecture in the Academy of Science (Videnskabs–Selskabet) in Kristiania (Oslo) on 3 May 1869; about 10 years after the first edition of *Origin* was published.

The book had already been bought by the University Library, with an English and a German edition in 1860. However, during the 10-year period 1860–1870, only 19 borrowers were registered. One of these was M. Sars, who borrowed the book on 25 April 1862. At this time he regarded Darwinism with the same distrust he had earlier felt for Lamarckism. In the following years, however, he changed his opinion.

Sar's lecture was the starting point of the coming debate around Darwinism, in which his two sons, the historian Ernst Sars and to an even greater extent the zoologist Georg Ossian Sars (1837–1927) would participate. The immediate effect of the lecture on the scientific community was however moderate. Two whole years would pass before the professor in medicine, E. F. Lochmann, tried to introduce a discussion on Darwinism during a meeting in the Academy of Science on 27 October 1871. In the minutes of the meeting one can read: "Lochmann tried to introduce a discussion on Darwinism. There was also some talk about it, for example by Faye and Esmark (professors in medicine and zoology), yet without going into details as one lacked a satisfactory account of the system of Darwinism to use as a basis. Monrad drew attention to this lack and urged one or another of them, who had studied Darwin's writings, to supplement this at the next meeting". Even this appeal did not lead anywhere and the professor in philosophy, M. J. Monrad, who later became one of the opponents of Darwinism in Norway, had to take the iniative himself. At the Academy of Science's annual meeting on 3 May 1873 he gave "a short account and criticism of Darwin's Teaching, to which Lochmann added some remarks".

This contribution was presumably a condensation of the lectures he held at the University during autumn 1872 and spring 1873, and which later formed the basis for the book *Tanke-retninger i den nyere Tid* (1874) (Ways of thought in recent times).

Even though little was known about Darwin within the natural sciences in the 1860s, several important cultural personalities had, nevertheless, acquired a knowledge of Darwinism early in this decade. In 1862 the social research-worker Eilert Sundt visited England and came into contact with the debate on Darwinism at close hand. This confrontation with Darwinism would have importance for Sundt's later scientific thought, not least for his theories on the development of the boat. He used the Norlands-boat and the Lista-boat as examples of Norwegian boat types, and tried to explain the boat forms as stages in a hundred year-old development towards better and better adaptation. The same applied also to the development of house-building over the centuries. In his investigations over these and similar problems, Sundt stood out as an alert cultural evolutionist.

The poet A. O. Vinje also visited England. In the spring of 1862 he started out on a journey which would last for more than a year. Like Sundt he came into contact with

Darwinism which later, among other things, is expressed in his views on the Franco-Prussian war (1870–1871), considered by him to be an example of "det Darwinske Princip for Naturens Udvalg af sit bedste Slag" (the Darwinian principle of Nature's selection of the fittest).

Several others also obtained firsthand knowledge of Darwin from visiting England. The historian L. K. Daa took part in the Manchester meeting of the British Association for the Advancement of Science in September 1861, where he also gave an address. Another historian, T. H. Aschehoug, visited England in 1850 and was introduced to Darwin himself.

Even then, Darwinian theories were nevertheless little discussed in the scientific milieu in Norway in the 1860s. Several reasons can be given for this. Two of the most important are that none of the cultural personalities mentioned belonged to the circle of the natural scientists and had therefore little opportunity to bring Darwin into that milieu, despite the knowledge of Darwinism they had obtained from personal visits to England. Secondly, Darwinism was often looked upon by the natural scientists themselves as a new interesting theory which could be discussed now and again, but which had little application within their individual research areas.

### THE NEW GENERATION OF SCIENTISTS

As a scientific theory, Darwinism first attained success with a new generation of natural scientists who began their research in the 1870s, and to a great or lesser extent based their scientific work on Darwin. Foremost among these should be mentioned the zoologist Georg Ossian Sars, the botanist Axel Blytt (1843–1898) and the geologist Waldemar Brøgger (1851–1940).

### Georg Ossian Sars

The first of these, Georg Ossian Sars, became a university researcher in 1870 and professor of zoology at the University of Kristiania (Oslo) in 1874. Sars founded his lectures in zoology on Darwinian theories from the very beginning. The lectures attracted attention and gathered large numbers of young students who here received their first introduction to modern ideas. The lectures contained comparatively radical ideas and it is not surprising that opposition was aroused among the more conservative university teachers. Sars was removed from *Anneneksamen* (the students first compulsory exam at the University) as it was felt that students at that level were too immature to withstand such an influence.

### Axel Blytt

Opposition to Sars soon became well-known, but that which met Axel Blytt is less known. Blytt's case is interesting as a typical example and we will therefore look a little more closely at his development as a Darwinist. He derived his earliest knowledge of the naturalist Darwin through his father, the botanist professor Matthias Blytt, who himself had contact with Darwin in the 1850s. Axel Blytt took his first trip abroad in 1866, to London, where he could follow the debate at close quarters. There he also became acquainted with the geogloist James Geikie, who later was to become an important contact between Darwin and himself. The importance of Darwinism as a scientific theory for Blytt is best expressed in his major work: *Forsøg til en Theori om Indvandringen af Norges Flora under vexlende regnfulde og tørre Tider* (1876).

"An English translation appeared in the spring (April–May) 1876 with the title: *Essay on the immigration of the Norwegian flora during alternating rainy and dry periods.* The major idea of the book was that different groups of Norwegian flora immigrated during periods of alternating dry and damp climates. The plant groups replaced one another and reached their maximum distribution in the corresponding period of time. Expressed in Darwinian terms this would mean that only the best adapted species had the possibility to evolve within each period of climate while those less well adapted succumbed. It was in other words natural selection.

Blytt's work received an enthusiastic reception among foreign scientists, including Darwin. Blytt had, before the book was officially published in Norway, sent Darwin a copy of his work and received 28 March 1876 a letter from Darwin, where he expressed his interest in the work: "I thank you sincerely for your kindness in having sent me your work on the Immigration of the Norwegian Flora, which has interested me in the highest degree. Your view, supported as it is by various facts, appears to me the most important contribution towards understanding the present distribution of plants, which has appeared since Forbes' essay on the effects of the Glacial Period". A month later, on 25 April, Geikie wrote in a letter to Blytt: "it may please you to know that Mr. Darwin has a very high opinion of your most interesting essay on the immigration of the Norwegian Flora". Darwin also wrote to A. R. Wallace on 25 June 1876: " . . . I may mention a capital essay which I received a few months ago from Axel Blytt on the distribution of the Plants of Scandinavia. . ."

He also expressed his approval on several later occasions as, for example, in a letter to Hooker dated 6 August 1881, where he again affirmed: "This seems to me a very important essay". Axel Blytt is most probably the only Norwegian scientist besides G. O. Sars, to be mentioned in any of Darwin's letters. Darwin's interest shows clearly the accordance between the fundamental ideas in Blytt's essay and Darwin's own views.

Although Blytt only reluctantly took part in the official debate around Darwinism and the theory of evolution, no one was in doubt as to his opinion. In 1876 a proposal was put forward by the Faculty of Mathematics and Natural Sciences that Parliament would grant money towards a new professorship in botany for Axel Blytt. During the treatment of this matter in Parliament a flow of rumours was started claiming that Blytt was a Darwinian. The aim of this was to hinder a supporter of Darwinism from obtaining such an influential position. Due to Blytt's professional reputation as a botanist, the campaign failed and Blytt became Professor of Botany at the University of Kristiania in 1880.

## W. C. Brøgger

The one scientist who perhaps more than anyone else was to become an official champion for Darwin in Norway was the geologist W. C. Brøgger. He originally studied zoology, but later went over to geology. In 1878 he became a university researcher and in 1882 a Professor in Mineralogy and Geology in Stockholm. In 1890 he returned to Norway as a Professor at the University.

His former teacher and predecessor there, Theodor Kjerulf, was averse to Darwinism; not so much from professional as from religious reasons. Brøgger, however, was to stand firm on Darwinism, both within his scientific field and as an enlightener and popularizer. Purely professionally this is best expressed in his major work: *Die silurischen Etagen 2 und 3 im Kristianiagebiet und auf Eker,* which was presented as a University programme (a series of scientific papers published by the University) in 1882.

## THE OPPOSITION

In many countries the main opposition to Darwin and his theories was to come from religious quarters; where Darwinism was looked upon as a direct contradiction to and renunciation of the Christian belief of creation. This was the case in England and one could perhaps expect the same in Norway. It is surprising therefore to see how little opposiiton came from the church to Darwinism at the end of the 1860s and beginning of the 1870s.

An important reason for this was that the Church had for many years taken a negative attitude towards natural scientific research, especially in those areas which could have an effect on matters pertaining to religion. It was therefore unprepared for the debate which was to come. Later, in the 1880s and 1890s, the Church's opposition would increase. If we consider the first decades of Darwinism in Norway, we will find that the main opposition did not essentially come from religious centres but from scientific circles within the University itself, especially from the philosophers. From a purely philosophical point of view Darwinism represents empiricism, which maintains that all true knowledge is based on experience. By experience is meant, within the scientific field methodically accomplished and experimentally controlled observations. One of the earliest representatives of this philosophical trend in Norway was the philosopher Niels Treschow (1751–1833), who is, by many scientists, considered as a pre-Darwinist due to his "evolutionary" thoughts in his most famous work: *Om Gud, Idee og Sandseverdenen samt de førstes Aabenbarelse i den sidste* (On God, Idea and the World of perception together with the revelation of the Former in the Latter), the first volume of which was published in 1831 when Treschow was 80 years old.

Treschow's philosophical views, however, did not gain approval at the University and Treschow himself was overshadowed and nearly forgotten. The main philosophical tradition at the University was to become Hegelism, which strongly opposed empiricism. In Norway Hegelism attained one of the strongest positions in the whole of the philosophical world; stronger perhaps than in Hegel's own homeland. The Hegelian professor Monrad was also to become the leading opponent to Darwinism.

In the autumn of 1882 The Academy of Science received an appeal from England to contribute to raise a monument to Darwin. Lochmann and Sars decided to form a committee to present the case to the Academy. Monrad, Lochmann, Sars and Blytt officiated as elected members. Both Monrad and Lochmann admired Darwin as a scientist, even if they opposed his theories.

As a result of the collection, announced by Sars at the annual meeting in 1884, Norway together with several other countries financed the portrait-statue of Darwin by Jospeh Boehms in the British Museum (Natural History).

By the time of Darwin's death in 1882, his ideas of organic evolution had been accepted within the scientific community of the University. The opposition still encountered came now mainly from groups concerned with the social and moral aspects of Darwinism, and from religious laity, which to a greater extent than before connected Darwinism with modern atheistic and materialistic ideas.


## THE TRANSLATION OF *THE ORIGIN OF SPECIES*

Although Darwin's own writings were relatively well known in scientific circles in Norway by the beginning of the 1880s, not one of these works had yet been translated and made available for the general public. In neighbouring countries there were already several translations. In Denmark *Artenes Oprindelse* (Origin of Species) came out in 1872 and *Menneskets Afstamning* (The Descent of Man) in 1874, both of which were translated by the Danish writer J. P. Jacobsen. In Sweden *Om arternas uppkomst* (Origin of Species) had already come out in 1871, translated by A. M. Selling.

The interested reader in Norway had more or less one-sided articles for and against Darwinism in newspapers and journals. The articles, however, were mainly biased contributions without factual information. The mycologist Olav Sopp (1860–1931) was aware of this situation and published in 1887 a small, popular introduction to Darwinism under the title *Udviklingslærens nuværende standpunkt* (The present state of the theory of evolution). It was published in the well-known book series *Bibliothek for de tusen hjem* (Library for every home), founded in the same year. This series, in reality one of the first Scandinavian series of low-priced books, had a strongly evolutionary character, and met with great opposition from conservative quarters. Sopp's book was divided into three main sections, the first dealing with Darwinism and the main outlines in *The Origin of Species*. Then follows a description of the conflict around Darwin abroad and finally Sopp discussed the question on the present position of the doctrine of evolution. The book was a powerful contribution in favour of Darwinism and Sopp several times repeats that "at the present time there is not one opponent of the theory of evolution among the more outstanding natural scientists".

During 1888–89 a Norwegian version in three volumes of *The Life and Letters of Charles Darwin* by his son Francis Darwin, was published. The books were translated by M. Søraas and Olav Sopp collaborated as scientific adviser.

Darwin's major work *The Origin of Species* was still not translated into Norwegian, but eventually in the years 1889–90, 30 years after it was first published in England and seven years after Darwin's death it was published under the title *Arternes oprindelse* and the subtitle *Gjennem naturlight udvalg eller de best skikkede formers bevarelse i striden for livert* (through natural selection or the preservation of the best favoured races in the struggle for life). The book was published in the series *Bibliothek for de tusen hjem* and was translated from the 6th edition of the original by Ingebret Suleng.

It is evident that the debate concerning Darwinism was more scientific towards the end of the century. Consequently the need increased for more pertinent information than that already available, both on a popular and at more professional level. A number of scientists realised this and attempted to remedy the situation. One of these was the botanist J. N. F. Wille (1858–1924), who was well informed regarding the theory of evolution. He was a prolific writer and took up the question of Darwinism in a great number of articles. Apart from a few articles on the most recent research results, Wille was more concerned with using Darwinism to explain social problems.

Several scientists were to follow him. Notable among these was the zoologist Kristine Bonnevie (1872–1948), the first Norwegian woman professor. In contrast to Wille, Bonnevie was more concerned with the scientific content and theoretical basis of the theory of evolution, which she, in her powerful professional position, had the best qualifications to take further.

As a scientific theory Darwinism, at the turn of the century was facing basic changes which were to lead towards the modern theory of evolution of today. Accordingly the scientific debate also changed. In the meantime, the moral and philosophical questions to which Darwinism had given birth remained the same and the debate around these issues would continue.

## REFERENCES

AMUNDSEN, L., 1957—1960 *Det norske Videnskabs-Akademi i Oslo.* Oslo 2 vol.

ARMAUER HANSEN, G., 1910 *Livserindringer og Betragtninger.* Kristiania.

ASBJØRNSEN, P. CHR., 1861 *Darwin's nye Skabningslære. Budstikken No 2—3:* 65—71.

BARRETT, P. H. (ed.), 1977 *The Collected Papers of Charles Darwin.* Univ. of Chicago Press.

BLYTT, A., 1876 *Forsøg til en Theori om Indvandringen af Norges Flora under vexlende regnfulde og tørre Tider. Nyt mag. f. naturv.* 21: 279—362.

BLYTT, A., 1876 *Essay on the Immigration of the Norwegian Flora during alternating rainy and dry Periods.* Christiania

BRØGGER, W. C., 1882 *Die Silurischen Etagen 2 und 3 im Kristianiagebiet und auf Eker.* Kristiania.

DARWIN, F. (ed.), 1887 *The Life and Letters of Charles Darwin.* London: John Murray.

DARWIN, F., and SEWARD, A. C. (eds.) 1903 *More Letters of Charles Darwin.* London: John Murray. *Det Kongelige Fredriks Universitet 1811—1911,* 1911 Christiania.

DAA, L., 1934—1952 *Dagbøker I—IV 1859—1893.* Oslo.

FREEMAN, R. B., 1978 *Charles Darwin A Companion.* Dawson.

JACOBSEN, J. P., 1872 *Artenes Oprindelse.* København.

JACOBSEN, J. P., 1847 *Menneskets Afstamning.* København.

LIE, THORE, 1981 *Axel Blytt og darwinismen. Blyttia* 39: 41—49.

MONRAD, M. J., 1874 *Tankeretninger i den nyere Tid.* Kristiania.

NORDGAARD, O., 1918 *Michael og Ossian Sars.* Kristiania.

NORDHAGEN, R., 1943 Axel Blytt. *En norsk og internasjonal forskerprofil (1843—1898). Blyttia* 1: 21—84.

SELLING, A. M., 1871 *Om arternas uppkomst.* Stockholm.

SOPP, OLAV, 1887 *Udviklingslærens nuværende standpunkt.* Kristiania.

(SULENG, INGEBRET) 1889—90 *Arternes Oprindelse.* Kristiania.

(SØRAAS, M.) 1888—89 *Charles Darwins Liv og Breve* 3 vol. Kristiania.

TRESCHOW, N., 1831—33 *Om Gud, Idee- og Sandseverdenen.* 3 vol. Christiania.

*Archives of Natural History* (1992) **19** (3): 407–410

# Darwin's *Archaeopteryx* prophecy

By GENE KRITSKY

Department of Biology,
College of Mount St Joseph,
Cincinnati, OH 45233–1670,
U.S.A.

The recent controversy about the British Museum's *Archaeopteryx* has brought renewed interest in the early history of the fossil and its importance (Wellnhofer, 1990). Discovered in 1861 in the Solenhoven quarries, this fossil "bird" flew into the storm surrounding Darwin's *Origin of Species*. After the discovery of the more complete Berlin specimen in 1877, *Archaeopteryx* was recognized by many to be a true transition between the classes Reptilia and Aves. Such a find should have been a great boost to Darwin and his theory of evolution by natural selection; however, he was publicly cautious. But an examination of Darwin's correspondence shows his excitement and pleasure with the fossil. Moreover, there is evidence that he predicted the discovery of an *Archaeopteryx*-like fossil over two years before its formal description.

The early history of the fossil is outlined by De Beer (1954) in his critical study of the British Museum's specimen. But details of the original discovery are irrelevant with respect to Darwin since he did not know or correspond with M. Haberlein, the original owner; M. Witte, who first examined the fossil; H. von Meyer, who published a short note naming *Archaeopteryx* in 1861; and A. Wagner, who published a description of the fossil in 1861. Wagner's paper was translated by W. S. Dallas and published in 1862 in the *Annals and Magazine of Natural History*, and this translation was likely Darwin's first information about the fossil since he had a subscription to the *Annals* . . . (Burkhardt and Smith, 1985; Burkhardt and Smith, 1990). But Wagner's conclusion that the fossil, which he named *Griphosaurus*, was of a "reptile with the simple tarsal bone of a bird, and with epidermic structures presenting a deceptive resemblance to birds' feathers" would not have given Darwin the impression that the fossil would be the long sought after evidence of evolution. Furthermore, Wagner leveled a charge against a "Darwinian interpretation" of the fossil. He challenged "Darwin and his adherents" to find the transitions between *Griphosaurus* and reptiles and *Griphosaurus* and birds (Wagner, 1862).

Later in 1862, the *Annals* . . . published a second paper on *Archaeopteryx*, the English translation of H. von Meyer's description of the isolated Jurassic feather. This paper described the feather as belonging to the species *Archaeopteryx lithographica*, and since the London specimen had feather impressions Meyer thought his isolated feather would have come from a similar animal. Meyer did not, however, deal with the Darwinian implications of the fossil (Meyer, 1862).

There is no evidence that Darwin took exceptional note of these papers. The different taxonomic names given to the fossils may have contributed to Darwin's not realizing that *Griphosaurus* was indeed a fossil bird. *Archaeopteryx* was originally named by Meyer in 1861 in a short note announcing the discovery of the British

Museum's specimen. Meyer's paper was published just weeks before Wagner's paper. Therefore, Wagner's *Griphosaurus* is a junior synonym of *Archaeopteryx* (De Beer, 1954).

During this time the fossil was being purchased by the British Museum and arrived in London on 1 October 1862 (De Beer, 1954). Richard Owen quickly prepared a study of the fossil and presented a paper to the Royal Society on 20 November 1862. In the audience was Hugh Falconer who wrote Darwin on 3 January 1863 about Owen's paper. Falconer was not impressed with Owen's research and he recognized the importance of *Archaeopteryx* to Darwin. Falconer's letter reads:

21 Park Crescent N.W.
3d Jan 63
My Dear Darwin,
. . . I was sorry to hear from your brother of the efflorescence which has been troubling you—and which he tells me is one of the reasons that has prevented you from coming to town. You were never more missed—at any rate by me—for there has been this grand Darwinian case of the *Archaeopteryx* for you and me to have a long jaw about. Had the Solenhofen quarries been commissioned—by august command—to turn out a strange being à la Darwin—it could not have executed the behest more handsomely—than with the *Archaeopteryx*. This is sober earnest—and that you should not have been in to town and see it and talk over it with me is a criminal proceeding. You are not to put your faith in the slip-shod and hasty account of it given to the Royal Society. It is a much more astonishing creature— than has entered into the conception of the describer—who compares it with the Raptors & Passeres & Gallinacea as a round winged (like the last) "bird of flight". It actually had at least two long free digits to the fore limb, and those digits bearing claws as long and strong as those on the hind leg. Couple this with the long tail—and other odd things—which I reserve for a jaw and you will have the gist of the mis-begotten bird creature—the dawn of an oncoming conception à la Darwin but I will not say more about it till you show yourself in town. A ludicrous event has turned up. John Evans appears to have hit upon the very obvious cast of the interior of the skull—undetected by the describer and before Owen's paper is out. We have Mr. Mackie describing the hemispheres and optic lobes of *Archaeopteryx*! Look to the geologists . . .
My dear Darwin
Yours ever sincerely,
H Falconer[1]

Darwin responded immediately (Darwin, 1897).

Down
Jan. 5th 1863
My dear Falconer,
. . . I particularly wish to hear about the wondrous Bird; the case has delighted me, because no group is so isolated as Birds. I much wish to hear when we meet which digits are developed; when examining birds two or three years ago, I distinctly remember writing to Lyell that some day a fossil bird would be found with end of wing cloven, i.e. the bastard wing and other part both well developed . . .
Ever my dear Falconer
Yours most truly
Ch. Darwin.

Did Darwin predict the discovery of an *Archaeopteryx*-like fossil? The mentioned letter to Lyell was written on 11 October 1859 and dealt in part with the differences between rudimentary and nascent organs.

Oct. 11th [1859]
My dear Lyell
. . . On theory of Nat. Select. there is a wide distinction between rudimentary organs & what you call germs of organs & what I call in my bigger book, "nascent" organs. An organ should not be called rudimentary unless it be useless,—as teeth which never cut through gums—the papilla representing the pistil in male flowers—wing of Apteryx,[2] or better, little wings under soldered elytra. These organs are

now plainly useless, & a fortiori they would be ueless in a less developed state. Natural selection acts exclusively by preserving successive slight, *useful* modifications, hence nat. select. cannot possibly make a useless or rudimentary organ. Such organs are solely due to inheritance . . . & plainly bespeak an ancestor having the organ in a useful condition.—They may be, & often have been worked in for other purposes; & then they are only rudimentary for the original function, which is sometimes plainly apparent. A nascent organ, though little developed, as it has to be developed, must be useful in every stage of development. As we cannot prophecy (*sic*) we cannot tell what organs are now nascent; and nascent organs will rarely have been handed down by certain members of a class from a remote period to present day, for beings with any important organ but little developed will generally have been supplanted by their descendants with the organ well developed . . . The small wing of Penguin, used only as a fin might be nascent as a wing; not that I think so; for whole stucture of bird is adapted for flight, & a penguin so closely resembles other birds that we may infer that its wings have probably been modified & reduced by nat. select. in accordance with its sub-aquatic habits. Analogy thus often serves as a guide in distinguishing whether an organ is rudimentary or nascent. . . The bastard-wing of birds is rudimentary digit; & I believe that if ever fossil birds are found very low down in series, they will be seen to have a double or bifurcated wing. Here is a bold prophecy! To admit prophetic germs is tantamount to rejecting theory of Natural Selection.—. . .
Yours most truly
C. Darwin (Burkhardt and Smith, 1991).

Darwin's prediction, while not dealing with the evolution of wings for flight, concerned the evolution of a rudimentary structure called the bastard wing, or the alula, which is formed from the much reduced first digit of the forelimb. *Archaeopteryx*, whose digits were well developed, was consistent with his prediction. Clearly well before the publication of the first German description of *Archaeopteryx* in 1861, the first English details about the fossil in 1862, and Owen's analysis in November 1862, Darwin predicted an important characteristic about the wing of a yet to be found fossil bird.

In subsequent editions of the *Origin of Species*, Darwin (1872) downplayed the importance of *Archaeopteryx* as a transitional fossil. Instead, he focused on how the fossil illustrated the incomplete nature of the fossil record. In the *Descent of Man* (Darwin, 1871), he refers to evidence by Huxley (1868) that *Archaeopteryx* is intermediate between the reptiles and the birds, but his presentation is limited to just one sentence rather than trumpeting the discovery.

In his private writings, Darwin was much more enthusiastic about *Archaeopteryx*. In a letter to Professor James Dana written just after he received Falconer's letter, Darwin wrote:

Jan 7th [1863] Down Bromley Kent
My dear Prof. Dana
. . . The fossil Bird with the long tail & fingers to its wings (I hear from Falconer that Owen has not done the work well) is by far the greatest fossil of recent times. This is a grand case for me; as no group is so isolated as Birds; & it shows how little we know what lived during former times . . .
Ch. Darwin[3]

## ACKNOWLEDGEMENTS

I thank Martin Levitt and Beth Carrol-Horrocks of the American Philosophical Society Library for their help in reviewing the Darwin/Lyell correspondence; Frederick Burkhardt for checking unpublished letters in his charge; the staff of the Lloyd Library in Cincinnati for help with obtaining the early papers on *Archaeopteryx*; and Greg McDonald for details on bird skeletal morphology. This research was completed while I was on sabbatical from the College of Mount St Joseph and on the staff at the Cincinnati Museum of Natural History.

## NOTES

[1] Hugh Falconer to Charles Darwin, 3 January 1863, American Philosophical Society Library. Courtesy American Philosophical Society Library.

[2] The *Apteryx* is New Zealand's kiwi.

[3] Charles Darwin to James Dana, 7 January 1863, American Philosophical Society Library. Courtesy American Philosophical Society Library.

## REFERENCES

BURKHARDT, F. and SMITH, S., 1985 *A calendar of the correspondence of Charles Darwin, 1821–1882*. New York, Garland Publishing, Inc.

BURKHARDT, F. and SMITH, S., 1990 *The correspondence of Charles Darwin, Vol. 6, 1856–1857.* Cambridge, Cambridge University Press.

BURKHARDT, F. and SMITH, S., 1991 *The correspondence of Charles Darwin, Vol. 7, 1858–1859.* Cambridge, Cambridge University Press.

DARWIN, CHARLES, 1871 *The Descent of man*. London, John Murray.

DARWIN, CHARLES, 1872 *The Origin of species*. London, John Murray.

DARWIN, FRANCIS, 1897 *The life and letters of Charles Darwin*. 2 vol. New York, Appelton and Co.

DE BEER, GAVIN, 1954 *Archaeopteryx lithographica*. London, British Museum (Natural History).

HUXLEY, T. H., 1868 On the animals which are most nearly intermediate between the birds and reptiles. *Annals and Magazine of Natural History* ser. 4, 2: 66–75.

MEYER, H. VON, 1862 On the *Archaeopteryx lithographica*, from the lithographic slate of Solenhofen. *Annals and Magazine of Natural History* ser. 3, 9: 366–370.

WAGNER, J. A., 1862 On a new fossil reptile supposed to be furnished with feathers. *Annals and Magazine of Natural History* ser. 3, 9: 261–267.

WELLNHOFER, PETER, 1990 *Archaeopteryx. Scientific American* 262 (5): 70–77.

(Accepted 9 March 1992.)

*J. Soc. Biblphy nat. Hist.* (1973) 6 (4): 293–295

# Short notes

OFFPRINTS OF DARWIN'S "CLIMBING PLANTS", 1865

The first edition of Charles Darwin's "On the movements and habits of climbing plants" was published as the first article in a double part, Nos 33 and 34, of Volume 9 of the *Journal of the Linnean Society of London, Botany.* The part is dated June 12, has a printed price of 4s, and was issued in green printed wrappers which list the contents and have the following imprint: LONDON:/SOLD AT THE SOCIETY'S APARTMENTS,/ BURLINGTON HOUSE;/ AND BY/ LONGMAN, GREEN, LONGMAN, ROBERTS, & GREEN,/ AND/ WILLIAMS AND NORGATE./ 1865.

Offprints of this paper occur in two distinct issues. Both have tipped in title leaves, both are in buff wrappers, and in both the title, the author, and his honours are the same. They differ in the note on the source of the original, in the imprint, and in one having the front wrapper printed from the same setting of type as the title page and in the other having it plain. In one the note reads [*From the* JOURNAL OF THE LINNEAN SOCIETY.], the imprint is the same as that of the part, except that it has been reset and there is no comma after Roberts, and the front wrapper is printed. In the other the note reads [*Being Nos.* 33 & 34 *of the 9th Volume of the* JOURNAL OF THE LINNEAN SOCIETY, *Section Botany.*], the imprint is LONDON:/ PRINTED BY TAYLOR AND FRANCIS,/ RED LION COURT, FLEET STREET./ 1865., and the wrappers are plain. In the first one there is no indication that Taylor and Francis were the printers.

The offprint is advertised in the *English Catalogue* Vol. II under the name of Longmans, at a price of 4s, and it is probable that what was offered for sale was the first of the two described. Copies that I have seen in libraries and in commerce have, all but one, been this, and the facsimile (1969) is the same. The Library of Congress *National Union Catalog. pre-1956 Imprints,* ND 0051169, records six copies all apparently of the commercial issue. I have seen only one copy of the second. It was lot 285 in Sotheby's sale of 6 June 1972, an author's presentation copy inscribed on the title page "From the Author" in Darwin's own hand, and with "W. B. Tegetmeier 1865" on the plain front wrapper, in the latter's hand. This issue is presumably the genuine author's offprint. The University Library, Cambridge, does not hold a family copy, but have a copyright one of the commercial issue; the Linnean Society does not hold the offprint in either form.

R. B. Freeman

EDITORIAL NOTE.

This note is complete; in the original printing other notes occupied pages 294 and 295.

*Archives of Natural History* (1999) **26** (2): 293–295

# Short note

### Note on the Fritz Müller—Charles Darwin correspondence

Over a period of nearly two decades, the German naturalist Fritz Müller (1822–97) and Charles Darwin (1809–82) corresponded regularly, their exchange being terminated only by Darwin's death. It was Müller's *Für Darwin* (1864) that led to the correspondence. In his book, Müller took the evolutionary ideas Darwin developed in the *Origin of species* (1859) and applied them to a specific zoological group, the crustaceans, in the hope of discovering affinities between various and very different species and of identifying common ancestral forms. If Darwin was right, Müller surmised, higher crustaceans passed through an embryonic stage similar to that of the nauplius larva of lower crustaceans (Winsor, 1976: 149). Müller's investigations subsequently confirmed this expectation, and *Für Darwin* provided a test case of Darwin's theories.

The book won Müller (who lived in comparative seclusion from the scholarly world in Brazil) international attention, not least of all that of Darwin, to whom it was read when he was unwell in the spring and early summer of 1865.[1] With the generosity so consistently emphasized by his biographers, Müller over the years supplied Darwin with countless zoological and botanical observations. Darwin in return bestowed on Müller the accolade of repeatedly referring to him as the "prince of observers" (Möller, 1920: 89), and he arranged for the publication of Müller's letters in journals, financially backed the translation into English of Müller's book (*Facts and arguments for Darwin*, trans. W.S. Dallas, 1869) and cited him extensively in his publications.

Except for the year 1881, the Müller-Darwin correspondence was at its liveliest during the 5 years after the publication of Müller's book, that is, between 1865–9, and it is with an aspect of this period that this communication is concerned.

The Müller-Darwin exchange is not preserved in full. Some of the letters that are missing from the known manuscript collections have appeared in print, however, in a work on Müller's life, labours and letters edited by a fellow naturalist and relative of Müller's, Alfred Möller (Möller, 1915–21).[2] Möller had spent almost 3 years (1890–3) carrying out mycological research under Müller's guidance in Blumenau, a German colony in Southern Brazil where Müller had emigrated in 1852. After Müller's death in May 1897, Möller asked his uncle's correspondents (or their surviving families) if he could borrow the letters his uncle had sent over the years, a request that was not once denied to him (Möller, 1921: iv). Möller had them copied and arranged for their return. He probably did this also in the case of the letters to Darwin, obtaining them through Darwin's son, Francis. But in this instance the letters appear to have gone missing, and it remains unclear to this day whether this happened in Germany, in transit, or at some later stage in England.[3]

In his letters, Müller frequently discussed details that also appeared in his published work, and since Möller was publishing the letters as well as the essay publications, and wanted to avoid overlaps, he was selective and at times tended to reproduce letters only in part (Möller, 1921: iv). Also, Müller corresponded with Darwin in

English, but in order to make these letters accessible to German audiences, Möller translated them back into Müller's native tongue. The only versions of this section of the Müller-Darwin correspondence that have been preserved are thus fragments in German translation.

One feature about these letters is most puzzling: assuming that the letters Möller obtained from Francis Darwin were kept together in one batch, if lost they would all be lost together, none of the letters published in Möller should therefore appear in the list of original letters known to be in existence. If one compares the Möller edition with the *Calendar of the correspondence of Charles Darwin, 1821–1882* (Burkhardt and Smith, 1994), however, it would seem that there are at least three letters in Möller's edition that also are part of the Darwin archive in the Cambridge University Library (CUL). The letters in question are listed in the *Calendar* as items 5226 (1/3 October 1866), 6140 (22 April 1868), and 6359 (9 September 1868).

A comparison of the Möller and CUL versions reveals that beyond the dates they have little in common (not even the species names tally). Irrespective of Möller's skills as a translator, it is inconceivable that as a trained biologist he would have mis-read the names. It seems plausible, then, that while bearing the same date, the Möller and CUL letters are not identical; i.e., the letters now in the CUL did not underlie the texts reproduced in Möller. This could mean that unless Möller mistook the date of the letters, Müller sometimes wrote to Darwin several times on the same day. Alternatively, since none of these pairs of letters is in fact in themselves complete, they could in each case be fragments of one and the same letter. This is the conclusion Professor David West of Virginia Polytechnic Institute and State University has arrived at, and it is obviously the one that makes most sense. But it leaves one question: upon receiving Möller's request, why did Francis Darwin not send him the complete set of Fritz Müller's letters to Charles Darwin, but in some cases chose to dispatch merely fragments of them instead?

Letters and portions of letters which contained information that Darwin deemed pertinent to his work were kept separate from other correspondence in portfolios that he assembled on particular topics, such as 'variation', 'classification', 'hybridism', etc. Since Darwin appreciated and often made use of Müller's observational skills, it seems likely that part of Müller's communications ended up in some of these portfolios. Now, the exact content of these portfolios is not known today, as they have been disassembled without inventories being made, but what we do know is that the papers from the portfolios were pasted into the bound volumes that constitute part of the Darwin archives in the CUL, and the two Müller letters dating from 1868 at least are in these bound volumes. What Francis Darwin sent Möller, it would seem, was the correspondence from the general collection of letters; of those in the portfolios he either did not know, or he did not want to disturb them.[4]

## NOTES

[1]  CD to FM, 10 August 1865, in Darwin (1888 **iii**: 37).

[2]  Möller was a first cousin once removed of Fritz Müller (his father's mother was the sister of Müller's mother), but Müller referred to him as his nephew (Möller, 1920: 138).

[3] David West, the leading authority on Fritz Müller, believes that Möller dealt expeditiously with the letters Francis Darwin had lent him. See memo enclosed in a letter from West, 23 September 1997, Fritz Müller Research File, Darwin Correspondence Project.

[4] Stephen Pocock in a note of 1987, Fritz Müller Research File, Darwin Correspondence Project.

## REFERENCES

BURKHARDT, F. and SMITH, S. (Eds), 1994 *Calendar: A calendar of the correspondence of Charles Darwin, 1821–1882. With supplement*. 2nd edition. Cambridge. Pp 690.

DARWIN, Francis (Ed.), 1888 *Life and letters of Charles Darwin*. London. Vol. 3, pp iv, 418.

MÖLLER, Alfred (Ed.), 1915–1921 *Fritz Müller. Werke, Briefe und Leben*. Jena. 3 vols.

WINSOR, Mary, 1976 *Starfish, jellyfish and the order of life: issues in nineteenth-century science*. Yale. Pp 228.

(Accepted 5 September 1998.)

ANNA-K. MAYER
Darwin Correspondence Project,
Cambridge University Library,
West Road,
Cambridge CB3 9DR.

*J. Soc. Biblphy nat. Hist.* (1975) 7 (3): 259–263

# Charles Darwin's *Queries about expression*

by R. B. FREEMAN and P. J. GAUTREY

University College London,
Gower Street, London W.C.1.
and
University Library,
West Road, Cambridge.

In 1972 we published (*Bull. Br. Mus. nat. Hist.* (hist. Ser.), Vol. 4, pp. 205–219) a facsimile of Charles Darwin's leaflet *Queries about expression,* and compared its text with that of the other two printed versions then known, one published by The Smithsonian Institution of Washington in 1868 and the other in all editions and issues of *The expression of the emotions in man and animals,* 1871.

The Smithsonian text is markedly different from that of the leaflet, whilst that in the book, though different, is clearly derived from the latter. The most important differences between the Smithsonian version and the leaflet are additions in the latter in queries, 2, 5, 10 and 13 and in the final remarks, and different wording in queries 7 and 9. We concluded that "the text of the queries printed by the Smithsonian Institution in 1868 was composed from an earlier version than that printed as a single leaf in 1867; though whether the copy was manuscript or printed is not apparent". We also noted that the word "printed" did not occur in any of the letters related to the queries.

Since then two further contemporary printings of the queries have come to light, as well as two further copies of the English printing, two more manuscript copies, and four relevant letters. These, taken with the documents recorded in our first paper, throw considerable doubt on the dating of the English leaflet.

The evidence for the first of the newly recorded printings is contained in a letter (Robin Darwin deposit, Cambridge University Library) from Asa Gray to Darwin, dated 26 March 1867, acknowledging one of 28 February. He writes "You see I have printed your queries – privately – 50 copies – as the best way of *putting* them where useful answers may be expected. Most of them will go into the hands of the Freedman's [sic] bureau, etc. – others to persons I or Wyman may know or rely on . . ." The "You see" at the beginning of this extract clearly indicates that he enclosed at least one copy of his printing with the letter. We know of no surviving copy of it, but, as Asa Gray was Fisher Professor of Natural History at Harvard University at the time, it was probably printed at Cambridge or Boston, Mass. It is possible that the version printed by the Smithsonian Institution was taken from a copy of Gray's version rather than Darwin's, and if so the change in title, the omission of the date, and the Americanisms may be due to Gray rather than to the editor of their *Annual Report.* In the letter from George Gibbs, at the Smithsonian, quoted in our first paper, the use of the word "circular" implies a printed document.

The Freedmen's Bureau, as it was popularly known, was set up by Congress in May, 1865, as the Bureau of Refugees, Freedmen, and Abandoned Lands, to handle the problems of the south derived from the civil war. Freed slaves were quite unsuitable to

194

Darwin's purpose, and he comments in *The expression of the emotions* (pp. 21–22) that "It would have been comparatively easy to have obtained information in regard to negro slaves in America; but as they have long associated with white men, such observations would have possessed little value". Asa Gray (1810–88) was the most distinguished American botanist of his day, and a frequent correspondent of Darwin's. He answered some of the queries from observations on Egyptians. Jeffries Wyman (1814–74) was a palaeontologist, mostly working on vertebrates. He did not answer the queries.

The second previously unrecorded printing came to our notice from a letter (Robin Darwin deposit) to Darwin from Robert Swinhoe, consular official and ornithologist, dated from the British Legation, Peking, 4 August 1868, referring to the publication of the queries in *Notes and queries on China and Japan,* a short-lived serial published in Hongkong. The queries appeared in Volume 1, No. 8, p. 105, dated 31 August 1867, in a note entitled "Signs of emotion among the Chinese" which is signed R.S. and dated Amoy, July 1867. Darwin's name does not appear; he is described as "a friend in England", and unfortunately Swinhoe does not make it clear whether the queries which he received were printed or in manuscript. The text of the queries is exactly the same as that printed by the Smithsonian, except for minor differences of punctuation and Americanisms. Darwin's letter to Swinhoe, which accompanied the queries is also in the Robin Darwin deposit. It is dated 27 February 1867.

The two manuscript copies of the queries are with some further replies to them which have recently been catalogued at Cambridge; both have one word answers appended. They are both in the same hand, that of R. Brough Smyth, of Australia, and bear the answers of Templeton Bunnett and H. B. Lane. Both are badly damaged by damp, but it is clear that, with minor differences of transcription, they are close to the text of the printed English leaflet. Both are however dated "Down, Bromley, Kent, Octr. 1867." The English leaflet has the same address, but only 1867 for date.

The four letters, two of which have copies of the English leaflet attached, are as follows:–

1. Darwin to G. H. K. Thwaites, dated only 31 January. "I enclose some printed copies of my queries on expression, with two of the more important ones a little amended. If you can stir up anyone to make a few observations on any race (tho' I well know how difficult it is to observe), I should be very obliged" (American Philosophical Society, Philadelphia). The word "printed" occurs for the first time in this letter. Letter 3 is almost certainly in reply to it, and the date would then be 1868. George Henry Kenrick Thwaites (1811–82) was director of the Royal Botanic Gardens at Peradenia, Ceylon. He did not answer the queries, but did put Darwin in touch with the Rev. Samuel Owen Glenie, then chaplain at Trincomalee, who wrote concerning the weeping of elephants.

2. Darwin to B. D. Walsh, dated 14 February 1868. The content of the letter is irrelevant, but written across the top is "This is a great hobby of mine, can you aid me?". A copy of the English printing of the leaflet, without manuscript alterations, is attached (Chicago Natural History Museum). "This" clearly refers to the enclosed leaflet. Benjamin Dann Walsh (1808–69) was State Entomologist of Illinois at the time. He replied on 25 March 1868 "The questions about Expression are altogether out of my line. We have no Red Indians here" (Robin Darwin deposit).

3. Thwaites to Darwin, dated 1 April 1868. "he has put C.D.'s printed list into the hands of several persons, but without any good results so far" (Cambridge University

Library). This seems to have been in reply to Letter 1, and again the word printed is used. Travelling time for a letter to Ceylon, at this date, was between 28 and 34 days.

4. Darwin to Thwaites, dated only 26 October, in Mrs Darwin's hand but signed by Darwin. "As you have been so very kind as to assist me about my queries, I enclose a few slightly corrected copies" (American Philosophical Society). Attached to this letter at the present time is a copy of the English printing of the leaflet, without any manuscript corrections. The word "few" would imply printed copies, and, taken in conjunction with Letters 1 and 3, we would suggest a date of 1868. If we have given the correct year for Letter 1 above, then some English version was in print by 31 January 1868, but we do not know which. Letter 2 shows that the only English version in leaflet form of which copies are known was in print by 14 February 1868.

The evidence for the existence of a printed English version as a pamphlet consists of Darwin's unequivocal statement in *The expression of the emotions* (p. 15) "Accordingly I circulated, early in the year 1867, the following printed queries . . .". The date 1867 printed at the foot of the surviving English printed text is indication of date of composition rather than of printing. There is plenty of evidence that Darwin was circulating queries, in some form, at least as early as 22 February to Fritz Müller, 27 February to Robert Swinhoe, 28 February to Asa Gray, 7 March to A. R. Wallace, and probably 28 February to Ferdinand Müller, and to others later in that year.

We are here suggesting that Darwin was mistaken in writing that these were printed queries; that they were all in manuscript, probably in Mrs Darwin's hand; that the manuscript, which had become much altered from its early form, was not sent to be printed until very late in 1867 or even early in January 1868; and that there was only one English printed version which reached Darwin from the printers not long before 31 January 1868. The evidence is circumstantial, but we feel that, taken together, it carries considerable weight. It can be disproved if a printed English version, in leaflet form and with the earlier text is found, or if a letter conclusively dated 1867 contains the word printed.

The following is a summary of the facts, and the conclusions which we draw from them:—

1. Asa Gray to Darwin, 26 March 1867. "You see I have printed your queries." This would seem a clear indication that what Darwin had sent was in manuscript.

2. Manuscript copy, dated 1867, with letter dated 28 February [?1867] to Ferdinand Müller of Australia, both in Mrs Darwin's hand. If the leaflet was in print in February 1867, why should Mrs Darwin transcribe it?

3. Robert Swinhoe in letter to Darwin dated 4 August 1868, in reply to Darwin's of 27 February 1867, say only "with its enclosure on human expression", with no mention of printed.

4. Copy in the hand of Dyson Lacy of Australia, undated. If he had received a printed leaflet, why should he both to transcribe it when Darwin had specifically stated that the number given to each query would suffice?

5. Letter to A. R. Wallace, dated 7 March [1867] "Please return these queries as it is my standard copy". If the queries were in print, why send his key copy? Also "You must not suppose the P.S. about memory has been lately inserted". There is nothing in the printed leaflet to indicate that the sentence about memory had been lately inserted; it comes eight lines from the bottom. On the other hand in the *Notes and queries on*

*China and Japan* and the Smithsonian versions the sentence on memory comes last. An examination of the whole Darwin-Wallace correspondance published in *Alfred Russel Wallace, letters and reminiscences,* (1916) shows certainly that this was written in 1867 and not 1860 as given. It should come between the letters Darwin-Wallace, 26 February 1867, and Wallace-Darwin, 11 March 1867.

6. Two copies in the hand of R. Brough Smyth of Australia, dated Down, Bromley, Kent, October 1867. If he had received copies of the leaflet, why should he bother to transcribe them? Even if he was down to his last copy and needed more, where did he get the date of October from, for the leaflet bears only the year, with no month? The text of these two October copies, which had presumably evolved almost to its final form by that time, is close to that of the printed leaflet.

7. Although we now have twenty manuscript copies and related letters, the first mention of the word printed in them is probably 31 January 1868, and there is a fully dated mention of 1 April 1868. The earliest known date of despatch of a printed copy is 14 February 1868. Of the five relevant letters of 1868, two use the word "printed" and another two have attached copies of the English printing. The fifth, to T. H. Huxley, dated 30 January 1868, merely says "give Mrs Huxley the enclosed". *More letters* states that the enclosed was a copy of the leaflet.

If these suggestions are accepted, the list of printings now stands as follows:

1. 1867. ?Undated, but 26 March or slightly before. ?Dropped title *Queries about expression for anthropological inquiry.* ?Cambridge or Boston, Mass. Printed for Asa Gray. 50 copies. This is the first edition, but it probably does not follow Darwin's manuscript exactly either in spelling or punctuation. It probably bears at top or bottom "By Charles Darwin, of Down, Bromley, Kent, England". No copy is known to survive. See also No. 4 below.

2. 1867. Dated 31 August 1867. *Notes and queries on China and Japan,* Vol. 1, No. 8, p. 105 title "Queries about expression". Printed with a covering note by R[obert] S[winhoe], but Darwin's name does not occur. Hongkong. The manuscript from which it was printed was posted in England on 27 February 1867. This is the nearest printed approach to the original text of the copies which Darwin sent out early in 1867, and it is the first English edition as opposed to American.

3. [?1868]. Undated by printer, but before 31 January 1868. Single sheet. Dropped title *Queries about expression.* "Down, Bromley, Kent, 1867" at foot. No printer or place. ?London. We believe this to be the only English printing in leaflet form in Darwin's lifetime. Copies at Cambridge University Library (2 and a corrected proof), American Philosophical Society (1), Chicago Natural History Museum (1); photograph in Freeman and Gautrey (1972).

4. 1868. *Ann. Rep. Smithson. Instn.* for 1867, p. [324]. Title "Queries about expression for anthropological enquiry." Below title "By Charles Darwin, of Down, Bromley, Kent, England." Text probably that of No. 1.

5. 1872. *The expressions of the emotions in man and animals,* pp. 15–16. Also in all later editions and printings, and in translations. Text is that of No. 3 with omissions.

We are grateful to the late Sir Robin [Robert] Darwin and to the Librarian of the University Library, Cambridge, for permission to quote from manuscript material at Cambridge. We are grateful also to the Librarians of the American Philosphical Society and of the Chicago Natural History Museum for permission to quote from letters in their archives. We are especially grateful to Mr P. Thomas Carroll, of the University of Pennsylvania, for drawing our attention to the documents in America which he was the first to record.

*Archives of Natural History* (1997) **24** (1): 145–148

# The early American printings of Darwin's
# *Descent of man* . . .

By JAMES W. VALENTINE

Museum of Paleontology and Department of Integrative Biology,
University of California,
Berkeley,
California 94720, U.S.A.

The earliest American issues of Darwin's *The Descent of Man, and Selection in Relation to Sex*, published by D. Appleton and Company, were based on British issues published by John Murray (Darwin 1871a,b). Freeman (1977) gave an account of the Murray issues, and listed and briefly discussed American issues that he had seen. In the course of preparing a handlist of American issues of Darwin's works I have been able to see and compare numbers of early Appleton and Murray issues of *The Descent* . . ., leading to a revision of the history of the early American printings.

## THE BRITISH ISSUES

Freeman's account of early British issues of *The Descent* . . . accurately describes the volumes I have seen, and only a few points need be reviewed here. Disregarding a variant known only from Darwin's own copy (Freeman 936), there are four issues dated 1871 (Darwin, 1871a , b, c, d), all of two volumes. In the first issue (Freeman 937) there are errata listed on the verso of the title page of Volume II, 17 for Volume I and eight for Volume II. Some copies have a note on a leaf [ix-x] tipped into Volume II that mentions an error affecting pages 197–299 in Volume I, and 161 and 237 in Volume II. In the second issue (Freeman 938: Darwin, 1871b) the errata have been corrected in both volumes and the list of errata is gone from the verso of the title page of Volume II, which carries a list of works by Darwin. Also, the leaf [xi-x] is missing, and the pages mentioned in the note have been corrected and reset. The third and fourth issues, called the seventh and eighth thousand, respectively (Freeman 939 and 940; Darwin, 1817 c, d), contain some textual changes but do not play a role in American printings.

## THE AMERICAN ISSUES

Five early American printings of *The Descent* . . . have been seen; all are of two volumes and are clearly based on British issues. However, these Appleton books are not precise reprintings but have been Americanized, so although they apparently follow their British models word for word the spellings and punctuations have sometimes been altered. For example, shew becomes show, whilst becomes while, colour becomes color, civilised becomes civilized, and so on; and semicolons sometimes become commas. Two of these Appleton printings are dated 1871 (Freeman 941 and 942; Darwin 1871 e, f), one is dated 1872 (Freeman 943; Darwin, 1872), one is dated

199

1873 (not in Freeman) and one is dated 1874 (not in Freeman). Freeman noted that for his number 941, the first volume is based on Volume I of Freeman 937 while the second volume is based on Volume II of Freeman 938. This curious arrangement is said by Freeman to change in the second issue, so that both volumes of Freeman 942 (and of 943 as well) are based on the volumes of Freeman 938, but this is not the case (see below).

The next Appleton issue listed in Freeman (his 946) is in one volume dated 1875, and is printed from stereotypes of the one-volume British edition of 1874 (Freeman 944), which was called the second edition. All subsequent Appleton texts are from stereotypes of British editions. The Americanized text of the earlier two-volume Appleton printings is not seen again.

## THE BRITISH SOURCES OF THE EARLY AMERICAN ISSUES

A simple way to determine which of the British issues, Freeman 937 or 938, has been used as a basis for the early American printings, is to examine the passages in the American volumes that refer to the pages that were reset in Freeman 938. Thus, in Freeman 937, Volume I, lines 1 and 2 of paragraph 3 on page 297 read: "All animals produce more offspring that can survive / to maturity; and we have every reason to believe that /". In Freeman 938, these lines read: "It is probable that young male animals have often / tended to vary in a manner which would not only have /". The text of all five of the early Appleton two-volume issues (wherein this passage occurs on page 288) agrees with the reading in Freeman 937. Freeman is thus correct about the basis of Volume I in the first American issue—Freeman 941 is based on Freeman 937. However, it is unlikely that Freeman 942 or any other Appleton issue has Volume I based on Freeman 938.

In Freeman 937, Volume II, a sentence contained on lines 14–18 of page 237 reads "Thus we can understand / how it is that variations arising late in life have chiefly / been preserved for the ornamentation and arming of the / males, the females and the young being left almost un- / affected, and therefore like each other." In Freeman 938, this sentence reads "Thus we can under- / stand how it is that variations arising late in life have / so often been preserved for the ornamentation of the males; and females and the young being left almost un- / affected, and therefore like each other." The text of all five of the early Appleton two-volume issues (wherein this sentence occurs on page 226) agrees with the reading in Freeman 938 (except that the semicolon is replaced by a comma); Freeman is correct here.

Thus all of these Appleton volumes have the same sources, Volume I from Freeman 937 and Volume II from Freeman 938. It seems unlikely that Appleton produced an issue with Volume I based on Freeman 938 and then reverted to using uncorrected text based on Freeman 937. However, two Appleton 1871 issues, that may be taken as corresponding to Freeman 941 and 942, can still be identified. In the earliest issue, all but one of the 17 errata found in Murray's Volume I are uncorrected (one was corrected, as Freeman notes, presumably when noticed by the compositor). A simple way to check this is to look at line 23 on page 26, where the word "kaola" is in error for "koala". In a second Appleton 1871 issue of Volume I koala is correctly spelled and the other errata are also corrected. Some Appleton 1871 issues of

Volume II list the 16 errata for Volume I on [vi], while in others that page is blank. (Incidentally, the erratum listed for "kaola" is spelled "kaolo", compounding the typographical error.) In all volumes, [vii] carries a note mentioning the errors on pages 297–299 of Volume I. As the Appleton Volume IIs are based on the second British issue, Freeman 938, the errata for Volume II, and the errors on pages 161 and 237, have already been corrected and are not listed in any Appleton issue.

Thus, there are two 1871 Appleton issues of *The Descent* . . . These volumes can be found in any combination; in some cases sets that appear to have been issued together, with both volumes in a similar condition, combine a Volume I with uncorrected errata with a Volume II that does not list errata, while others combine a Volume I with corrected errata with a Volume II that lists errata. Clearly, both 1871 variants of each volume were available at the same time, at least for a while. It seems likely that the presence of an 1871 Appleton issue of Volume I with corrected errata misled Freeman into supposing that it was based on the second British issue which also has corrected errata, and into supposing that the subsequent early Appleton issue that he saw was also so based.

## ACKNOWLEDGEMENTS

Thanks are due to Eric Korn, bookseller of London, for pointing out that there is confusion over the early Appleton issues, and for his expertise in kaolo.

## REFERENCES

DARWIN, C., 1871a *The Descent of Man, and Selection in Relation to Sex.* 2 vols. John Murray, London. (Title verso of Vol. 2 lists 17 errata in Vol. 1 and 8 in Vol. 2, all uncorrected in text. Freeman 937.)

DARWIN, C., 1871b *The Descent of Man, and Selection in Relation to Sex.* 2 vols. John Murray, London. (Title verso of Vol. 2 lists Darwin works, errata of 1871a corrected in text. Freeman 938.)

DARWIN, C., 1871c *The Descent of Man, and Selection in Relation to Sex.* 2 vols. John Murray, London. (Title page states "Seventh thousand". Freeman 939.)

DARWIN, C., 1871d *The Descent of Man, and Selection in Relation to Sex.* 2 vols. John Murray, London. (Title page states "Eighth thousand". Freeman 940.)

DARWIN. C., 1871e *The Descent of Man, and Selection in Relation to Sex.* 2 vols. D. Appleton and Co., New York. (Text Americanized, text of Vol. 1 based on Murray's 1871a (Freeman 937) with all but one of the errata uncorrected, remaining errata listed on [vi] of Vol. 2. Text of Vol. 2 based on Darwin 1871b (Freeman 938). Freeman 941.)

DARWIN. C., 1871f *The Descent of Man, and Selection in Relation to Sex.* 2 vols. D. Appleton and Co., New York. (Text Americanized, text of Vol. 1 based on Murray's 1871a (Freeman 937) with all errata corrected, no errata listed in Vol. 2. Text of Vol. 2 based on Darwin 1871b (Freeman 938). Advertisements different from Appleton's 1871e. Evidently Freeman 942.)

DARWIN, C., 1872 *The Descent of Man, and Selection in Relation to Sex.* 2 vols. D. Appleton and Co., New York. (As Darwin 1871f (but unique mixture of advertisements). Freeman 943.)

DARWIN, C., 1873 *The Descent of Man, and Selection in Relation to Sex.* 2 vols. D. Appleton and Co., New York. (As Darwin 1871f (but unique mixture of advertisements). Not in Freeman, held by the Biological Sciences Library, University of California, Berkeley.)

DARWIN, C., 1874a *The Descent of Man, and Selection in Relation to Sex.* 2 vols. D. Appleton and Co., New York. (As Darwin 1871f (but unique mixture of advertisements). Not in Freeman, held by the Biological Sciences Library, University of California, Berkeley.)

DARWIN, C., 1874b *The Descent of Man, and Selection in Relation to Sex*. 1 vol. John Murray, London. (Title page states "Second edition" and "Tenth thousand". Freeman 944.)

DARWIN, C., 1875 *The Descent of Man, and Selection in Relation to Sex*. 1 vol. D. Appleton and Co., New York. (From stereotypes of Darwin 1874b (Freeman 944). Freeman 946.)

FREEMAN, R.B., 1977 *The Works of Charles Darwin, an Annotated Bibliographic Handlist*, Second Edition. Dawson, Folkestone, and Archon Books, Hamden CT.

(Accepted 14 May 1996.)

*J. Soc. Biblphy nat. Hist.* (1976) 8 (1): 78–82

# Darwin's 'American' Neighbour

By KENNETH G. V. SMITH and R. E. DIMICK

Department of Entomology,
British Museum (Natural History),
London, S.W.7.

and

Oregon State University,
Corvallis, Oregon 97331,
USA.

For about four years (1873–77) Charles Darwin had as neighbours a barrister and his wife, Mr. & Mrs. Wallis Nash. They took up residence in "The Rookery", which stands about one-quarter of a mile north of the village of Down in Kent.

Wallis Nash recorded his impressions of Charles, other Darwins and their neighbours in chapter XIV 'Charles Darwin – a personal sketch' of his book *A Lawyer's Life on Two Continents* (Nash, 1919). This book is now a rarity and as far as we can trace the only copy in a British library is that at Cambridge University. Our notes are compiled from a copy in the possession of one of us (R.E.D.). His wife, Louisa A. Nash, had earlier recorded her impressions in an article in the *Overland Monthly* (Nash, 1890).

Both of these accounts, although somewhat anecdotal, give an interesting view of family life in the Darwin household, of the village life of Down and the part played in it by the Darwins, and of the character of Charles Darwin himself.

Of Mrs. Nash's article Francis Darwin (1892) said:

"some pleasant recollections of my father's life at Down, written by our friend and neighbour, Mrs. Wallis Nash, has been published in the *Overland Monthly* (San Francisco), October 1890".

Wallis Nash is not mentioned in either of Francis Darwin's biographies (1887, 1892) or the autobiography (Barlow, 1958).

On page 132 of Mr. Nash's book we read:

"Mr. Darwin was the treasurer and manager of the village Friendly Society,[1] taking charge of their little savings, keeping their accounts. I think he showed more simple pride in the successful balancing of the yearly statements of his poorer neighbors' provision against sickness and old age than in receiving some degree from a well-known foreign university",

and on page 134 Darwin is quoted as saying:

"Trollope, in one of his novels, gives as a maxim of constant use by a bricklayer, "It's dogged as does it!" and I have often and often thought this is a motto for every scientific observer",

and again:

"As far as my experience goes, what one expects rarely happens".

The well-known Elliot and Fry (1881) photograph of Darwin is included as a plate.

Apart from Darwin, there are comments on other eminent naturalists such as Sir John Lubbock, Herbert Spencer and H. N. Moseley. We learn that it was Darwin who recommended Moseley, recently free from his post as the *Challenger* naturalist, as a suitable expert to accompany Nash on a trip to Oregon with a view to assessing its natural resources. As a lawyer Nash had been consulted by a firm of Jewish bankers who were interested in land grants in Oregon and wished to enlist British capital in the enterprise. Moseley subsequently (1878) published a book on Oregon and Nash (1878–1919) published several in addition to becoming an editorial writer on the *Oregon Journal*.

In Nash's books there is much of interest to the natural historian. In *A Lawyer's life* . . . (p.146) we read:

"a couple of Indians came in out of the dark, one carrying slung over his shoulder, some long, dark beast, which he jerked on the counter before the store-keeper. Moseley pricked up his ears and came to take notice. From nose tip to tail the animal was about four or four and a half feet long, plainly of the otter type – the fur dark brown and glossy: but the feet were webbed. "I have never met this before," said Moseley to me. "It is the sea-otter of the Pacific." The Indian began to dicker with "Bush" [Hammond, the store-keeper] for the hide: the bidding started at two hundred dollars, and Moseley's face fell, for, by slow degrees it went to four hundred, and changed hands at that. The price was too high for him, and he had to content himself with the skeleton, which we arranged to have cleaned by the ants at a neighboring ant-heap in the wood. In due time that skeleton followed him to Oxford, and took its unique place in the Museum of Natural History. Even then these sea-otters were rare – now they are all but extinct."

The sea-otter (*Enhydra lutris* L.) was exterminated on the Oregon coast many years ago and has only recently been re-introduced. The reference to Oxford of course, refers to Moseley's occupation of the Linacre Chair of Human and Comparative Anatomy in 1881. We have made enquiries to see if this skeleton still survives in the University Museum at Oxford, but Messrs. J. Hall and E. Taylor inform us (*in litt.*) that the only sea-otter skeleton in the Zoology Department is without data or locality but was apparently purchased by an individual called Clark of the Museum of Comparative Anatomy, Cambridge from a French dealer Edouard Verreaux of Paris and the receipt dated 27th July 1869 still exists. Clark subsequently sold this skeleton to Rolleston (Linacre Professor), probably in the same year. The specimen is positively identifiable by some missing pieces mentioned at the time of purchase. The fate of Moseley's specimen is therefore unknown.

In *Two Years in Oregon* (1882) he discusses deer, elk, land otters, waterfowl and other birds. He described the Columbia River salmon industry and the kinds of salmon found in the Yaquina and Siletz areas. He also clearly described the symptoms of "salmon poisoning" in dogs and wolves and the immunity developed by the occasional dog.

Wallis Nash (figure 1) was born near London in 1837, educated at Mill Hill School and New College University of London and finally moved to Oregon in 1879 with "a number of boys to whom he taught farming." He played a large part in the development of Oregon. He was one of the builders of the Oregon Pacific Railroad; helped frame the constitution of the Oregon Agricultural College and was secretary and later President of the Board of Regents; he also practised law, farming and was an editorial writer for the *Oregon Journal* as well as being the author of several books.[2] Nashville (Lincoln County) and Nash Crater in the Cascade Range were named in his honour. He died in 1926. One of the Nash sons was named L. Darwin Nash, who, with two other brothers, attended Oregon State College and was an early graduate, eventually becoming an Oregon legislator.

In 1973 The Oregon State Board of Higher Education designated a new Bioscience building on the Oregon State University campus to be named 'Nash Hall'. This new modern six-storey building houses the Departments of Fisheries and Wildlife and the

Figure 1. Wallis Nash (1837–1926).

Department of Microbiology. The dedication of Nash Hall took place on April 27th 1974 and there were 30 descendants present, (grand and great-grandsons and daughters) originating from the families of the four deceased sons. The family provided a very fine display of objects having historical interest, including the several books written by Wallis Nash, a signed copy of Darwin's *The Movement and Habits of Climbing Plants*, copies of three letters from Darwin to Nash, several paintings, an outstanding portrait of Charles Darwin in brushed indian ink by Mrs. Wallis (Louisa) Nash (probably made at Down during Mr. Nash's first absence in Oregon), a photograph of Emma Darwin and other objects relative to the establishment and development of the Oregon State University. The portrait and Darwin letters are owned by Mrs. Lester Barton (née Louise Nash) and Miss Kathryn Nash (great-granddaughters of Wallis and Louisa Nash), who have very kindly allowed us to publish two of the letters here for the first time, the third being only an unimportant brief note.

'May 29, 1878

'My Dear Mr. Nash –

    'I thank you sincerely for your most Kind Dedication and I heartily wish that I deserved it more fully than is the case.

    'Emma began reading aloud your volume[3] to us yesterday evening and we like it much. Though not as yet arrived at new ground it tells several new things. There is not a word in it superfluous and this, as far as my experience goes with books of travels, is an extremely rare virtue.

205

'Pray thank Mrs. Nash for her note. We often and often regret your and Mrs. Nash's absence and say if you were here we would talk about this and that point.

'We ought not to regret you more for Beckenham[4] is a larger field for the unbounded goodness of both of you.

'Believe me,

'Yours ever very sincerely,

(signed) Charles Darwin

'Frank has gone to Cambridge for some electrical work with plants, otherwise he would have sent a message to Mrs. Nash.'

'Feb. 1, 80

'My dear Mr. Nash,

'I thank you cordially for your long and very interesting letter. Your life sounds very prosperous and I am delighted to hear that you are all well and happy. We heard some time ago with much alarm of your illness but I trust it was not as bad as it sounded.

'I can well understand your new life, for in old days I well remember thinking that a colonist's lot, with children, was a happy one. I remember especially this in regard to Tasmania. Frank will tell you what little news there is to be told about this quiet place. But I must send my kindest remembrances to Mrs. Nash. You will both be a heavy loss here, believe me, my dear Mr. Nash.

'Yours very sincerely,

(signed) Charles Darwin'

## ACKNOWLEDGEMENTS

Our thanks are offered to Messrs. J. Hall and E. Taylor of the University Museum, Oxford, to the Custodian of Down House, Down, Kent, to Mr. P. J. Gautrey of the Cambridge University Library and to the Nash family for their kindness in supplying information and allowing us to publish the Darwin letters. Mr Gautrey and Dr. Sydney Smith are thanked for their help in transcribing parts of the Darwin letters.

## NOTES AND REFERENCES

1 There is a brief mention of these activities in *Life and Letters*, vol.1, pp.142–3 and in the University College Library at Cambridge there is the only known copy of an original printed letter 'To the members of the Down Friendly Club'. I thank Mr. R. B. Freeman and P. J. Gautrey for a copy of this letter produced by xerography.

2 Always interested in new ventures, Nash secured Alexander Graham Bell's patent rights to the telephone for England and the first telephone in England was in his office and the first message passed from there to Queen Victoria, at Osborne House.

3 *Oregon: there and back in 1877* (1878) London. The dedication referred to is a printed one reading: "To Charles Darwin in token of a friendship wherein his gentle courtesy has almost induced forgetfulness of his greatness, this book is, by his permission, dedicated".

4 Apparently the Nash family moved back to their old home before finally deciding to emigrate to Oregon.

BARLOW, N. (Editor). 1958. The autobiography of Charles Darwin. Collins, London. pp.253.

CORNING, H. McK. (Editor). 1956. Dictionary of Oregon History. Bindford & Mort, Oregon. pp.281.

DARWIN, F. 1887. The Life and Letters of Charles Darwin. John Murray, London. 3 vols.

DARWIN, F. 1892. Charles Darwin: his life told in an autobiographical chapter and in a selected series of his published letters. John Murray, London. pp.356.

MOSELEY, H. N. 1878. Oregon: its resources, climate, people and productions. Edward Stanford, London. pp.125.

NASH, L. A. 1890. Some memories of Charles Darwin. *Overland Monthly* (San Francisco) October 1890: 404–408.

NASH, W. 1878. Oregon, there and back in 1877. McMillan, London. pp.285.

NASH, W. 1882. Two years in Oregon. Appleton & Co., New York. pp.311.

NASH, W. 1904. The Settler's Handbook. J. K. Gill, Oregon. pp.191.

NASH, W. 1905. The Farm, Ranch and Range in Oregon [Lewis and Clark expedition pamphlet]. R. Whitney, Oregon. pp.35.

NASH, W. 1919. A lawyer's life on two continents. The Gorham Press, Boston. pp.212.

*J. Soc. Biblphy nat. Hist.* (1974) 7 (1): 93–105

# Samuel Butler, Darwin and Darwinism

By BRIAN COLEMAN

4640 West 10th Avenue, Apt.204,
Vancouver, B.C.,
Canada.

Nora Barlow, grand daughter of Charles Darwin, in her re-issue of *The autobiography of Charles Darwin*[1] gives the fullest documentation yet of the controversy between her grandfather and Samuel Butler. As an early enthusiast for Darwin, Butler exchanged letters with him on the meanings of *The Origin of Species*. Darwin, in his turn, encouraged Butler in his early writings. Through a series of misunderstandings, however, their relations became strained at different points, until Butler chose to single Darwin out as his principal opponent in his fight against what he took to be the Establishment of science. After the publication of *Erewhon*, Darwin wrote to Butler to assure himself that the "Book of the Machines" was not a caricature of *The Origin of Species*. Butler's reassurance continued their good relations and he used to visit Darwin at Down. Although in his first evolutionary book, *Life and Habit*, his loyalty to Darwinism was already on the wane in favour of Darwin's predecessors, Buffon, Lamarck, and Erasmus Darwin, the unfavorable review of it by scientists put the enemy in clearer view. As he moved further from natural selection as a basis for evolution, Butler felt himself more isolated from the good opinion of Darwin and others, whose approval he originally sought, and more threatened by them.

Nora Barlow writes of the final break between her grandfather and Butler:

> On Charles Darwin's seventieth birthday in February 1879, there was issued in Germany a congratulatory number of the German periodical *Kosmos* (II, Jahrg, Heft 11), containing an article by Dr. E. Krause on Dr. Erasmus Darwin's contribution towards the history of the Descent-theory. In May, 1879, Butler published *Evolution Old and New, or the Theories of Buffon, Dr. Erasmus Darwin and Lamarck compared with that of Mr. C. Darwin*, without being aware of Krause's article in 169 *Kosmos*. Meanwhile Krause was enlarging his essay for translation; it formed the second part of Charles Darwin's *Life of Erasmus Darwin*, published in November of the same year. Whilst Krause had been engaged in this collaboration, Charles had sent him a copy of Butler's work, and some of Krause's additions consisted of disparaging references to Butler's ideas . . ..
>
> Unfortunately Charles Darwin's Preface to his *Life of Erasmus Darwin* omitted to state that Krause's original essay had been altered – exactly how this happened is explained later. Butler soon compared the supposed correct translation with a copy of the original, and the differences le him to conclude that the unacknowledged alterations formed a covert attack against himself; the public would think his views had been condemned, even before the publication of *Evolution Old and New*, and by an independent German scholar. (Nora Barlow, pp. 168–69).[1]

Butler clamored for a personal and public apology from Darwin, and might well have been satisfied with some form of explanation from him. Darwin, as *The Autobiography* reveals, was hurt, and was alarmed by Butler's caustic wit. Darwin wanted to make some reply to Butler, but in consultation with family and friends, he was dissuaded, which added more pain to Butler's already jaundiced view of Darwin and Darwinism. Neither ever learned of the hurt caused to the other, and it was only in 1911 that Butler's biographer, Henry Festing Jones, in collaboration with Francis Darwin, published an exchange of letters[2] between the two men on this and on other topics that showed the acrimony in clearer perspective. It was partly because of this background and partly because of his own ventures into the origin of species that Butler chose to discredit Darwin and natural selection in the history of evolution.

Ostensibly, Butler's complaint against Darwin was twofold: 1) Darwin had neglected to give due credit or had ignored altogether the contributions of his predecessors in evolution, principally Buffon, Lamarck, and Erasmus Darwin, and 2) Darwin had taken mind out of the universe in his supposed dismissal of design in creation and his substitution of chance through natural selection. The history of the theory of evolution until Darwin published *The Origin of Species* in 1859 was, in a sense, a cumulative effort. Without the work of Lyell in geology, of the classification of plants and animals, and of the educated guesses of earlier investigators, like Linnaeus and Buffon, Darwinism would probably not have been. In the same way, one can say that without the metaphysics of the Schoolmen that formed an important basis of Renaissance science Newtonian physics would probably not have been. But just as it took the genius of Newton to see the world the way he did, so it took the genius of Darwin to see the world the way he did. This does not mean to say that only Newton or Darwin had talents that made them what they were, (after all, Wallace was co-discoverer with Darwin of natural selection), but, nonetheless, it was Newton in his varied activities and Darwin in his who accomplished what they did.

Evolution, as an alternative to the theory of special creation, had its beginnings in modern times in the mid-eighteenth century. Before then, although special creation was always assumed, it had not been formed into a 'doctrine of immutability' until the seventeenth century when investigators began to appraise creation differently. Its popularizers were: "Milton, Ray and Linnaeus – and especially Milton, whose description of the creatures emerging fully formed from 'the earth' had so captured his readers' imaginations as to be accepted as authoritative."[3]  An historian of science gives this account of the beginnings of the theory of evolution:

> In this year [1749] Diderot published in London his *Lettres sur les Aveugles,* in which . . . he makes use of the ancient Empedoclean idea of the origin of living things. In the same year Scheidt printed, for the first time, the *Protogaea* of Leibniz, in which Leibniz at least hints at the possibility of transformation of species . . . it was in the same year that Buffon brought out the first three volumes of his great *Histoire Naturelle* . . . He [Buffon] attempted a natural explanation of the origin of the earth in general, and of sedimentary rocks in particular, and thus did for Lamarck much what Lyell did for Darwin: he fostered the idea of a gradual temporal development due to natural causes. He directed attention to the problems associated with the theory of evolution, and he was, as far as I am aware, the first to appreciate that the geographical distribution of animals contained an essential part of the evidence relating to the origin of species. Finally, he recognized that many species of animals had become extinct before the appearance of man on earth, and that there were, consequently, causes of extinction which owed nothing to human agencies.[4]

But, unlike Lamarck, Buffon cannot be said to have developed or even to have subscribed to a theory of evolution. Although Lamarck's distinction between vertebrate and invertebrate animals continues to give him renown, he was, as an evolutionist, one of the "last of the subjective and deductive philosophers in direct line from ancient Greece."[5]  His vitalism is more a response to organic mechanism, that started mainly with Harvey, than a serious attempt to accord theory with organic development. "In no serious sense, therefore, is Lamarck's theory of evolution to be taken as the scientific prelude to Darwin's. Rather, as epilogue to an attempt to save the science of chemistry for the world of the organic continuum, it was one of the most explicit examples of the counter-offensive of romantic biology against the doom of physics."[6]  Erasmus Darwin in *Zoonomia,* published in 1794–6, "put forward a theory of evolution based on David Hartley's psychological theory of associationism,"[7] though he is at least as well known for being the grandfather of Charles.  And so the stage was set for a wider study and acceptance of the theory of evolution.  More significantly, still, was the work of two little known evolutionists, William Wells and Patrick Mathew, who came as close as possible to an approximation of what was later to be Darwin's theory of natural selection.

> William Wells delivered before the Royal Society of London in 1813 a paper which contains an almost complete anticipation of Darwin's major thesis, natural selection . . . There is no record that the paper [*An Account of a White Female, Part of Whose Skin Resembles that of a Negro*] aroused any particular attention and it was published again until 1818 after the author's death.[8]
>
> In 1831 an obscure Scotch botanical writer, Patrick Mathew by name, published a book entitled *On Naval Timber and Arboriculture.* Although Mathew was a contemporary of Darwin nothing seems to be known of his life or of his birth and death dates.  This is unfortunate because Patrick Mathew is the first clear and complete anticipator among the progressionists of the Darwinian theory of evolution. (Eiseley, p. 125)[8]

The many-sidedness of Darwinism (biological, moral, theological, economic) was part of a larger tradition of discussion that preceded 1859 and continued after that date.  Geology, in the early nineteenth century, had only recently lost its theological connections, though, as part of natural theology in the writings of Paley and others, it and nature study in general continued to be seen in a religious perspective.  It was as part of this view that Robert Chambers published his *Vestiges of Creation* in 1844.  Chambers presented an account of evolution, on the basis largely of intuition, that came close enough to Darwin's to encourage Darwin to begin to write to do justice to himself as well as to the imperfect and somewhat fictionalized account of evolution that Chambers gave.  Chambers was writing theology as much as he was guessing at natural history: "that tradition of amateurism in science, by no means averse to hypothesis, largely turned to religious ends and inclined to apply religious criteria, that was even then besetting the general mind with doubt and challenge and defense of contemporary geology; the same in which that geology itself had had its gradual remote beginnings."[9]  It was this same many-sided discussion of the implications of pre-Darwinism evolution which educated the British public for 1859,[10] and which, in the writings of Malthus, for example, aided Darwin to conceive of natural selection.  "Darwinism was an extension of Laissez-faire economic theory from social science to biology." (Young, pp. 3–4).[7]  For these reasons, when the *Origin* did come its impact appears to have been immediate:

Summary of the publishing history of the *Origin of Species,*[11] English editions corrected by Darwin, 1859–1890.

| Edition | Date | | | Printing |
|---------|------|---|---|----------|
| First | Nov. | 26, | 1859 | 1,250 |
| Second | Dec. | 26, | 1859 | 3,000 |
| Third | By Apr. | 26, | 1861 | 2,000 |
| Fourth | Dec. | 15, | 1866 | 1,500 |
| Fifth | Aug. | 7, | 1869 | 2,000 |
| Sixth | Feb. | 19, | 1872 | 3,000 |
| | | | 1873 | 2,000 |
| | | | 1875 | 1,500 |

| Sixth, corrected or | 1876 | 1,250 |
|---|---|---|
| | 1878 | 2,000 |
| | 1880 | 2,000 |
| | 1882 | 2,000 |
| | 1882 | 2,000 |
| | 1883 | 2,000 |
| | 1885 | 2,000 |
| | 1886 | 2,000 |
| | 1887 | 3,500 |
| | 1889 | 2,000 |
| | 1890 | 2,000 |
| | | 39,000 |

By 1898 the total profits for Murray's were £2,709/19/8; for the Darwin family, £5,385/8/8, or £11/6/6 more than the stipulated two-thirds . . . From 1872 to 1890 the average sale was 1,561 copies a year, a sales rate of 130 copies a month. It was that publisher's delight, a bread-and-butter book. In the first decade the rate was 65, in the second 98, and in the third 112. Forty-two per-cent of the total was sold in the first fifteen years, 58 per cent in the second, to which should be added the several thousand of the expensive two-volume edition. The real triumph of Darwin's book came after his death. The profits of the American pirates must have been enormous. It would be interesting to know the sales in foreign languages. But now that the book is no longer protected by copyright, it would be as hopeless a task to search out all the reprints as it would be to discover those of its great-and almost as shattering-coeval, The Rubaiyat of Omar Khayyam. (Peckham's *A Variorum Text*, p. 25)[11]

Nor was it as absent of acknowledgement to earlier evolutionists and to the theory of design in the universe, as Butler suggests it is. In order to answer the theory of special creation, Darwin was obliged to state the mechanics of organic development without re-course to super-natural intervention at every turn, but, still, he does refer several times throughout *The Origin* to the glory of creation and to God as its creator. Nor does he repudiate entirely the contributions of Lamarck and other evolutionists, though he did not set out to add the labours of an historian of science to his other works.

The early and cordial exchanges between Butler and Darwin and the later grudging approval of Darwin's contribution to evolution are scant. When Butler does speak of Darwin, he is out to blame, not to praise. And his later and rare approval of Darwin's place in evolutionary theory is more in the nature of subterfuge, to conceal his shrill abuse of Darwin, while he gets on with his real business of anti-Darwinism. Darwin writes to Butler in 1865 of his opinions on *The Evidence For the Resurrection of Jesus Christ:* "It seems to me written with much force, vigour, and clearness; and the main argument to me is quite new. I particularly agree with all you say in your preface." (*Canterbury*, vol. 1. p. 185).[12] In *Life and Habit*, Butler can still speak of Darwin as "a great man," (*Life.*, col. 4, p. 212). Even in the 1882 Preface to *Evolution, Old and New,* he can write: "I have partaken of his hospitality, and have had too much experience of the charming sim-plicity of his manner not to be among the readiest to at once admire and envy it." (*Evolution*, vol. 5, p. XIII). In arguing his complaint against Darwin in public, he gives a history of his former admiration and good relations with Darwin: "[while in New Zealand] I became one of Mr. Darwin's many enthusiastic admirers." (*Unconscious*, vol. 6, p. 12). The snub still rankled: "I sometimes allowed myself to hope that *Life and Habit* was going to be an adjunct to Darwinism which no one would welcome more gladly than Mr. Darwin himself." (*Unconscious*, vol. 6, p. 23). Canon Butler, who was with Darwin at Shrews-bury and later at Cambridge, could speak of Darwin's works more objectively than his son: "I have always thought Darwin's evolution cruelly wanted gaps filled up, but his voyage of the Beagle Coral Islands and fertilization of Orchids will make him live." (*Family*, p. 208; Canon Butler to his son, 9 May 1882).[13] The best that his son could say in his later writings of Darwin's work is: "Mr. Darwin will have a crown sufficient for any ordinary brow remaining in the achievement of having done more than any other writer, living or dead, to popularize evolution. This much may be ungrudgingly conceded to him, but more than this those who have his scientific position most at heart will be well advised if they cease henceforth to demand." (*Luck*, vol. 8, p. 220).

Butler's true business of attacking Darwin, his person and his works, tells of a hurt and a dislike that he shows nowhere else, except in his feelings towards his father. The same disillusionment in his father's good faith, that Ernest speaks of in *The Way of All Flesh*, is echoed here in the good faith that Butler formerly had in Darwin. Beginning with the smouldering tones of rancour in *Life and Habit* and increasing, towards the end of Darwin's life and after, to the shrill abuse in *Luck, or Cunning*, Butler never doubts the righteousness of his relentless warfare with the powers of darkness, Darwin and scientists. At the beginning, "No one can tell a story so charmingly as Mr. Darwin" (*Life*, vol. 4, p. 222), and to make unmistakably clear that he is speaking in irony and not in earnest, he adds for good measure: "he has almost ostentatiously blindfolded us at every step of the journey." (*Life*, vol. 4, p. 226). In *Evolution, Old and New*, both in the 1882 Preface and in the original edition, the smouldering begins to flame up: "Mr. Darwin's manner . . . must go for what it is worth." (*Evolution*, vol. 5, p. 332), and, of Darwin's works: "the confusion and inaccuracy of thought." (*Evolution*, vol. 5, p. 316). But in the 1882 Preface, before he gets down to his true business, he cushions the blows that are to come, perhaps to make his own part in the debacle appear open to reason: "It is unfortunately true that I believe Mr. Darwin to have behaved badly to me; this is too notorious to be denied; but at the same time I cannot be blind to the fact that no man can be judge in his own case, and that after all Mr. Darwin may have been right, and I wrong." (*Evolution*, vol. 5, p. XIII). *Unconscious Memory* begins to get down to brass tacks: "there was one evolution, and . . . Mr. Darwin was its prophet." (*Unconscious*, vol. 6, p. 7). Referring to his two-sided complaint against Darwin, the wrongs to himself as well as to the older evolutionists, he writes "my indignation has been mainly aroused, as when I wrote *Evolution, Old and New*, before Mr. Darwin had given me personal ground of complaint against him, by the wrongs he has inflicted on dead men." (*Unconscious*, vol. 6, p. 56). *Luck, or Cunning*, which appeared four years after Darwin's death, follows Darwin to the grave in its tones of abuse. Here, Butler attacks once again his competence, and his integrity as well: "Mr. Darwin wanted to hedge" (*Luck*, vol. 8, p. 48): "Mr. Darwin was the Gladstone of biology, and so old a scientific hand was not going to make things unnecessarily clear unless it suited his convenience." (*Luck*, vol. 8. p. 73); "It was not Mr. Darwin's manner to put his reader on his guard." (*Luck*, vol. 8, p. 136); "poor simpleminded Mr. Darwin." (*Luck*, vol. 8, p. 148); Darwin was "pretending to take us into his confidence" (*Luck*, vol. 8, p. 181); "Mr. Darwin played for his own generation" (*Luck*, vol. 8, p. 215), and so on. Even by association with Darwin one is found guilty: "Mr. Romanes is just the person whom the late Mr. Darwin would select to carry on his work." (*Luck*, vol. 8, p. 49). Butler writes of his disloyalty to his grandfather, Erasmus Darwin: Darwin's: "over-anxiety to appear to be differing from his grandfather." (*Luck*, vol. 8, p. 47). And, not without jealousy, Butler says of his foe: "It is ill for any man's fame that he should be praised so extravagantly. Nobody was ever as good as Mr. Darwin looked . . . I heartily hope I may never be what is commonly called successful in my own lifetime." (*Luck,* vol. 8, p. 196). Further comment is added concerning the discrepancy between things as they seem to be and as they are: "whom can we trust? What is the use of science at all if the conclusions of a man as competent as I readily admit Mr. Darwin to have been, on the evidence laid before him from countless sources, is to be set aside lightly and without giving the clearest and most cogent explanation of the why and wherefore?" ('The Deadlock in Darwinism,' vol. 19, p. 39). In his correspondence and notebooks, Butler seeks support and further justification for his position: "I am to meet young Darwin [Francis] . . . on the 5th at the rooms of a common friend . . . . I am sure he means to try and come round me in some way." (Savage, pp. 171–72; Butler to Miss Savage, 26 January 1878).[14] His own explanation of the Darwins' silence is: "I have kept well within the facts –– which doubtless is the reason why no answer has ever appeared either from Mr. Charles Darwin or any of his sons." (BM Add Ms 44032, p. 140; Butler to Hyde Clarke, Esq., 2 December 1887).[15] In an ironic indictment of Darwin and scientists in general: "when I think over the names of our more prominent men of science from Huxley downwards there is not one in whose uprightness and amiability of character I have more confidence than I have in Mr. Darwin's." (BM Add Ms 44048, p. 101; 15 October 1891).[15] And with the velvet glove off: "if Charles Darwin is the best thing science can turn out what can be expected from the rank & file?" (BM Add Ms 44048, p. 102; 15 October 1891).[15]

At the same time that Butler carries out his *ad hominem* arguments, he seeks to discredit Darwin's writings as well. Until Butler collated the different editions of *The Origin,* in spite of criticism of Darwin by others, "my faith in him, though somewhat shaken, was far too great to be destroyed by a few days' course of Professor Mivart." (*Unconscious,* vol. 6, p. 25). Just as Butler had lost 'faith' in the credibility of the Gospels by a collation of them, so he repeats his performance with the editions of *The Origin,* forgetting that Darwin's word is not 'revealed' and that time may alter a naturalist's point of view as it may that of anyone else. But for Butler, if anyone should do as he did with Darwin, "he will know what Newton meant by saying he felt like a child playing with pebbles upon the seashore." (*Unconscious,* vol. 6, p. 34).

Butler deserved credit in his efforts to collate the *Origin.* He may have been the first to have done so, and he may even deserve, in some degree, being called "the first careful historian of the evolutionary movement."[16] Butler, as later evidence shows, recognized important changes in the different editions of the *Origin:*

> In my recent variorum text of the work are to be found between fifteen and twenty thousand individual variants, or well over seven thousand sentence variants, a sentence variant ranging from the exclusion or addition of a sentence to a minor and apparently non-semantic change in punctuation. The majority of these are probably of not very great importance, but there are thousands of sentence variants of crucial significance.[17]

But Butler failed to understand Darwin's reasons for these changes and he never made any realistic efforts to understand them. Instead, he saw them as further evidence of intrigue on Darwin's part. C. D. Darlington attempts to unravel the complications of Darwin's thinking in the different editions of the *Origin:*

> 'My theory' appears forty-five times in his first edition of the *Origin of Species.* In successive editions 'my' is gradually deleted. By the sixth edition of 1872, it becomes 'the theory' in forty-four of the forty-five cases . . . Nobody noticed it, before Butler, nobody cared. It was (and this is the strangest circumstance in the story) a confusion which his friends and his enemies obviously agreed in wanting. How could this be so? Darwin's scientific friends wished to imagine that only

Mr. Darwin's new illumination had convinced them of the absurdity of the creation theory which, like their opponents, they had so long and so fondly held. And Darwin's enemies were glad enough to believe that Mr. Darwin was unique and original in his own absurd and dangerous opinions. And neither side (including Darwin himself) knew what to do about the technical problem of heredity and variation which had to be dealt with in asking how evolution had occurred.

Thus in public the problem of whether evolution had occurred (which should not have been Darwin's problem) gradually pushed into the background the problem of how it had occurred (which should have been Darwin's problem). And Darwinism came to mean evolution not natural selection.

This was fortunate for Darwin because, while the change was taking place, he was engaged in retreating from the theory of natural selection. 'The theory' was coming to mean something different.[18]

Darwin, as Butler perceived, had indeed learned his lessons of evolution from the earlier evolutionists. Whether Darwin fully realized the influence of Lamarck and others on his thinking is another matter:

Darwin was essentially a transitional figure standing between the eighteenth century and the modern world. He had never entirely escaped certain of the Lamarckian ideas of his youth, whether they came by way of Lyell, or independently from his grandfather, or, as is more likely, from both. As a consequence it is not surprising that in a time of stress he grew doubtful that natural selection contained the full answer to the sallies of his critics. He fell back, therefore, toward ideas he had never totally repudiated but which, in the first edition of the *Origin*, had been allowed to remain in the background, masked, in a sense, while the major emphasis had been placed upon natural selection. (Eiseley, pp. 245–46)[8]

But it was through his own efforts of synthesis and observation that Darwin achieved the perception of nature that he did:

great acts of scientific synthesis are not performed in a vacuum. The influences, the books, the personalities surrounding a youthful genius are always of the utmost interest in terms of the way his own intellectual appetites come to be molded. Darwin's impact upon biology was destined to be so profound that much of what he absorbed from others was remembered as totally his own achievement. This happened because many of the biological works written before the *Origin of Species* became old-fashioned and ceased to be read . . . It would be easy to get the impression from the first edition of the *Origin of Species* that Darwin conceived of the evolutionary theory solely by field observation in South America. That his belief in its possible truth had been strengthened in this manner is likely enough, but it does not negate the fact that he went aboard the Beagle already aware of an existing hypothesis which he might have the opportunity of testing in the field. His genius lay in the fact that he was willing to test it; no preconceived emotional revulsion hindered him, no appetite for any exciting evolutionary theory prevented his development of a more satisfactory mechanism by which to explain its effects. (Eiseley, pp. 155–56)[8]

*Luck, or Cunning,* once again, closes his sights on the enemy and the problem at hand: "the pitchforking, in fact, of mind out of the universe, or at rate its exclusion from all share worth talking about in the process of organic development, this was the pill Mr. Darwin had given us to swallow." (*Luck,* vol. 8, p. 6). In favour of the older evolutionists and against Darwin: "He was heir to a discredited truth; he left behind him an accredited fallacy." (*Luck,* vol. 8, p. 45). In so far as Butler objects to what Darwin says and not merely to Darwin himself, the heart of the problem is his supposed denial of design[19] that, in Butler's eyes, has tended to dehumanize man and to replace the harmony of nature with a well-greased machine: "how largely this theory is responsible for the fatuous developments in connection alike with protoplasm and automatism." (*Luck,* vol. 8, p. 104). Just as Butler overlooks the reasonableness of Darwin's position for the sake of his eighteenth-century biology, so he also minimizes the reasons for Darwin's ready acceptance by those who took pains to evaluate the completeness and cogency of his arguments. "If the *Times* and a few other leading papers had not backed him, he might have written till now and made very little impression." (BM Add Ms 44045, p. 190).[15] But it was not, as Butler suggests, a conspiracy of vested interests in Darwin's favour, but his compelling logic which has then and since proved the essential merits of his case. Lyell, for example, until his last years would not change the thinking habits of a lifetime and assent to Darwin's theory, but the cogency and mounting evidence were so strongly in favour of the theory that as an old man he too became convinced of its basic reasonableness.

In place of Darwin, however, Butler turned backwards to Buffon, Lamarck, and Erasmus Darwin for his evolutionary direction and, at least, with dead men, he was less likely to have to bear the cross purposes of differing points of view. Although he was much akin to Lamarck in a similar emphasis on a 'vital principle,' that urges us to act and to become a certain way, he was more critical of him than he was of any of the older evolutionists: "He attempted to go too fast and too far." (*Evolution*, vol. 5, p. 223). And, although outside of evolution, Paley, whose efforts were to establish design in the universe, was after Butler's own heart: "I know few writers whom I would willingly quote more largely, or from whom I find it harder to leave off quoting when I have once begun." (*Evolution*, vol. 5, pp. 13–14). In contrast with what Butler says above of Darwin, he refers to the "shrewd and homely mind" (*Evolution*, vol. 5, p. 100) of Darwin's grandfather, which is suggestive of the primitive wisdom that Ernest Pontifex found in his great-grandparents. But the main theory of evolution he attributes to Buffon, "to whom it indisputably belongs." ('The Deadlock in Darwinism,' vol. 19, p. 5). His trusting to Buffon was not only because of Butler's sympathy for his point of view, but also because Buffon was able to retain the affection of his son, and these "are the only people whom it is worth while to look to and study from." (*Evolution*, vol. 5, p. 66). On the one hand, Butler is seemingly looking for people whom he can trust in and who cannot talk back to him, and on the other, he is expressing: "a moral resentment which wants more out of nature than science finds there." (Gillespie, p. 345).[6]

In retrospect, historians of science have enlarged the areas of Darwin's influence on the habits of men's thinking, then and since. And these changes in our modes of thinking may have been at the heart of Butler's opposition to what he understood by Darwinism:

One of the criticisms most commonly made of the Darwinian theory in the years after the publication of the *Origin* was that it was not inductive: it was based on assumptions instead of facts. Darwin, it was said, had deserted the true British scientific tradition, inaugurated by Bacon and brought to fruition by Newton . . . Broadly speaking these questions were answered in two different ways. One was an empiricist answer, the other an idealist one, and in the Darwinian controversy Darwin's supporters quite consistently sided with the empiricists, while his opponents almost equally consistently took the idealist line. The Darwinians found philosophical support in the writings of J. S. Mill, and the long British empiricist tradition, while the anti-Darwinians found theirs in the idealistic philosophical tradition from Plato onwards, and in the writings of the foremost philosopher of science of the age, William Whewell . . . If the periodical press can be taken as evidence, the idealistic, anti-Darwinian view of scientific philosophy was favoured by the majority of the educated public in the 1860's. The explanation is probably not far to seek. Idealistic philosophy was traditionally the philosophy of the Church, which had always been suspicious of thoroughgoing empiricism . . . like the term induction, the word hypothesis became a key word in the Darwinian controversy. Induction was the good, positive ideal which Darwin had deserted for its bad, negative counterpart, variously named hypothesis, theory, speculation, assumption, imagination, fancy, guess, and the like. (Ellegard, pp. 174, 175, 176, 189.)[19]

A vital aspect of the Darwinian revolution consists in Darwin's contribution to the method and philosophy of science. The nature of the Darwinian revolution is the revolution of man's conception of nature. Copernicus brought astronomy within the domain of science. Darwin brought zoology within the domain of science. Darwin made the biological sciences objective, experimental, phenomenal, and empirical. As Darwin viewed the world of Reality, Being possessed no antecedent priority. Being, if being there was, was to be found in the process of natural exploration. Darwin contributed to reverse the order of Being and process. Being and permanence were conceived, not as ultimate coordinates of temporal change within an antecedent transcendental system. They were conceived as logical coordinates emergent in the study of natural parts. Historically, the question is not whether Parmenides was wrong and Darwin and his successors right. Significant is the fact that Darwin separated the study of process from the concept of Being. The impact of Darwin is the impact of scientific method on the life sciences and on the sciences of human life.[20]

The deceit that Butler finds in Darwin, his person and works, he also finds in the life and works of most contemporary scientists. Butler is sensitive to discrepancy between what seems to be and what is, particularly if he feels that any one or any group of interests has caused him injury. In *God the Known and God the Unknown,* he pursues his thinking on the nature of God, more forthrightly "than modern science is prepared to admit." (*God,* vol. 18, p. 12). In his later writings, he sees an analogy between the conniving of the professional in religion and the professional in science: "our scientific elders or chief priests." (*Evolution,* vol. 5, p. 54). Both are playing tricks on us with the unknown. "Men of science are of like passions even with the other holy ones who have set themselves up in all ages as the pastors and prophets of mankind." (*Evolution,* vol. 5, p. 223). The spuriousness of the two is identical, except for a change of name: "that new orthodoxy which is clamouring for endowment, and which would step into the Pope's shoes to-morrow, if we would only let it." (*Evolution,* vol. 5, p. 317). In seeking perhaps an explanation for his own lack of recognition in science, Butler points to Lamarck's lack of recognition in his lifetime: "Science is not a kingdom into which a poor man can enter easily, if he happens to differ from a philosopher who gives good dinners, and has 'his sisters and his cousins and his aunts' to play the part of chorus to him." (*Evolution,* vol. 5, p. 223). Scientists, like the clergy, as Butler shows them in caricature elsewhere, claim to have gifts of understanding apart from the common sense of others, which irritates Butler: "So-called men of science will not allow that they are men of like intellectual infirmities with the common run of mankind." (*Life, vol. 2,* vol. 18, p. 116). Butler learned his lessons of distrust of his elders in the school of hard knocks of his family, and he never forgot these lessons in any of his later dealings with those in authority, as long as he was, in some sense, under authority of religion, of the academicians and of scientists: "they mean turning the screw upon us" (*Life, vol. 2,* vol. 18, p. 134). But the scientists, of all the others, perhaps because they were themselves still struggling to have an equal say in areas of influence with those of the still traditional culture of classical studies, did not let him have his own way. And he matched their resistance with all that he could give, in journals which would air his usually shrill tones and in his own books: "we have given life pensions to some of the most notable of these [Darwinians] biologists, I suppose in order to reward them for having hoodwinked us so much to our satisfaction." (*Luck,* vol. 8, p. 6). All was fair in the stakes for evolution: "men like Professor Huxley do not serve protoplasm for naught." (*Luck,* vol. 8, p. 114). "If terror reigns anywhere among scientific men, I should say it reigned among those who have staked imprudently on Mr. Darwin's reputation as a philosopher." (*Luck,* vol. 8, p. 124). Central to his complaint against science, here as elsewhere, is the tension in it between appearance and reality: "Whatever faith or science the world at large bows down to will in its letter be tainted with the world that worships it. Whoever clings to the spirit that underlies all the science obtaining among civilized peoples will assuredly find that he cannot serve God and Mammon." (*Ex Voto,* vol. 9, pp. 213—14). Although he makes this same complaint against religion as he does against science, he considers the first less sinning than the second: "I, who distrust the doctrinaire in science even more than the doctrinaire in religion, should view with dismay the abolition of the Church of England, as knowing that a blatant bastard science would instantly step into her shoes." ('A Medieval Girl School,' vol. 19, p. 215). For Butler, at any rate, the science of his day was not so much on-the-make, as it was the leading oracle of the times: "Science is a mode of revelation, and so far as we can see the only mode that is now vouchsafed to us." (*Dr. Butler,* vol. 2, vol. 11, p. 446). A banner at the opening of the Temple in *Erewhon Revisited* reads: "Science as well as Sunchildism" (*Erewhon Revisited,* vol. 16, p. 140), one falsehood alongside another. 'Science' offers him the opportunity to deflate its assured reputation and to hit Darwin at one and the same time. As his good fight progresses, he sees the shoe that Darwin once wore on the other foot: "Heretofore it was I who wanted to draw them & cd not get them to break silence--now they have been well drawn & it is I who am holding my tongue, & the more I hold it the more they will try & draw me, & the more the really influential people will incline towards me." (*May,* p. 121; Butler to May, 9 April 1884).[21] At different points in his fight with science and scientists, he occasionally sees victory: "there is nothing but their skill in hoodwinking the public which can save them from disaster." (BM Add Ms 44033, p. 315; Butler to Marcus Hartog, 6 November 1891).[15] *The Times,* too, because of its past association with Darwinian interests, is suspect: "It had a swagger article on M. Pasteur's supposed discovery of a cure for hydrophobia. It was very plausible, but so are the prospectuses of many companies which never pay a dividend." (BM Add Ms 44047, p. 14; June 1887).[15]

If Butler could not speak the language of the scientists and if they would not listen to him speaking his own language, then he would judge them on the basis of style, which he and others of his humanistic training would understand, even if they should not agree with his judgements. Commenting on a sentence[22] from *The Origin,* he writes: "which yet upon examination proves to be as nearly meaningless as a sentence can be." (*Life,* vol. 4, p. 213). Butler asks the reader to excuse any faults in his style which may be attributed to association, for: "if one has gone for some time through a wood full of burrs, some of them are bound to stick." (*Life,* vol. 4, p. 245). Nor do the older evolutionists get away scot free from his literary probings. Of Buffon he writes "a single hasty passage in so great a writer may well be pardoned under such circumstances." (*Evolution,* vol. 5, p. 150); and of Erasmus Darwin: "Dr. Darwin's use of the word 'thence' here is clearly a slip, and nothing else; but it is one which brings him for the moment into the very error into which his grandson has fallen more disastrously." (*Evolution,* vol. 5, p. 201). But, on the positive side he refers to: "the un-theory ridden language of Dr. Darwin." (*Evolution,* vol. 5, p. 224). After his reading in the Darwinians he exclaims; "How refreshing to turn to the simple straightforward language of Lamarck." (*Evolution,* vol. 5, p. 332). But this is just so much preamble to an examination of Darwin as a writer: "Mr. Darwin is not a clear writer." (*Unconscious,* vol. 6, p. 41). It is not only Darwin's writing which, in Butler's eyes, shows the hollowness of his meaning, but his self-consciousness of accuracy: "I ought to have suspected inaccuracy where I found so much consciousness of accuracy, but I did not." (*Unconscious,* vol. 6, p. 43). Again, if there is any fault in his own style, Butler affects to put the blame on those whom he has been reading: "as one who has caught a bad habit from the company he has been lately keeping." (*Unconscious,* vol. 6, p. 198). Of himself and of anyone else, whose business it is to watch clear style and thinking, he says: "It is not the ratcatcher's interest to catch all the rats." (*Unconscious,* vol. 6, p. 198). Butler, who is inclined in his writings to identify analogy and metaphor, and who makes his own rules of science, says of Darwin: "the expression natural selection must be always more or less objectionable, as too highly charged with metaphor for purposes of science." (*Luck,* vol. 8, p. 50). But in the end, as Butler thinks he recognizes, it is "Mr. Darwin with one of the worst styles imaginable [who] did all that the clearest, tersest writer could have done." (*Luck,* vol. 8, p. 217). In his literary criticism of the Darwinians, Butler sees himself as: "a fly in a window-pane. I see the sunshine and freedom beyond, and buzz up and down their pages, ever hopeful to get through them to the fresh air without, but ever kept back by a mysterious something which I feel but cannot either grasp or see." ('The Deadlock in Darwinism,' vol. 19, p. 39).

But in earlier and in more guarded moments, Butler's feelings for science were not wholly hostile. At these times, he can sound like many a Victorian in his admiration and respect for science. *A First Year in Canterbury Settlement,* 1863, his earliest published book, speaks this way: "Being hungry, far from home, and without meat, we ate the interesting creature [a kind of kaka], but made a note of it for the benefit of science." (*Canterbury,* vol. 1, p. 162). In 'A Dialogue,' from the *Press,* Christchurch, New Zealand, 20 December 1862, Butler imagines a dialogue on the values of science: " 'I believe in Christianity, and I believe in Darwin. The two appear irreconcilable. My answer to those who accuse me of inconsistency is, that both being undoubtedly true, the one must be reconcilable with the other.' " ('A Dialogue,' vol. 1, pp. 193–94). *The Fair Haven* pays a back-handed compliment to science, when it compares the completeness of evidence in scientific judgement with the absence of evidence in religious judgement: "What should we say if we had found Newton, Adam Smith, or Darwin, arguing for their opinions thus?" (*Fair,* vol. 3, p. 152). In the same vein: "Can any step be pointed to as though either Church [Rome or Canterbury] wished to make things easier for men holding the opinions held by the late Mr. Darwin, or by Mr. Herbert Spencer and Professor Huxley?" ('A Medieval Girl School,' vol. 19, p. 214). And not to resist a play on ideas: "Do you--does any man of science believe that the present orthodox faith can descend many generations longer without modification?" (BM Add Ms 44032, p. 83; Butler to Mr Blunt, 5 July 1887).[15] Just as Butler would sometimes stay at monasteries when he visited Italy, so, he writes: "It is a great grief to me that there is no place where I can go among Mr. Darwin, Professors Huxley, Tyndall, and Ray Lankester, Miss Buckley, Mr. Romanes, Mr. Grant Allen, and others whom I cannot call to mind at this moment, as I can go among the Italian priests." (*Alps,* vol. 7, pp. 50–1). When some Italian friends said that they preferred scientists to clerics, Butler shares the opinion of the former, though he chooses to continue on good terms with the latter, "with [whom] I suppose I am on the whole more sympathetic." (*Alps,* vol. 7, p. 48). Writing to his sister May, on recent hypotheses in astronomy, he speaks with the 'holy abandon' of the science enthusiast: "I don't quite understand it, but no doubt it is all right." (*May,* p. 141, 15 April 1885).[21]

But the science he is talking about, in his later years at any rate, is part of the same idealism, that elsewhere makes it impossible for him to reconcile discord between the way he fancies things to be and the way things are:

Science, after all, should form a kingdom which is more or less not of this world. The ideal scientist should know neither self nor friend nor foe–he should be able to hob-nob with those whom he most vehemently attacks, and to fly at the scientific throat of those to whom he is personally most attached; he should be neither grateful for a favourable review nor displeased at a hostile one; his literary and scientific life should be something as far apart as possible from his social; it is thus, at least, alone that anyone will be able to keep his eye single for facts, and their legitimate inferences. (*Luck*, vol. 8, p. 218).

It may be that this is as much an apology for his own life of science as a model for others, who did not accept him on his terms. In this way, when he defends Buffon from the abuse of his critics, he may be speaking on his own behalf: "that he was more or less of an elegant trifler with science, who cared rather about the language in which his ideas were clothed than about the ideas themselves, and that he did not hold the same opinions for long together; but the accusation of instability has been made in such high quarters that it is necessary to refute it still more completely." (*Evolution*, vol. 5, p. 84). He himself wanted to be taken seriously at the beginning of his evolutionary theorizing, and never quite lost the hope of belated recognition: "I got the *Athenaeum* last week to announce my new book *Luck or Cunning?* which they did, I was pleased to see that they put it well up among their Science news, not among the Literary events." (*Family*, p. 273, Butler to his sister, Harriet, 15 March 1886).[13] Nor did he lose faith in the worth of his theorizing: "One great complaint against me is that I am not a professional scientist. It is not shown that what I say is unsound."[23] But his terms of science are outside the objective terms of science, as they are generally understood.

Ostensibly, Butler is fighting the good fight for wrongs caused to dead men, to the harmony of creation and to his own person. But if seen in terms of the whole of his life and writings, his battle with science and scientists is one more battle in the warfare between fathers and sons. At the beginning of his relations with Darwin, there was no provocation on Darwin's part to cause injury, on the contrary, Darwin took some pains to encourage Butler, as the younger man, in his developing interests as a writer. But Butler, seemingly, could not for long accept Darwin as a great man without an irresistible inclination to debunk him. Perhaps when he first attempted this in 'The Book of the Machines' he simply could not help himself, for this was how his imagination worked. But later he may have come to believe in his own unflattering picture of Darwin, for in *Life and Habit* he makes it his business in part to rebuke Darwin and scientists on a number of points. When Krause seemingly impugned some of the ideas expressed in *Evolution, Old and New*, Butler may have considered this as sufficient justification for his increasing ill-will against Darwin, his person and works. Because of the set pattern of his imagination, Butler could not trust to goodness or greatness and take Darwin and others on the basis of their reputations. Butler came to believe in some degree that he was persecuted as he came to believe in some degree in his own theory of evolution.

Although at the beginning of his evolutionary writings, in *Life and Habit*, he may have taken himself seriously as one man of science among others, his later battles with scientists are fought not so much to do himself credit as to discredit them. Though he published three more evolutionary books, they were largely a repetition of the thinking in *Life and Habit*. He was repeating the arm-chair science of a Robert Chambers at a time when such hit-and-miss efforts had shown themselves to be wanting in comparison with the results of those who worked within the scientific method. Butler was playing tricks with a chimaera of science, which he himself had created.

## NOTES AND REFERENCES

[1] Darwin, Charles; *The autobiography of Charles Darwin 1809–1882*, ed. Nora Barlow. London, 1958.

[2] Jones, H. F.; *Charles Darwin and Samuel Butler a step towards reconciliation.* London, 1911.

[3] Willey, Basil; *Darwin and Butler two versions of evolution.* London, 1960. p. 39.

[4] Wilkie, J. S.; Buffon, Lamarck and Darwin: the originality of Darwin's theory of evolution, from *Darwin's biological work some aspects reconsidered*, ed. P. R. Bell. Cambridge, 1959. pp. 270–287.

[5] Simpson, George Gaylord; Lamarck, Darwin and Butler three approaches to evolution. *American Scholar.* **30** (2), Spring 1961: 239.

[6] Gillispie, Charles Coulston; *The edge of objectivity.* Princeton, 1960. p. 276.

[7] Young, Robert M.; *The impact of Darwin on conventional thought.* A talk delivered to the Victorian Society at the National Portrait Gallery. 8 February 1968. p. 2.

[8] Eiseley, Loren; *Darwin's century.* London 1959. p. 120.

[9] Millhauser, Milton; *Just before Darwin Robert Chambers and Vestiges.* Middletown, Connecticut, 1959. p. 57.

[10] "Chambers' *Vestiges of Creation* . . . sold over 25,000 copies in Britain before 1860, and the phrenological work of George Combe on which it drew for its views on man sold 50,000 copies between 1835 and 1838 and was selling at the rate of 2,500 copies a year in 1843. By the end of the century, according to A. R. Wallace, it had sold 100,000." (Young, p. 5). See also footnote 7.

[11] Peckham, Morse; Introduction to his edition of *The origin of species. A variorum text.* Philadelphia, 1959. p. 24.

[12] Unless otherwise stated, all quotations from Butler's writings are taken from the *Shrewsbury Edition*, edited by Henry Festing Jones and A. J. Bartholomew. London, 1923–1926.

[13] *The family letters of Samuel Butler 1841–1886*, edited by Arnold Silver. London, 1962.

[14] *Letters between Samuel Butler and Miss E. M. A. Savage, 1871–1885*, edited by Geoffrey Keynes and Brian Hill. London, 1955.

[15] *Samuel Butler papers.* British Museum Add Ms 44027–44054.

[16] Barzun, Jacques; *Darwin, Marx, Wagner.* Boston, 1941. p. 118.

[17] Peckham, Morse; Darwinism and Darwinisticism. *Victorian Studies*, **3** (1), September 1959: 21.

[18] Darlington, C. D,; *Darwin's place in history.* Oxford, 1959. pp. 35–36.

[19] "The Design argument is an argument from analogy. From the fact that some processes and phenomena in our ordinary experience are regularly explained as due to human design and forethought, is drawn the conclusion that similar processes in the history of the world, before the entry of man on the scene, are due to Divine Design and Providence." (Alvar Ellegard, 1958, Darwin and the general reader. *Goteborgs Universitets Arsskrift*, **64**: 117).

[20] Loewenberg, Bert James; The mosaic of Darwinian thought. *Victorian Studies*, **3** (1) September 1959: 18.

[21] *The correspondence of Samuel Butler with his sister May*, edited by Daniel E. Howard. Berkeley and Los Angeles, 1962.

[22] "I can no more believe in this." Charles Darwin, *Origin of species.* Sixth edition. London, 1876. p. 171.

[23] *Further extracts from the note-books of Samuel Butler*, edited by A. J. Bartholomew. London, 1934. p. 56.

*Archives of Natural History* (1982) **10** (3): 509–514

# Charles Darwin at Glenridding House, Ullswater, Cumbria

By H. P. MOON

Professor Emeritus,
Department of Zoology,
University of Leicester,
Leicester.

In 1881 Charles Darwin and some of his family had a holiday in the Patterdale district (Darwin, 1887, **3**: 356) and in a letter to Hooker dated 15 June 1881, he gave his address as Glenrhydding House, Patterdale, Penrith (Darwin & Seward, 1903, **2**: 433), Glenrhydding being an unusual spelling of Glenridding.

Against the whole background of Darwin's life and work, this particular episode is not of great significance but from the point of view of local interest, it is worthy of a record as it is not generally known. The two visits which Darwin made to the Lake District, one to Coniston[1] and the other to Ullswater being additions to the long list of distinguished men and women who, in some way or another, have associations with the district. It seemed worth while to give some account of this holiday and the various Darwins who stayed at Glenridding House, as Charles Darwin was one of the few men in the whole of history whose work has been the basis for a fundamental change in human thought. It is therefore, of no mean interest that he and some of his family should have an association with Glenridding House, and that one can, to a certain extent, retrace and visualise his holiday.

Glenridding House is a large family house one quarter of a mile to the north of Glenridding village in Cumbria on the shores of Ullswater; the house is built between the lake and the main road. Glenridding is in the Chapelry of Patterdale, hence the reference to Patterdale, meaning the district of Patterdale. The house stands on a level strip of ground a few yards back from the lake, between the grounds of the Ullswater Hotel and Cannon Crag, the first of the rocky headlands of the rugged and wooded shoreline stretching north to Glencoin. It is a square-built two storey stone building, the stone being covered over by a light coloured stucco; the corner stones, windows and doorways are picked out in a light greenish blue colour. A conspicuous feature is the fine but simple wrought ironwork of the balconies. On the south and west sides, a continuous balcony runs round the house, supported on cast iron pillars from ground level. The spaces between the pillars are spanned by slender arches, the gap between the arches and the balcony above being filled in with cast iron lattice work. On the east side, two end rooms have individual balconies built over the projection of ground floor bay windows. At ground level, there is a walk-way on the south and west sides, sheltered from the weather by the balcony above and bounded by a low cast iron trellised fence. At the back, there are buildings in local stone. The rooms are high and spacious with large windows giving fine views of the lake and across to Place Fell, Blowick, and Silver Point.

The house with its blue door and window edgings and corners, the green of the grass strip sloping to the lake, which is so blue in the sunshine, is reminiscent of the type of house to be seen beside Swiss or north Italian lakes.

The house was probably built by the Askews, a well known local family of former days, in about 1832, but sold in about 1860 to the Greenside Mining Company, and soon afterwards to Robert Bownass who had recently built the Ullswater Hotel (Morris, 1903: 53–54; 59). Robert Bownass may have intended to use the house as additional accommodation for guests at the Ullswater Hotel, whose grounds adjoin the garden of Glenridding House. His son, Thomas Bownass, certainly used the house as an annexe to the hotel, and it has always been referred to locally as the annexe, which may explain why this very interesting house has not attracted the attention that it deserves.

One of the party with Darwin at Glenridding House in 1881 was his married daughter, Henrietta Litchfield, whose book *Emma Darwin – A Century of Family Letters*,[2] refers to the Patterdale holiday (Litchfield, 1904). This book with its letters and notes is a rich source of information about the Darwins, and contains detailed pedigrees of the Allens, Darwins, and Wedgwoods.

Referring to the Patterdale holiday, Henrietta has the following note—

On June 2nd, we all went to a house at Patterdale taken for a month. I think that this second visit to the Lake Country was nearly as full of enjoyment as the first. It was an especial happiness to my mother for the rest of her life to remember her little strolls with my father by the side of the lake. I have a vivid picture in my mind of the two often setting off alone together for a certain favourite walk by the edge of some fine rocks going sheer down into the lake. (Litchfield, 1904, 2: 316–7).

This is followed by an extract from one of Emma Darwin's letters dated Sunday (June 1881)

The day has turned out even more beautiful than the first Sunday. We all, but F, went in the boat first up and then coasting downwards as far as How Town landing place, where we got out. Bernard was with us dabbling his hand in the water and very quiet and happy. It was very charming up amongst the junipers and rocks. William was much delighted but is rather troubled by wishing for Sara. F. got up to his beloved rock this morning but, just then a fit of his dazzling came on and he came down. I mean to get there soon. (Litchfield, 1904, 2: 317).

In later years, the family often referred to Darwin as F. for father.

The situation of the house as one sees it today, agrees with Henrietta Litchfield's note. There is the level ground between the house and the lake where Charles and Emma would have strolled, while a few yards away to the north is Cannon Crag, a conspicuous glacially rounded rock headland going sheer and smooth into deep water. There is a way to the top of the Crag by means of a gate close by the house, and the view from the Crag, twenty-five feet above the lake, is not only impressive but of great interest.

At the present time, the path from Glenridding House up to Cannon Crag is boarded off and rather over-grown, but there is a good path leaving the main road at Stybarrow Crag, running along the lake edge and over Cannon Crag to re-join the main road behind the Crag. Until recently, the path continued past Glenridding House to the Ullswater Hotel grounds. This path is still known to the older people in Glenridding as 'Askews Walk' and presumably was part of the Glenridding House estate and may well have been the path along which Emma and Charles went for their "favourite walk".

Cannon Crag would be the "favourite rock" to which Darwin evidently so often went. The view from the top of the crag has many features of geological interest, which would not have been missed by Darwin and such a place would have become for him, "a beloved rock". A practical point is that Cannon Crag would be the nearest place of general interest that would be within his capacity, but even so, as related in Emma's letter, this short distance was sometimes too much for the ailing Darwin who was to die in the following year.

The boat excursion described in the extract from Emma's letter, quoted earlier, included some interesting members of the Darwin family. Emma says, "we all but F., went in the boat", so we know who else was with Darwin at Glenridding House apart from Emma herself and Henrietta Litchfield.

Henrietta Emma Litchfield born in 1843 was Darwin's only daughter to survive, two girls dying as babies and a third at the age of ten. Henrietta, who as she grew older, became the eccentric Aunt Etty of Gwen Raverat's charming *Period Piece* (1952) and was a person of ability and character. Darwin was greatly indebted to Henrietta for her excellent suggestions and corrections when she read the proofs of his book on *The Descent of Man* (1871), and asked her to accept a memorial from him of £25–£30 in recognition of her help (Litchfield, 1904, 2: 229–230, 241). She was evidently well thought of by no less a person than Thomas Henry Huxley who nicknamed her *Rhadamanthus Minor*.[3] He also sent her a geological cartoon illustrating the evolution of the 'Ur-Hund' which referred to Polly one of the well remembered Darwin dogs (Litchfield, 1904, 2: 232, 252).

There is no reference to her husband being one of the party at Glenridding House, but it is worth saying something about Richard Buckley Litchfield. Despite Gwen Raverat's rather comic picture of Uncle Richard, he was a well read and able man. Richard Litchfield was an authority on music and taught the subject at the London Working Men's College, of which he was one of the Founders (Litchfield, 1904, 2: 251). He used to take his music class on walking parties that on several occasions ended at Down House to be welcomed by Charles and Emma Darwin. Their visits were a great success and would include singing in the garden under the lime trees (Litchfield, 1904, 2: 264).

Darwin had a high opinion of Richard Litchfield and in Chapter 4 of his work on *The Expression of the Emotions in Man and Animals* (1872), he quotes him at length when discussing expression in music, and mentions this in a letter to Henrietta (Litchfield, 1904, 2: 255–256). Richard Litchfield also sketched out the draft of a Parliamentary Bill on vivisection – a question about which Darwin and a number of leading scientists were very concerned (Darwin, 1887, 3: 204). He also arranged the letters and wrote many of the notes in *The Century of Family Letters*, but his death prevented the completion of this task (Litchfield, 1904, 1: viii).

The William of the boat party was William Erasmus Darwin, Charles Darwin's eldest son, born in 1839 and married in 1877 to Sara Sedgwick, the Sara referred to by Emma in her letter. There is a nice account of William by Gwen Raverat in her Chapter on the five Uncles. William was at Rugby and Cambridge where he occupied the rooms in Christ's College that had been his father's (Litchfield, 1915, 2: 166). He became a banker in Southampton, and involved himself in various public duties, for example, the County Council, the Southampton Water Works, and re-organising the town's charity and medical relief system. Southampton, with the exception of Bristol, was the most pauperised town in England. (Litchfield, 1904, 2: 199, 275).

His most important interest was the building up of the Hartley Institute to University College status which lead finally to the University of Southampton. He was Treasurer of the Hartley Council and later of the University College for forty-one years until his death in 1914. It was largely due to William Darwin that the Institute achieved University College status in 1902 (Litchfield, 1904, 2: 287; Patterson, 1962).

William Darwin was well read with an interest in botany and geology, and his observations on the orchid *Epipactis palustris* in the Isle of Wight are quoted by Charles Darwin in his book, *The various contrivances by which Orchids are fertilised by Insects* (1862). William was also much respected for a speech he made at a Cambridge dinner given in memory of his father (Litchfield, 1915, **2**: 167–171) from which Henrietta gives extensive and interesting quotations.

The Bernard of the boat party was the child of Francis, Darwin's third son, and at the time of the Patterdale holiday only five years old. Francis had married Amy Ruck in 1874 but in 1876 she died when Bernard was born (Litchfield, 1904, **2**: 280), and the motherless Bernard was the centre of special affection in the Darwin family – at any time a family which seemed to take children to itself. Henrietta tells how Emma Darwin, then sixty-eight years of age "took up the old nursery cares as if she was a young woman"; but it greatly changed her life. Emma, in a letter to Henrietta about Bernard writes that, "Your father is taking a good deal to the Baby. We think he (the Baby) is a sort of Grand Lama, he is so solemn" (Litchfield, 1904, **2**: 280).

As a little boy of two years old sitting on Darwin's knee, Bernard was fascinated by the "bright spots" – the spots of light reflected from Darwin's pocket magnifying glass (Litchfield, 1904, **2**: 288). Darwin, however, could be stern with Bernard when he was a little older, for Emma writes in a letter to Bernard's father Francis, that Darwin admonished Bernard for being over-bearing with little Alice, "B was astonished, but it quite answered" (Litchfield, 1904, **2**: 297). One of Charles Darwin's characteristics, of course, was his detestation of any form of oppression or injustice. In after years, Bernard became one of the world's most famous golfers and wrote extensively on the game.

Down House was a haven for children. Henrietta records that her elder cousin, Julia Wedgwood, said that, "the only place in my father's and mother's house where you might be sure of not meeting a child, was the nursery." Henrietta tells how many a time, even during her father's working hours, a sick child was tucked up on his sofa to be quiet and safe and soothed by his presence (Litchfield, 1904, **1**: 468).

The last member of the boat party was Darwin's wife Emma, whose personality is so well revealed by the letters and notes in her daughter's book.

Emma was one of those remarkable Victorian women of firm faith, high principles and a courageous sense of duty, who despite endless child-bearing and anxieties, maintained her character and intellect. Emma was always interested in the world around her to the end of her life, and was a person of great sympathy and wise counsel to whom people could unburden themselves. She had a keen sense of humour and a delight in the small and simple things of life.

Emma created a warm and friendly home at Down where Darwin could work and think and feel secure in his frequent periods of ill health and depression; the world owes more to Emma Darwin than is generally realised.

Finally, there was Charles Darwin himself who climbed up to Cannon Crag, his favourite rock, while the boat party rowed to Howtown. He was an old man and not in good health, yet Henrietta's note suggests that he enjoyed the holiday, which was apparently nearly as successful as the first one to the lakes at Coniston. None the less, he does seem to have been affected by the despondency that so often afflicted him, judging by his remarks in a letter to Hooker from Glenridding House on 15 June 1881.

> I am rather despondent about myself, and my troubles are of an exactly opposite nature to yours, for idleness is downright misery to me, as I find here, as I cannot forget my discomfort for an hour. I have not the heart or strength at my age to begin any investigation lasting years, which is the only thing

which I enjoy; and I have no little jobs which I can do. So I must look forward to Down graveyard as the sweetest place on earth. This place is magnificently beautiful, and I enjoy the scenery though weary of it, and the weather has been very cold and almost hazy. (Darwin and Seward, 1903: 433).

This sad letter shows how the years of struggle with depression and poor health were taking toll of the ageing Darwin. On 12 July 1881, he wrote to Wallace from Down;

We have just returned home after spending five weeks on Ullswater. The scenery is quite charming, but I cannot walk and everything tires me, even seeing scenery, talking with anyone or reading much. What I shall do with my few remaining years of life, I can hardly tell. I have everything to make me happy and contented, but life has become very wearisome to me. (Marchant, 1916, 1: 319).

It is clear that Darwin's enjoyment of the Ullswater holiday was overshadowed and he was sufficiently ill at one time for a doctor to be called (Colp, 1977: 92–93). According to Colp there is no mention of this in any of the Darwin letters, journals or private diaries and Henrietta Litchfield makes no mention of a visit by a doctor.

The distinguished physcian, Sir Arthur Salusbury MacNalty, in an article on the ill health of Charles Darwin, recalls that his son, Dr F. C. MacNalty, told him that he visited Darwin at Glenridding House and diagnosed angina pectoris. He considered that Darwin's health was precarious as there were signs of myocardial degeneration (MacNalty, 1964). Dr MacNalty was the local doctor at the time, the *Medical Directory* for 1881 giving his address as Patterdale, Westmorland.

The holiday at Ullswater was Darwin's last for he died less than a year later and was buried in Westminster Abbey, close to the grave of Sir Isaac Newton. He did not rest in the sweetness of Down churchyard, but perhaps the thought that his countrymen had conferred so great an honour would have pleased this modest and reserved man. It is strange to stand on top of Cannon Crag or look out through the big front windows of Glenridding House and realise how little the lake and fells have changed since Darwin saw these same views for the last time.

NOTES

1 Darwin's first visit to the lakes was to Coniston in 1879 which he enjoyed, and Ruskin drove over from Brantwood to visit him (Litchfield, 1904, 2: 299). They had met many years ago at Dean Buckland's house in Oxford, as young men with a common interest in geology. There is no known record of where Darwin stayed at Coniston.

Although the anticipation of a journey always worried Darwin, he enjoyed the journey to Coniston and was not in the least put out when they missed the connection at Foxfield (Darwin, 1887, 1: 129); Litchfield, 1904, 2: 299). There is, of course, no railway now from Foxfield to Coniston.

2 There are two editions of *Emma Darwin. A Century of Family Letters.* The 1904 edition was for private circulation; the 1915 edition was published because many people who read the 1904 edition felt that it would be of interest to a wider circle of readers. The two editions differ slightly in content as Henrietta Litchfield made some alterations in the 1915 edition.

John Murray have a voluminous file of correspondence with Henrietta concerning the 1915 edition. This correspondence would make fascinating reading and I greatly regret that I have never been able to avail myself of John Murray's kind offer to let me look at this correspondence if ever I was in London.

3 *Rhadamanthus* is a former generic name for the lily.

ACKNOWLEDGEMENTS

In the course of writing this account, I was helped by a number of people. I owe a very real debt of gratitude to Miss Sheila Kay of Penrith, Cumbria, who for many years, has been studying the problem of Darwin's health. Miss Kay replied to an appeal I made in the local press for any information about Darwin at Ullswater and provided me with important references.

I also acknowledge the help of my friends, Mr James Cooper, Mr Lane, Mr Harold Oglethorpe, the Reverend John Rogers, and my sister, Miss Barbara Moon.

I also express appreciation to my wife for assistance in typing the manuscript.

REFERENCES

COLP, R. Jr, 1977 *To be an invalid. The Illness of Charles Darwin.* Chicago: University Press.

DARWIN, F. (ed.), 1887 *The Life and Letters of Charles Darwin.* London: John Murray.

DARWIN, F. and SEWARD, A. C. (eds.), 1903 *More letters of Charles Darwin.* London: John Murray.

LITCHFIELD, H. E., 1904 *Emma Darwin. A Century of Family Letters.* 2 vols. Cambridge: University Press.

LITCHFIELD, H. E., 1915 *Emma Darwin. A Century of Family Letters.* London: John Murray.

MACNALTY, A. S., 1964 The ill health of Charles Darwin. *Nursing Mirror*, i–ii.

MARCHANT, J., 1916 *Alfred Russel Wallace Letters and Reminiscences.* London: Cassell.

MORRIS, W. P., 1903 *The Records of Patterdale.* Kendal: Titus Wilson.

PATTERSON, A. T., 1962 *The University of Southampton* Southampton: University Press.

RAVERAT, G., 1952 *Period Piece.* London: Faber and Faber.

*Archives of Natural History* (1986) **13** (1): 19–24.

# Darwin in Chinese

By R. B. FREEMAN

Department of Zoology and Comparative Anatomy,
University College London,
London,
WC1E 6BT

There are only eight entries for translations of any of Darwin's works into Chinese in the second edition of my *The works of Charles Darwin* (1977) and all eight concern *On the origin of species*. I did enquire from Peking at the time, but was unable to trace any other translations. The first five of these entries concern the translation by Dr Ma Chun-wu of Shanghai which appeared bit by bit, from 1903 onwards, the whole not being completed until 1920. This same translation was used in 1957 for an edition produced in Taiwan: the publishers of this state, in error, that the original had first appeared in 1918. The other two entries are for an entirely new translation which first appeared in Peking in 1955 and was reprinted in 1972.

In April 1981, Mr Chou Pang-li of Shanghai wrote to ask me if I could send him a copy of Nora Barlow's (1958) transcription of her grandfather's *Autobiography* with the original omissions, which Francis Darwin had made in 1887, restored. He asked also for Sir Gavin de Beer's transcription (1974) which had been checked against the manuscript, explaining that Chinese libraries had not, for many years, been buying books in English and no copies were available to him. Chou had already translated five of Darwin's works into Chinese from Russian texts, including in 1958 the independent transcription of Darwin's *Journal* by S. L. Sobol' which had appeared in 1957, before the English version of de Beer in 1959. He told me that he was putting together a Chronicle of Darwin's life and for the next year I was able to help him over points of difficulty particularly about people and places. In exchange, he sent me notes on all the Chinese translations of which he was aware, as well as copies of such as he could find. Chou, was born in 1918, died on 31st August 1982, after his *Chronicle* had appeared, but whilst the *Autobiography* was still at the press. Mdme Ku Yuan, his widow, very kindly sent me copies of it when it appeared in December.

The following list is made up from the notes which Chou sent me, arranged in order of titles given in the second edition of my *Handlist*, with a few additions not noticed by him. I have retained the spellings as Chou gave them, rather than try to alter them to the recent transcriptions: for example, I have retained Peking throughout rather than Beijing. I have noted the whereabouts of those few copies which I have seen in British libraries, including those on my own shelves. It will be seen that sixteen works have been available in Chinese in fifty five different editions or printings. Four recent biographies have been added to the list, as well as the exhibition catalogue of the Peking centenary celebrations.

## Journal of researches

1. 1941 Chungking, The Commercial Press. 2 vols. Translated by Huang Su-feng. Issued in February. Based on Second English edition (1845). Chou Pang-li comments that the text has many errors of translation.

2. 1955 Chungking, The Commercial Press. As No. 1, but second edition.

3. 1957 Peking, Science Press, 676 pp., portr., 13 maps, text figs. Translated with preface by Chou Pang-li. Preface and introduction by S. L. Sobol'. Translated from Russian F241, the English original being Oxford University Press, World's Classics 1930, F136. Copy R. B. F.

## On the tendency of species to form varieties

4. 1983 in *Some Materials of Biology for Reference*. Vol. 18, pp. 1–10. Science Press Peking, October, 4,000 copies. Translated by Chou Pang-li from F1557 = F359. Contains Darwin's paper, pp. 4–7, and Letter to Asa Gray, pp. 8–10.

5. 1983, the same, second printing, 4,000 copies, late October. Copy R.B.F.

## On the origin of species

6. 1902 Yokohama, *Hsin-min tsung Pao*, No. 8, pp. 9–18, 22 May. Translated by Dr Ma Chun-wu. This consists of the historical sketch only. F634 part.

7. 1903 Shanghai, Chung-hwa Press. Translated by Dr Ma Chun-wu. This consists of Chapters 3 and 4 only, either as a single pamphlet or as two separate pamphlets. F634 part.

8. 1904 Shanghai, Chung-hwa Press. Vol. I. Translated by Dr Ma Chun-wu. This consists of the historical sketch and Chapters I–V. It appeared in the spring. F635.

9. 1920 Shanghai, Chung-hwa Press. 4 vols. Translated by Dr Ma Chun-wu. The first 5 chapters revised from No. 8; the rest is new. This is the first complete translation. No. 14 dates this ?1918 = F637. F638.

10. 1922 Shanghai, Chung-hwa Press. All as No. 9, but Second impression.

11. 1947 Peking, Sheng-huo Bookshop. Vol. I. Translated by Chou Kien-ren. All published, about half the text.

12. 1954–55 Peking, Sheng-huo, Tou-shu, Hsin-chih, Three Cooperative Press. 3 vols. Translated by Chou Kien-ren, Yeh Tu-chuang and Fang Tsung-hsi. This is a completion of No. 11. Copy R.B.F.

13. 1955 Peking, Science Press, Translated by Shih Yun-cheng. Edited by Wu Hsien-wen, Chen Shih-shiang and Chu Hung-fu. This is an edition in one volume from Russian F762 and with preface of that edition by K. A. Timiryazev. Darwin's *Autobiography*, No. 50, in the version of Francis Darwin is added. F639. Copy British Museum (Natural History), London.

14. (1957) Taipei, Taiwan, T'ai-wan Chung-Hua She-Chu. 2 vols. Translated by Dr Ma Chun-wu. Dated 46th year, 2nd month. Copies British Museum (Natural History), London; Down House, London.

15. 1963 Peking, The Commercial Press, 625 pp., reprint of No. 12 in one vol.

16. 1972 Peking, Science Press. Translated Shih Yun-cheng, Chen Shih-shiang, Wang Ping-yuan, Cheng Tso-hsin and Chen Pao-shan. Edited by Wu Hsien-wen, Chen Shih-shiang and Chu Hung-fu. This is a revised edition of No. 13. F640a. Copy Zoological Society of London.

17. 1974 Shanghai, Renmin Press, iii, 330 pp., 7 parts in one vol. This is a new edition of No. 15 with the same translators etc.

18. 1981 Peking, The Commercial Press, 190 pp., portr. A reprint of No. 5. Copy R.B.F.

### Fertilisation of orchids

19. 1965 Peking, Science Press, x, 229, pp. Translated by Tang Tsin, Wang Fa-tsuan, Cheng Hsin-chi and Hu Chang-hsu. May, 2450 copies. Translated from the Second edition, 1890, F810. Copy R.B.F.

### On the movements and habits of climbing plants

20. 1957 Peking, Science Press, x, 102 pp. Translated by Chang Chao-chien. Edited by Lou Cheng-hou. October, 1580 copies. Translated from First edition, 1865, F833.

21. 1959 Peking, Science Press. Second printing, as No. 20 but Nos 1581–2381.

### Variation under domestication

22. 1957–58 Peking, Science Press. 2 vols. Vol. I translated by Yeh Tu-chuang and Fan Tsung-hsi. Vol. II translated by Yeh Tu-chuang alone. Vol. I September 1957, 1800 copies. Vol. II December 1958, 1720 copies.

23. ?date Peking, Science Press. 2 vols. Second impression. Date not traced, assumed from existence of No. 25.

24. ?date Peking, Science Press. 2 vols. Third impression. Date not traced, assumed from existence of No. 25.

25. 1973 Peking, Science Press. 2 vols. Fourth impression. All as No. 22. November.

### The descent of man and selection in relation to sex

26. 1930 Shanghai, Commercial Press. 9 vols. Translated by Dr Ma Chun-wu. April.

27. 1932 Shanghai, Commercial Press. As No. 26, but Second impression.

28. 1939 Shanghai, Commercial Press. As No. 27, but Second edition.

29. 1959 Shanghai, Commercial Press. As No. 28, but Third edition.

30. 1981 Peking, Commercial Press. Translation by Pan Kuang-tan and Hu Shou-wen. New edition.

31. 1982 Peking, Science Press. Translated by Yeh Tu-chuang. May.

**Expression of the emotions in man and animals**

32. 1939 Shanghai, Commercial Press. 2 vols. Translated by Chou Kien-hou.

33. 1958 Peking, Science Press, ii–3–254 pp. Translated by Chou Pang-li. Introduction by S. G. Gellerstein. Translated from Russian F1213. December, 2174 copies. Contains *Biographical sketch of an infant*, pp. 235–243, No. 38. Copy R.B.F.

34. 1959 Peking, Science Press. As No. 33, but Second impression. September, Nos 2175–4174 copies.

**Insectivorous plants**

35. ?1983 Peking, Science Press. Based on 1882 edition, New York. This edition was planned, but had not appeared by the end of 1983.

**Cross and self fertilisation in the vegetable kingdom**

36. 1959 Peking, Science Press. 254 pp. Translated by Siao Fu, Chi Tao-fan and Liu Tsu-tung. February, 2080 copies. Based on Second edition F1251.

37. 1963 Peking, Science Press. As No. 36, but Second edition. April, copies 2081–3080.

**Biographical sketch of an infant**

38. 1958 Peking, Science Press. Pp. 235–243 in No. 33 above. Copy R.B.F.

39. 1959 Peking, Science Press. As No. 38, in No. 34.

**The power of movement in plants**

40. 1982 Peking, Science Press. Translated by Lou Cheng-heu. Edited and revised by Chou Pang-li. Based on Third thousand, 1882. F1328.

**The formation of vegetable mould, through the action of worms**

41. 1954 Peking, Chung-hwa Press. 149 pp. Translated by Shu I-shang. December, 2800 copies.

**Life and letters and autobiography**

42. 1917 Peking, *Student Magazine*, Nos 1, 3 & 7. Translated by Chou Tai-hsuan. *Autobiography* only, based on Francis Darwin's version.

43. 1935 Peiping, Chung-shan Bookshop. Translated by Chang Meng-wen. *Autobiography* only, based on Francis Darwin's version.

44. 1935 Shanghai, Shih-chieh Press. Translated by Chou Yun-to. *Autobiography* only, based on Francis Darwin's version.

45. Shanghai, The Commercial Press. Translated by Chuan Chu-sun. *Autobiography* only, based on Francis Darwin's version.

46. 1947 Peking, Sheng-huo Bookshop. Translated by Su Chia-o. *Autobiography* only, based on Francis Darwin's version.

47. 1948 Peking, Sheng-huo Bookshop. As No. 46, but Second edition.

48. 1949 Peking, Sheng-huo Bookshop. As No. 47, but Third edition.

49. 1950 Peking, Sheng-huo, Tou-shu, Hsin-chih, Three Cooperative Press. As No. 48, but Fourth edition.

50. 1955 Peking, Science Press. As No. 49, but Fifth edition, revised.

51. 1957 Peking, Sheng-huo, Tou-shu, Hsin-chih, Three cooperative Press. 2 vols. Translated by Yeh Tu-chuang and Meng Kuang-yu. Vol. I, May; Vol. II, October, 20,000 copies. This is an edition of the whole work.

52. 1963 Peking, The Commercial Press. 2 vols. As No. 51, but Second printing.

53. 1982 Peking, The Commercial Press, viii, 189 pp. Translated by Bi Li [pseudonym of Chou Pang-li]. December, 2270 copies. *Autobiography* only, based on Barlow F1479, with the original omissions restored and compared with that of de Beer F1508 and Sobol' F1541. Copies Down House, R.B.F.

### Charles Darwin and the voyage of the Beagle

54. 1958 Peking, Science Press, xv, 296 pp. Translated by Chou Pang-li. April, 1660 copies. Translated from Russian F1586, from Nora Lady Barlow's book F1594.

### Charles Darwin's diary of the voyage of H.M.S. Beagle

55. 1958 Peking, Science Press, 350 pp. Translated by Chou Pang-li. August, 1109 copies. Translated from Russian F1585, from Nora Lady Barlow's book F1566.

### Biographies

56. 1981 Li Kueng-yu [*The founder of scientific evolution — Darwin*]. 185 mm, 55 pp., portr., 8 text figs, Peking, The Commercial Press. July, 6000 copies. Foreign History Series. Copy R.B.F.

57. 1982 Chang Ping-lun and Chen Tu-sheng [*Charles Darwin*]. 184 mm, 296 pp., 8 portrs on 4 leaves, 1 map, 2 pedigrees, Peking, Chinese Youth Press. March.

58. 1982 Chou Pang-li [*The chronicle of Charles Darwin's life*]. 200 mm, vi, 356 pp., 2 portrs, 52 text figs, Peking, Science Press. March, 6,000 copies, 1–3100 in cloth, 1–2900 in paper. Copies Cambridge University Library, Down House, R.B.F.

59. 1982 Korsinskaya, Vila [*The great naturalist Charles Darwin*]. 185 mm, 2,205 pp., 12 portrs, 119 text figs, Shanghai, Science and Technical Press. April. Translated by Sun Chao-kun. Revised by Chou Pang-li. From Russian text, Leningrad 1956. Copy R.B.F.

60. 1983 Chou Pang-li. As No. 58, but corrected edition. 2,508 copies. Copy R.B.F.

**Exhibition catalogue**

61. 1982 Beijing Natural History Museum [*Charles Darwin: a guide to the exhibition of Darwin's life . . . for the commemoration of the death of Charles Darwin*]. 180 mm, [12] pp. with text figs, Peking, Natural History Museum. April 19. Copy R.B.F.

*Archives of Natural History* (1988) **15** (1): 61–62

# Darwin in Chinese: some additions

By P. J. P. WHITEHEAD

Zoology Department,
British Museum (Natural History),
London SW7 5BD

Hearing of my intended visit to China in October 1986, the late Richard Freeman asked me to look out for any additions to his list of Darwin works or Darwiniana translated into Chinese. In his second edition of *The works of Charles Darwin* (Freeman, 1977) he had located only eight titles (all concerning *On the origin of species*). With the help of Mr Chou Pangli (of Shanghai) and later his widow he was subsequently able to document sixteen Darwin or Darwinian works in Chinese in fifty-five editions or printings (Freeman, 1986—sixty-one items, six being already catalogued).

Since I was pressed for time in Beijing (Peking) and since Shanghai was presumably scoured by Mr Chou, I sampled only the library of Nanjing (Nanking) University, using the 1985 printing *General catalogue of books in natural science in China* (edited and published by Liaoning University Library); this includes all books published in China to the end of 1984. Unfortunately, there was time to examine only some of the volumes, either there or in the library of the Department of Biology.

It was not until I returned to London that I learned of Richard Freeman's death. Without his help the following notes and additions are necessarily brief, but they are offered as a tribute to his scholarship and his profound knowledge of Darwinian bibliography.

The numbers prefixed FC are those of his 1986 listing of Chinese works, to which I have added 'a' for the three new titles. NUL is the Nanjing University Library and NUBD is Nanjing University Biology Department. New titles are indicated by an asterisk.

**On the origin of species**

FC 9, FC 12 and FC 16 seen at NUBD. The dating of FC 9 is implied as 1918 by FC 14; in fact it was published in 1920 (no Freeman copy and none indicated in his notes).

**The descent of man and selection in relation to sex**

*FC 28a, being a reprint of the 1939 2nd edition (FC 28). Copy not been, but it is indicated in the back of the 1959 third edition (FC 29), which I examined at NUBD. *FC 31a, being a reprint of 1983 by the Commercial Press, Beijing, 987 pp., of the 1981 new edition translated by Pan Kuang-tan and Hu Shou-wen (FC 30).

**Life and letters and autobiography**

FC 51 and FC 52 were seen at NUL, volumes 1 only of the first, volume 2 only of the second. However, since FC 52 (1963) is merely a reprinting of FC 51 (1957), the pages for both can now be given as 569 and 618 pp.

*FC 53a, being *The life and letters of Charles Darwin* edited by Francis Darwin, translated by Ye Du-zhuang and Ye Xiao, published by the Science Press, Beijing, 398 pp. Not seen, data from Liaoning catalogue.

### Charles Darwin and the voyage of the Beagle

FC 54, seen at NUBD.

### Charles Darwin's diary of the voyage of HMS Beagle

FC 55, in NUL, not seen.

This adds only three titles to Richard Freeman's list, which suggests that Mr Chou's searches were rather thorough. I am greatly indebted to Dr Chen Jian-xiu of the Biology Department, Nanjing University, for his help in searching out the volumes and translating for me.

## REFERENCES

FREEMAN, R. B., 1977 *The works of Charles Darwin. An annotated bibliographical handlist*, 2nd revised and enlarged edition. Dawson, Folkestone and Archon Books, Shoe String Press, Connecticut. Pp 235.

FREEMAN, R. B., 1986 Darwin in Chinese. *Archives of Natural History* **13**: 19-24.

*J. Soc. Biblphy nat. Hist* (1978) 8 (4): 351—366

# The Charles Darwin — Joseph Hooker correspondence: an analysis of manuscript resources and their use in biography

By JANET BROWNE

Department of History of Science,
Imperial College,
London S.W.7

The Charles Darwin (1809—1882) — Joseph Hooker (1817—1911) correspondence spans the most tumultuous and eventful years of the nineteenth century, when natural history forced itself into scientific and public affairs through the debate over evolution and natural selection. Yet, ironically, the letters between these two eminent protagonists for selection reflect only a small part of the debate. From 1843 to Darwin's death in 1882 they spontaneously confined discussion to purely scientific matters in which botanical questions and problems of plant distribution predominated. The point is made not to denigrate botany nor to criticise the quality of the correspondence, but rather to stress the importance of plant studies to the history of natural selection and evolution theory.

The correspondence totals some 1300 items: about 800 letters from Darwin to Hooker, and about 500 the other way. Not as large as many manuscript collections, this is certainly the largest body of Darwin letters to one individual that also remains intact with its replies. It makes up for what it lacks in volume by sheer intellectual weight: singly, and as a collection, the letters make significant contributions to Darwinian and Hookerian studies.

But why use the manuscripts — the autograph letters — when there is such a wealth of published sources? A dip into the *Life and Letters of Charles Darwin*[1] could lead through a library of comparable works: *More Letters of Charles Darwin;*[2] *Emma Darwin: A Century of Family Letters;* the *Lives and Letters* of Sir Joseph Hooker, Charles Lyell, Thomas Henry Huxley, A. R. Wallace, Asa Gray, and perhaps on to those of Richard Owen, R. I. Murchison, A. Sedgwick, L. Agassiz, Edward Forbes, and so on.[3] These are all universally regarded as authoritative sources for nineteenth century naturalists.

Yet a re-evaluation of the *Life and Letters* type of biography is essential. When the manuscripts are extant it is important and necessary to compare them with the printed works: the latter not only contain errors of transcription and dating, but they also suffer from questionable methods of composition. The authors were high Victorians and invariably closely connected with their subjects through family, personal, or intellectual ties. The mere fact of living one generation removed from the great man gave biographers the necessary detachment whilst ensuring that they had to rely heavily on friends and relatives for information. The end-product consequently offers a selection of hazy recollection, polemic and evasion, anecdote and family tradition, disguised under an air of 'tell it like it was'.

Because the genre explicitly includes letters the books tend to be treated as primary sources. Here they will be examined as historical documents within their own significant context, and the *Lives and Letters* of Darwin and Hooker will be compared with the manuscript collections of their letters, not to characterise the one as 'bad' evidence against the 'good' but to show how a complementary study can significantly aid our interpretations of biography, and perhaps alter the received widsom about these two eminent naturalists. Such an approach is commonplace among critics of literature and hence the study of bio-graphy is well advanced,[4] but few attempts have been made to place scientific biography in this tradition and apparently none for the *Life and Letters* group. Dr. Outram's work on the biographies of Cuvier is, however, a noteworthy exception.[5]

What is known of biographers and biography in the nineteenth century? The basic infor-mation in *Lives and Letters* was often presented so that the subject was moulded to fit the ideals and conventions of his memorialist, and so that myth and dogma were perpetuated. More important however, is the fact that the great man frequently cast himself in a particular role, or roles, during his lifetime and a famous figure could thus create his own biography. In the scientific world Huxley and Murchison[6] immediately spring to mind. A dutiful bio-grapher would do nothing to dispel or modify such images and, indeed, *Lives and Letters* as a literary style were not designed to cope with informed assessments of scientific achievement. They were purely a vehicle for the revelation of the 'inner man' and an exemplary character study: hence biographers relied on the subject's own estimate of his scientific worth and were uncritically laudatory.

Most people dismiss such 'Victorian' memorials for their lack of artistic quality and their over-solid pragmatic approach, but it is useful to remember that it is this prosaic element that encourages us to use them as sources for history. Over the nineteenth century the num-ber of commemorative volumes increased exponentially and were rapidly standardised into the two volume, crown octavo sets that are so familiar today. The desire to read of the famous ensured a stable market for the publisher and, by mid-century, it was expected (though not always endorsed) that death was followed by a resurrection in print. Carlyle,[7] with his belief that history was the lives of great men, did more to encourage the genre than perhaps even Boswell, and towards the end of the century *Lives* were being produced, and read, in enormous quantities. Uniform sets such as the *Men of Letters* and *Men of Science* series attempted to fill the gap left by personal memorabilia for some serious evaluation of 'lives and works' but even these were couched in eulogistic terms. Typical of the era and symptomatic of the popularity and saleability of biography was, of course, Leslie Stephen's tribute to his nation, the *Dictionary of National Biography*.

The universal justification was its didactic worth and elevating message, and much of the limited discussion over the aims of biography was dominated by the question of ethics. Since Boswell had revealed Dr. Johnson 'warts and all' public opinion had swung fiercely towards reticence, moral evaluation and enlightenment in its reading matter.[8] The Evangeli-cal movement, although it could hardly have caused this development, nevertheless fostered and furthered such an approach to the extent that even biographers with little or no sym-pathy for religious enthusiasm adopted the style. The second generation pushed it to the excesses that we now characterize as 'Victorian' and which, at the end of the century, pro-voked such violent rebuttals as Gosse's *Father and Son* and Strachey's *Eminent Victorians.*

Exemplary biographies depended for their merit not on the writer but on the moral quality of their subject. The biographer was modestly supposed to be only the compiler and organiser of documents that displayed in their proper order virtues and suitably stigmatised vices. The power to make or break a hero must have indirectly encouraged the author to waive his own judgement (or at least to say so) and present the evidence so that the reader could draw his own conclusions; by allowing the great man to tell his own story the contemporary goal of objectivity was assured. The use of letters went hand-in-hand with this 'documentary' ideal and the quasi-scientific desire to let facts speak for themselves, increasingly bearing the burden of the book as impartiality, reticence and sensibility became the norm. As there already was in existence a strong alternative tradition of epistolary novels and collected letters of literary figures, late Victorians must have felt that adding letters to a *Life* in some way increased its adherence to the truth. Although it is difficult to imagine what a biographer could produce without correspondence as his raw material, authors in the last third of the nineteenth century relied solely on the published letter – in its entirety – to fill up the pages of their books. And fill them up they did: not only were vast numbers of letters written, but almost as many were preserved, so that biographers suffered under an *embarras de richesses.*

With this in mind it is easy to dismiss Francis Darwin's (1848–1925) *Life and Letters* and *More Letters* of his father Charles as typical of the genre. Credit must, however, be given to Francis for realising that the best way to portray his father was through his large and varied correspondence. During a secluded life Darwin's main contact with the outside world was via the written word and that was how many of his colleagues came to know and respond to him. As he lived through letters, so letters reflect his life – more accurately perhaps than Hooker's. Both the *Life and Letters* and *More Letters* of Charles Darwin were widely acclaimed when first published in 1887 and 1903 respectively, and remain highly praised not just for the mass of information they contain but for genuine literary merit and charm. Over and above its expected sales the *Life and Letters* proved to be a best-seller and, about one month after publication, John Murray could boast that he had disposed of 4000 copies, and the following week that he was printing the sixth thousand.[9] However the success of Francis Darwin's collections of Darwin's letters should not obscure the case that they contain many editorial errors; errors which might appear to be trivial but which hide valuable historical information.

For my own work on the development of plant geography it has been important to know precisely when Darwin familiarised himself with Edward Forbes' (1815–1854) famous paper published in the first memoir of the Geological Survey, 1846: *On the connexion between the distribution of the existing fauna and flora of the British Isles, and the geological changes which have affected their area.*[10] This is conventionally regarded as the first formal merger between geology and botanicozoological distribution: in other words, Forbes here proposed a series of invading floras and faunas moving over the British Isles as and when land movements permitted. Darwin too had been working on similar ideas and it might reasonably be expected that he would have expressed immediate interest in Forbes' theory.

However, in Darwin's correspondence with Hooker such a discussion occurs only before the event – in 1845 – and some ten years later in 1855 and 1856. The comments form the basis of an interesting psychological study of Darwin and his reaction to being 'forestalled' by Forbes, which adds a significant factor to Darwin's rejection of continental extensions as an explanation for distribution. In his *Autobiography;* in the long manuscript version of his book *Natural selection;* in *On the origin of species* and in letters to Hooker[11] Darwin insisted that Forbes had anticipated his own theory on the origins of alpine and arctic geographical distribution patterns. But if the two published presentations are compared there are fundamental differences which Darwin would have noticed had he fully examined Forbes' account. The comments arising in Darwin's correspondence of 1845 are based only on verbal and informal written communications about Forbes' theory as put forward to the British Association of that year, whereas those arising at the later time of 1855 are clearly based on a recent reading or re-reading of his formal and fullest statement to the Geological Survey of 1846. The major piece of evidence in favour of Darwin reading Forbes' long article on publication is a letter from him to Hooker, published in *More Letters of Charles Darwin*, and dated '1845 or 1846'.[12] The editor, Francis Darwin, stated that "Sir Joseph Hooker's letters to Mr. Darwin seem to fix the date as 1845, while the reference to Forbes' paper indicates 1846".[13]

Going through the preceding letters to Hooker in the autograph collection it becomes clear that Darwin senior was not referring to this, 1846, paper, nor even to the similarly worded notice of 1845, but instead to a completely different report on the fauna of the Aegean seas.[14] So there is no evidence that Darwin did read Forbes on publication and, although this does not prove that he did *not* read it, a source of confusion has been eliminated. The question would not have arisen had not Francis Darwin introduced a spurious connection between the letter and this particular paper. The letter itself can now be dated to the summer of 1845.[15]

Not only dates can be recovered from studying the manuscripts in conjunction with the published works. A simple mistake introduced unwittingly into Darwin's *Life and Letters* by Joseph Hooker, has led to scholarly ignorance of a long and interesting, unpublished, letter. It deserves attention as the first to be sent from Hooker to Darwin, and because it invalidates the published 'first' letter.[16] In his reminiscences for the *Life and Letters of Charles Darwin* Hooker stated that he received his first letter from Darwin in December 1843. Francis Darwin printed the text of the letter using Hooker's account as the 'lead-in', although he dated the letter itself with only the year. By the time that Hooker's *Life and Letters* was published in 1918 this date – December 1843 – was an unquestioned part of the Darwin legend. Hooker's biographer, his godson Leonard Huxley (1860–1933), would however have done well to question it, for his concentration on December caused him to overlook Hooker's first reply dated November of the same year. Leonard Huxley compounded the error by publishing another letter (actually the second reply) in Hooker's *Life* and calling that the first response to Darwin.[17] So the first Darwin to Hooker letter is printed in Darwin's *Life* with a date that has only illusory exactitude, and the mistake in Hooker's reminiscences led to the publication of another letter – the supposed reply – in Hooker's own *Life* thirty years later. The genuine reply according to Leonard Huxley could not exist, nor does it exist for historians today unless they have consulted the manuscript collection.[18]

Leonard Huxley was drawn into error by not questioning the accuracy of Francis Darwin's editorial work and through ignorance that the date had, in the first instance, been derived from Hooker. This case, and that which preceded it, are salutory examples of the benefits of using manuscript material. But consultation of autograph letters is always essential if dates and readings are to possess any real authority. Unfortunately, the *Life and Letters* treatment can contain more fundamental and far-reaching faults than these. However, before presenting a critique of *Life and Letters* it is both logical and useful to digress to its counterpart and to stress the intrinsic value of original resources, and, in particular, the Darwin – Hooker correspondence. It is a measure of the quality of this collection that, despite exhaustive examination by Darwinists, it can and does provide such fresh and stimulating source material.

The letters written before the *Origin of Species*, 1859, can be seen as a dialectic between botany and geology, which seems to have been Hooker's and Darwin's own characterisation of their positions. Hooker over some twenty years called Darwin a geologist in the face of ample contradictory evidence for Darwin as a zoologist, botanist, entomologist, or any of the 'ists' pertaining to nature study. Darwin, on the other hand, always referred to 'you Botanists' when writing to Hooker. Equally, they describe themselves in similar terms. Taking up such positions was not strictly necessary, as both men were competent over an enormous range of contemporary science, and it perhaps reflected a desire to polarise issues and to reduce mutual problems to their simplest terms by asking 'What would a non-botanist or non-geologist say to this?' Consciously or unconsciously Darwin projected a certain image of himself as a geologist and no doubt Hooker reciprocated. Writing in August 1856 Hooker told Darwin that:

> "The difference between us lies in this: that you are more of a geologist than botanist & feel the real weight of the geological objections – I am no geologist at all & enjoy the aforesaid freedom of motion . . ."[19]

In reply Darwin stated that Hooker's geological letters were of more value to him than even Lyell's, and implied that the advantage lay in Hooker standing outside geology. This was an attribute that Hooker correspondingly recognised in Darwin, as he wrote to John Tyndall:

> "Darwin is the only one of them [geologists] I know who has any accurate conception of the right or wrong of a botanical problem . . ."[20]

The correspondence presents an opportunity to see not only what Darwin and Hooker did, and what they thought that they were doing, but also helps us to recover their self-images. The historical necessity for such a recovery can hardly be over-stated nor can its corollary, the several shades of truth behind self-perception and projection, be ignored.

Immediately related to these questions is the utilisation that each man made of his colleague's scientific knowledge. Darwin deliberately took botanical concepts from Hooker and introduced them into his geology, whilst Hooker followed up several geological problems during the course of his studies that are essentially derived from Darwin. If this exchange had been less self-conscious nothing more could have been drawn from the fact but, from their obvious awareness of their 'roles' considerable historical weight can be laid on it. Darwin, in particular, extracted methods, concepts, and techniques from allied and subsidiary disciplines, firstly to provide a wide range of evidence for his theories, and secondly, to see if these additions and cross-fertilisations offered anything to his methodology. In botany he found a new and exciting field opening out in front of him from which he could pick and choose information and interpretations like flowers.

These general relations between the sciences – botany and geology – and their physical expression in the correspondence between Hooker and Darwin take definite direction if the history of the development of biogeography is superimposed. Biogeography is, after all, the merging of geology with botany and zoology to form a science of plant and animal distribution in time and space. Reducing it to dangerously simple terms, biogeography was the child of geology. But it did not spring fully formed from Zeus' head: it required shaping and continuing inter-penetration of the various factors involved, in the manner suggested by Darwin's and Hooker's correspondence. Naturalists who were familiar with spatial distribution such as Hooker, Brown, H. C. Watson, Gray, Bentham and De Candolle, had to co-ordinate with geologists and palaeontologists such as Darwin, Edward Forbes, Murchison and Agassiz, who knew about fossil distribution in time, before biogeography became a discipline in its own right. Contrary to expectation, Darwin's work, which gave such continuity and unity to the natural world, did not effect this fusion. Biogeography was already established, admittedly in diffuse form, before Darwin imposed evolution theory and thereby gave it direction. The prevailing methodology of this nascent science dictated that concepts from one discipline should be transposed to others. The Darwin-Hooker correspondence presents an actual example of a conscious attempt at cross-fertilisation through Darwin's use of botanical statistics under Hooker's guidance.

In close collaboration with Hooker through the 1850s Darwin tabulated some dozen floras and one or two monographs from the animal kingdom to show the relative proportions of certain taxa in large and small groups.[21] Statistics were, to naturalists, more of a study of proportions than anything else, and closely resembled population surveys and actuarial tables. In botany the proportions were worked out of species to their families or genera; the major taxonomic groups to the whole flora; or the proportion of endemic forms to commoners, first in absolute terms and then converted to decimals or expressed as numerators of 1000. Augustin De Candolle was the first to popularise such arithmetical analysis and it had been rapidly integrated into the botanist's repertoire, and rather more slowly, via Lyell, into geology. Lyell had made the study of proportions an original and productive technique for estimating the relative ages of strata by the comparison of extant to extinct species.[22] Yet Darwin could not make the intellectual leap from statistical geology to statistical botany. He needed the active intervention of Hooker. Hooker reinforced Darwin's high opinion of botanical methodology and convinced him that he could profitably use a technique that can hardly have been unfamiliar to him within the geological context.

The results of Darwin's calculations demonstrated that more varieties were produced in numerically large species which also occupied a wide area. Using various methods he showed that these were also the varieties most likely to turn into species ('incipient species') and that the advantages which had made the parent successful would equally contribute to the success of the offspring. He became preoccupied with large and dominant species, spread over wide ranges, with many individuals, and which belonged to similarly potent genera. Darwin affirmed that these were the source of most new species. This idea of 'winner takes all' is very different from his first attempts at explaining the origin of species in which he relied heavily on geographical isolation.[23] Now, with widespread and dominant species in his theory he had to find some mechanism other than isolation.

He did this by inventing a 'principle of divergence' that was to be used in tandem with natural selection.

(Darwin's thinking can be followed this far through his correspondence: indirectly the same correspondence throws new light on his published works and in particular offers an explanation for the series of revisions and amendments in his long, unpublished manuscript from which *On the origin of species* was abstracted.[24] The sequence of composition and interpolation that has been brought out by Stauffer in his editorial notes to this manuscript can, for the first eight chapters, be attributed to Darwin's discovery of the principle of divergence.)

Darwin's explanation of divergence is confusing and his use of it appears to be methodologically unsound, but he was emphatic about its importance, arguing that it was crucial to the action of selection. The principle expressed the tendency for natural selection to favour the most different variety out of any continuum, and that this extreme form was most likely to occupy a different niche, or place, in nature. A continuous selection for difference in the one direction would eventually make the variety a species in its own right. Darwin's use of divergence is therefore closely linked with arguments about the evolutionary potential of widespread and dominant species. His letters to Hooker show the botanist to be his mentor for the latter, and by implication an important factor in the discovery of the former.

It is customary when tracing historical influences at this point to insert a caveat and state that once exerted the influence loses status because it is absorbed and transformed by the recipient. However, in the case of botanical statistics it is curious to see Hooker not only providing the basic information, but also regulating the result – with Hooker's continuing guidance and criticism Darwin had substantially modified and improved his theories about the kind of species that contributed most to evolution. Over and above this achievement Darwin had virtually simultaneously uncovered a conceptual gap and, happily, thought of something – divergence – to fill it.

When Darwin in his *Autobiography* retrospectively analysed the difference between his first essay on species and the *Origin* of 1859 he stated that the earlier works lacked only one thing: his principle of divergence. Substantiating evidence comes from the fact that he felt it necessary, when under great personal distress, to add a description of divergence to his submission to the Linnean Society in July 1858. This took the form of a copy of a letter to Asa Gray (1810–1888), the American botanist, dated 5 September 1857.[25] Much has been made of Darwin's effortless abstraction of his theory for Gray in September, compared with his excuse that it was 'really impossible' to do so for A. R. Wallace some few months beforehand.[26] But in May 1857 it surely was impossible for Darwin to precis his work for he had not thought of divergence; by September he had. The principle possessed such heuristic power that, once discovered, he could reduce a mass of interrelated propositions and probabilities to manageable proportions, and even welcome an opportunity to present an abstract.

Study of the autograph letters has thus revealed noteworthy ideas and problems that deserve further attention: the roles in which Hooker and Darwin cast themselves; their attempts to effect a cross-fertilisation between the various sciences; Darwin's faith in this principle as exemplified by his use of botanical statistics; the part that botanical statistics played in the sophistication of evolution and natural selection theories, and the subsequent discovery of a remarkable conceptual aid in divergence. None of these could have come from an examination of the appropriate *Lives and Letters*.

To return to the more general historiographic problems concerned with *Lives and Letters*, it can be seen that the extant manuscript collections provide important clues to the writing of official biographies.

After Darwin's death in April 1882 Hooker offered the Darwin autograph letters to his eldest son William and they were gratefully accepted.[27] But the letters were not permanently handed over until 1889[28] although they were actively used by the Darwin family during the intervening years for the compilation of the *Life and Letters*. Even after publication they returned at least once to Hooker for his advice on the sequel volumes of *More Letters*. Francis Darwin had, at first, no intention of drawing up a life of his father, especially since the role of biographer had immediately been assumed by others. Darwin had died on April 19th and within three days proposals were made to publish some kind of extended scientific memoir. Romanes (1848–1894), the sub-editor of *Nature*, wrote simultaneously to both Hooker and Huxley asking for a notice, and after some initial confusion Hooker gladly relinquished the task to Huxley.[29] He did however suggest that a longer article might follow under his own direction and in early May outlined his plans to Francis Darwin – the article was to assess Darwin's place in the honorium of science and to include a selection of his letters to Hooker. But by the 6th May Romanes who, no doubt, wanted a more tangible part in the creation of Darwin's memorial, had proposed to Hooker that the article should become a book with several contributors, himself among them. It is interest-

ing to see that Romanes conducted his enquiries through Hooker and received answers from
the Darwin family along the same channel: apart from a sensitivity towards the bereaved,
Romanes probably thought that the intervention of Hooker would materially aid his cause.
His proposal, meeting with Hooker's approval, was forwarded to Francis Darwin. At this
point Francis stepped in and quashed everything: "It seems to me to be always a mistake",
he wrote,

> "that anything like a biography of anyone should be written immediately after his death . . . the fact
> that a good part of the work would fall into other hands does cause us some apprehension. I know
> and fully appreciated Romanes' regard for my father but I have but small confidence either in his tact
> or literary taste, and I think he is especially likely to go wrong when writing under the influence of
> strong recent feeling."[30]

Romanes compromised and brought out a five part memoir in *Nature* through May and June
1882 in which each section was by 'other hands'.[31] Francis Darwin lost no time in inserting
his own notice in *Nature*,[32] and later the newspapers and the *Athenaeum,* asking for gifts of
or copies of his father's letters to be kept for an official biography.[33] He told Hooker that
he hoped this would deter others yet that he was not prepared to write it himself.

However, in September 1882 Francis was pushed into declaring himself an author through
an unfortunate incident with the German naturalist Haeckel. Haeckel had contravened the
family's instructions by publishing one of Darwin's rare letters relating to Christianity and
his lack of faith in the established Church.[34] The 'Haeckel business' caused them a certain
amount of distress and a great deal of displeasure so that Francis was virtually obliged by the
norms of the day to produce a *Life,* as an official commemoration.

It is instructive to compare this apparent unwillingness with the easy path to literary production that Leonard Huxley enjoyed. As T. H. Huxley's son he ascended to the role of his
biographer through the determined intervention of his mother Henrietta with Macmillan. A
potential scientific biographer, Michael Foster, was pushed aside in favour of Leonard, who
was, according to Henrietta, "too retiring . . . too modest to take the initiative".[35] In
August 1895 Leonard Huxley told Macmillan:

> "At my mother's particular wish I propose to write a Life of my father. Prof. Michael Foster and Ray
> Lankester have offered to help me in all the technical scientific part . . . "[36]

After the success of the *Life and Letters of T. H. Huxley,* 1900, Leonard Huxley proceeded
to edit a selection of educational writings by his wife's uncle, Matthew Arnold,[37] before
acquiring the commission from Lady Hyacinthe Hooker to write the *Life and Letters* of Sir
Joseph.

The general conclusions to be drawn from only two examples indicate that the choice of
biographer was firmly under the control of the family and that, wherever possible, a member
of that family should take on the work. By this emphasis on the personal rather than on
scientific qualifications the contents of the ensuing biography were almost pre-determined
and any evaluation of the great man's contributions to science would certainly be derivative.
Although there is no reason to believe that filial feeling was incompatible with a knowledge
of contemporary science and its problems, it seems to be the case that *Lives and Letters* were
thought to be inappropriate places for an examination and analysis of science. Henrietta
Huxley persuaded Michael Foster and Ray Lankester to give way to Leonard Huxley yet
maintained their interests by allowing them to contribute to the scientific parts.[38] Similarly,
Hyacinthe Hooker bypassed Hooker's son-in-law Thiselton-Dyer, who as Assistant-Director
of Kew Gardens was well qualified to undertake a botanical biography, to give Leonard

Huxley the work. Even Francis Darwin's botanical interests and scientific background, which gave him a unique sympathy with his father's aims and methods, appear not to have been the decisive element in his taking up the task of a biography. It is fortunate though that Francis Darwin was so familiar with plant studies, as it is this background knowledge that adds substantially to the charm and authenticity of his *Life and Letters* and *More Letters*. But generally speaking the science in the *Life and Letter* genre is either derivative, actually contributed by some-one else, or, in the rare case of a scientific author, attenuated. Proof sheets were circulated amongst the subject's friends for comments and advice was solicited. Whole chapters, such as Huxley's on the reception of the *Origin of species,*[39] or Hooker's reminiscences of Darwin, were included in the various biographies. And, as will be seen, accounts and characterisations of events and personalities become stereotyped into the mould given them by the biographer or even by the great man himself during his lifetime.

In his notice in *Nature* Francis Darwin had called for gifts or copies of his father's letters, and during the five years that he worked on the *Life and Letters* and the subsequent years on *More Letters* he had transcribed for him the larger part of Darwin's correspondence. He wrote to Huxley:

> "Will you whenever you have time (there is no sort of hurry) look out any letters of my father's to you that I might see and make copies of. Of course your wishes or slightest feeling about what should & should not be published shall be law . . . "[40]

But, although Francis Darwin was later to receive editorial aid and major contributions from Huxley, he never relied on him to the extent that he relied on Hooker for advice at every stage, nor did he send Huxley copies of letters – as he did to Hooker – for comments. Apart from Francis Darwin, Hooker was the only one of Charles Darwin's friends to use the material at all stages of composition: the original manuscripts; the transcripts; and eventually the proof sheets. It appears that during the first few years after Darwin's death in 1882 Francis Darwin kept all the original Darwin to Hooker letters, effecting some preliminary sorting and dating, and having transcripts made. In July 1885 he wrote to Hooker hinting that he might like to add comments to the proof enclosed:

> "I wonder if you would read over this chapter and give me any suggestions that occur to you – I feel as if it were scrappy – & has a poor sort of feel about it from the arrangement I think. You will see that Fitton, Boot (or Boott) Lonsdale etc are only casually mentioned so that the briefest note would do . . . "[41]

On hearing that Hooker was, in fact, drawing up considerable notes about Charles Darwin, Francis offered to return the original letters:

> "I am extremely glad that you are making notes about my father – shall I send the earlier of the two vol[umes] now? or do you not at all care for it till August."[42]

Hooker used the earlier manuscript letters to write up the reminiscences that are featured in the introductory chapters of the *Life and Letters.*[43] If he had not already done so (i.e. before sending the originals to Francis in the first instance) this was probably the time that Hooker added the many dates to the earlier part of the collection, which, by being printed in the *Life and Letters,* and subsequently used again for *More Letters,* are of great historical significance. Even checking against an original does not necessarily verify a date when dealing with Darwin letters.

In addition to his enthusiastic proof reading Hooker tossed out ideas and information at an astonishing rate for Francis Darwin to incorporate at will: whilst proof reading Chapter IX Hooker suggested:

"Is it worth a passing allusion to his having taken into his service a sailor from the Beagle, or was this before he went to Down. I thought I remembered one. Passibly [sic] an allusion to Parslow may be lugged in – he was so integral a part of the family and known to all visitors as such."[44]

This point was repeated verbatim by Francis Darwin in the *Life and Letters*,[45] as were many others. With few exceptions all Hooker's contributions were incorporated into the published work: he had not only corrected the text of the letters, but also added notes about chronology, events and personalities, and reminiscences to the editorial sections that linked items. Francis Darwin obviously valued these additions and therefore called again on Hooker to help with the composition of *More Letters*.

For *More Letters* he sent Hooker the originals of his own letters to Darwin (which Hooker was delighted to find preserved) and copies of all Darwin's letters in reply. Hooker made a number of pencil corrections to both copies and originals which Francis Darwin acknowledged gratefully, and, no doubt, it was this continuing interest of Hooker's that encouraged Francis to send him the proof 'slips' of the two geographical distribution chapters. Again, Hooker made corrections and footnotes that appear in print.

The transcribed correspondence seems to have led as active a life as the genuine letters, and it deserves a brief digression. There are two sets of copies in addition to the real letters at the University Library, Cambridge (Darwin to Hooker; Hooker to Darwin – this last is now incomplete),[46] and a further shorter set of Darwin to Hooker copies at the Royal Botanic Gardens Kew. These were made at Francis Darwin's request in 1889 to acknowledge Hooker's gift of the originals.[47] The copyist was H. W. Rutherford (?–1924) a member of the Cambridge Library staff and collaborator with Francis Darwin on the catalogue of Charles Darwin's books left to the Botany School there. Francis Darwin made a particular effort to help Rutherford, financially and otherwise, and employed him in his spare time for many years in copying Darwin's papers.[48] Rutherford received 3d. per page of transcript, apparently 1d. over the usual rate, as explained by Lady Hooker to her friend and anxious authoress, Marianne North:

"Mr. Fr. Darwin paid for copying his father's letters; this I find was 3d. per page or copy (i.e. *side* of paper) . . . but this seems a large sum and I did not tell the Miss Nobles what Mr. Darwin had paid his copyists; and I find that a *usual* charge is 2d. a page for copy."[49]

The general historiographical point to this analysis of the composition of one particular *Life and Letters* is that such processes should always be taken into account. Most of the ground work was conducted from copies of letters, often by someone other than the author, and the proof sheets were hawked around for notes and emendments, because strong feelings of nepotism and regard for suitability often meant that the authors were not scientifically capable. In Hooker, Francis Darwin had a fairly reliable associate, but the potential for abuse in such situations was enormous. Equally, the autograph letters have a history that by itself should invite caution about the biographies: from original to transcript, from transcript to proof, there were many opportunities for error. Added to this there are problems concerned with unacknowledged contributions. Hooker's all-pervasive influence over the published Darwin collections leaves room for serious distortion, a distortion which is magnified by the fact that his role in constructing the Darwin story is largely anonymous.

This last point is one that deserves, and even demands, expansion. Hooker's contributions were, through modesty or strategy, mostly concealed in the Darwin biographies. Because passages and notes were not published as his opinions, the reader is invited to believe them as the simple truth. Hence a tentative thought can turn into dogma and tradition which is nurtured by continuing use of various *Lives and Letters* as primary historical sources. The stereotyped account of the famous Oxford meeting of the British Association in 1860 is an excellent example of this 'invitation to believe', and well illustrates the conjoined power of anonymity and the official biography.

The event hardly needs more than a reminder that the meeting provided a platform for statements about evolution, and that Thomas Henry Huxley exchanged insults with the Bishop of Oxford, Samuel Wilberforce (1805–1873). The main points of the debate are entrenched in Victorian scientific 'mythology'. Even the most recent and scholarly of papers on this meeting typifies a whole series of reconstructions by accrediting Leonard Huxley in the *Life and Letters* of his father, with the most complete body of information available to historians,[50] and follows the popular account. Yet what is this history based on? Contemporaries were agreed that they could not remember Huxley's or the Bishop's exact words, and even Hooker, as one of Huxley's closest friends and an active participant in the debate, could not remember such a crucial point as which of Huxley's female relatives Wilberforce had besmirched.[51] Huxley left no account of his own except for a letter to Dyster[52] and another to Francis Darwin at a later date.[53] Not only was there contemporaneous disagreement as to Huxley's words, there was also some controversy over which of the 'combatants' had triumphed: Balfour Stewart (1828–1887) wrote on the 4th July 1860 that he thought that "the Bishop had the best of it".[54] There was even contemporaneous distortion, lucidly ridiculed by Monckton Milnes in the House of Commons, and chronicled by Grant Duff in his *Notes from a Diary, 1851–1872:*[55]

> "The British Association is meeting this year at Oxford, and there has been a great scene between the Bishop of Oxford and Huxley, of which Monckton Milnes has brought to the House of Commons a comic version. According to him, Huxley asserted "that the blood of guinea pigs crystallises in rhombohedrons". Thereupon the Bishop sprang to his feet and declared that "such notions led directly to Atheism!" "[56]

There is in the literature far more to the story than that which is transmitted through Leonard Huxley's biography. But historians continue to use the published accounts in the *Lives and Letters* of Darwin, Huxley, and Hooker, and their reconstructions therefore conform to the evidence which is selectively put forward for an 'official' version. As published, the story is a composite one ultimately based on the 'eyewitness' account in the *Life and Letters of Charles Darwin.* And this was written by Hooker some twenty-five years after the event.

Hooker had volunteered a sketch of the Oxford meeting merely to add colour to what he supposed to be Francis Darwin's account:

> "Have you any account of the Oxford meeting? if not, I will, if you like, see what I can do towards vivifying it (and vivisecting the Bishop) for you. I had utterly forgotten that letter of mine, and am amused to find that it recalls the scene so clearly. I am perhaps not a fair judge of Huxley's contribution . . . "[57]

Francis Darwin seized the opportunity to write to Huxley and asked him for his own recollections:

> "I wish I had some account of the celebrated Oxford meeting. It is a thousand pities that my father destroyed his letters [,] there seems to have been one from you – and I think from Hooker – I should be glad of any account of it even if slightly Tichbornian."[58]

but had to be satisfied with his promise to read the proof sheets. Hooker, on the other hand provided a mass of entertaining material that was eventually published as the 'eyewitness' account. At the time of submission Hooker was careful to point out that "it is impossible to be sure of what one heard, or of impressions formed, after nearly 30 years of active life",[59] and asked that Huxley should vet it:

> "Here is my screed. I do not like it altogether, but can do no better. I should like Huxley to see it if you put it in print. Pray Anglicise it where necessary. Huxley may not like the bit about himself: if so it must come out. I have been driven wild formulating it from memory"[60]

It is at present impossible to say if Huxley did edit Hooker's account, but he certainly saw the page proofs which contained Francis Darwin's presentation, and thus, by implication, approved this account.[61] The *Life and Letters of Charles Darwin* therefore contains as 'official' a version as could be devised.

Five years after the *Life and Letters,* in 1892, Francis Darwin printed a one volume abbreviated *Life* of his father[62] and for this added to Hooker's story some minor details from other sources and two important contributions: one from the Rev. W. H. Freemantle; and the other from Huxley himself.[63] The latter is the famous letter in which he remembered crying 'The Lord has delivered him into my hands'. It is sufficient to point out that Huxley had had thirty-one years to remember this point, and five years to decide what, if anything, he desired to add to the record.

When Leonard Huxley came to prepare a separate rendering of the meeting to honour his late father he naturally used the Francis Darwin one volume *Life* as his source material. In 1898 – three years after Huxley had died – Francis Darwin replied to Leonard's query:

> "The eyewitness was Hooker – I suppose he told me not to publish his name tho' I can't imagine why. I am pretty certain it was written for the *Life* i.e. in the 80s. It is most curious how different accounts vary. You will see your father's account of "The Lord has delivered" etc in the one volume version of the *Life* "Charles Darwin, his life told etc" 1892 p.240 . . Your father's letter to me is by a misprint dated 1861 instead 1891."[64]

In the event Leonard Huxley paraphrased Francis Darwin's second, more complete, description for his *Life and Letters of T. H. Huxley,* 1900, to which he added some embellishments from other sources.[65] Then, even though the 'eyewitness' account had gone through three biographies picking up extra details in each one, Leonard Huxley revamped it for a fourth time for the *Life and Letters of Sir Joseph Hooker* 1918.[66] This biography, published in the same year as Lytton Strachey's *Eminent Victorians,* represents everything that is typical of the *Life and Letters* genre, and Leonard Huxley's conservative use of evidence aptly came full circle when Hooker's account passed through three biographies only to return to his own. By repeating and reaffirming the same story the younger Darwin and Huxley, abetted by Hooker in his old age and endorsed by Huxley before his death, had forged a legend that remains in the same conventional format today. The source material without Hooker's 'vivifying and vivisecting' recollections, and without the four biographies, is inadequate and anecdotal at worst, and unexplored at best. Surely it is fruitless to continue to reconstruct the Oxford meeting when the evidence cannot, at the moment, provide the answers, and the history of the transmission of even that evidence is ignored.

In drawing a general conclusion it is only necessary to state that in these instances, and possibly many others, manuscripts and autograph letters are much more than a simple alternative to the published version; such material transcends the 'facts' and legends that are passed on to modern historians through the medium of *Lives and Letters.* No-one would, however, deny the value of these biographies and it is unrealistic (not to mention unnecessary) to demand that all research should be conducted from archival resources. But *Lives and Letters* should be acknowledged as historical documents in their own right, with all their concomittant problems, and be considered as limited sources for the history of Victorian science.

ACKNOWLEDGEMENTS

This paper was undertaken with a Keddey-Fletcher Warr scholarship from the University of London. I am indebted to the Syndics of the Cambridge University Library, the Governors of Imperial College of Science & Technology, the Royal Botanic Gardens Kew, and the British Library, for permission to quote from their manuscript collections, and particularly to their respective librarians and archivists, Peter Gautrey, Mrs. Jeanne Pingree and V. T. H. Parry, for their help and interest. Professor A. R. Hall and Dr. Marie Boas Hall of the Department of History of Science, Imperial College, brought their combined critical faculties to bear on the several drafts of this paper, and I am grateful to them and Sydney Smith of St. Catherine's College, Cambridge for their encouragement and knowledge. Professor F. H. Burckhardt kindly read and commented on the draft during a busy trip to Cambridge. David Roos of Northwestern University, Illinois, advised me on many points and I am especially grateful for his heuristic recommendations.

## NOTES

1 DARWIN, F. 1887. *The life and letters of Charles Darwin including an autobiographical chapter edited by his son.* London, Murray.

2 DARWIN, F. and SEWARD, A. C. 1903. *More letters of Charles Darwin. A record of his work in a series of hitherto unpublished letters.* London, Murray.

3 LITCHFIELD, H. 1904 and 1915. *Emma Darwin: a century of family letters 1792–1896.* London, privately printed and Murray.
HUXLEY, L. 1918. *Life and letters of Sir Joseph Dalton Hooker based on materials collected and arranged by Lady Hooker.* London, Murray.
HUXLEY, L. 1900. *Life and letters of Thomas Henry Huxley.* London, Macmillan.
LYELL, K. M. 1881. *Life, letters and journals of Sir Charles Lyell.* London, Murray.

4 The most useful works on biography and its trends that I have found are:
REED, J. W. 1966. *English biography in the early nineteenth century 1801–1838.* Yale University Press.
ATTICK, R. D. 1966. *Lives and letters. A history of literary biography in England and America.* New York.
CLIFFORD, J. L. 1970. *From puzzles to portraits, problems of a literary biographer.* North Carolina.
GARRATY, J. A. 1957. *The nature of biography.* New York.
NICOLSON, H. 1968. *The development of English biography.* Oxford.
I am grateful to David Ross, Fulbright scholar at Imperial College, London (now at Northwestern University, Illinois) for his advice and enthusiasm about this section, and for his more general criticisms of the paper as a whole.

5 OUTRAM, D. 1976. Cuvier and scientific biography. *Hist. Sci.* 14, 24: 107–142. Dr. Outram makes the point that biography was probably the most important medium for the diffusion of the history of the life sciences in the nineteenth century.

6 Murchison had told Geike that he would leave him suitable financial incentives for the preparation of a biography long before his death. It is easy to suppose that Geike immediately began collecting material whilst Murchison was still alive. I am grateful to John Thackray of the Institute of Geological Sciences for this information.

7 ANNAN, N. 1952. Historians reconsidered: Carlyle. *History Today* 2: 659–665.

8 CLIFFORD, J. L. 1962. *Biography as an art. Selected criticism 1560–1960.* London & Oxford.

[9] *Life and letters C.D.* appeared in the review columns on 19 November 1887. Dr. Richard Freeman has kindly suggested to me a publication date in late October or early November, to catch the Christmas market. Joseph Hooker wrote to Francis Darwin on 19 and 26 December 1887 with the information about Murray: Royal Botanic Gardens Kew, J. D. H. outgoing letters, vol. 3, 164–165.

[10] *Memoirs of the Geological Survey of Great Britain* I: 336–403.

[11] *Autobiography* as printed in *Life and letters C.D.* I, 88.

STAUFFER, R. C. (ed.) 1975. *Charles Darwin's Natural Selection being the second part of his big species book written from 1856–1858.* London, Cambridge University Press. p. 535.

DARWIN, C. R. 1859. *On the origin of species by means of natural selection or the preservation of favoured races in the struggle for life.* London, Murray, p.366.

Darwin to Hooker, 29 June 1845, Cambridge University Library Darwin MS 104, 35, also printed in *More Letters C.D.* I, 408–409: "Forbes is doing apparently very good work about the introduction and distribution of plants. He has forestalled me in what I had hoped would have been an interesting discussion . ."

[12] *More Letters C.D.* I, 51 and also C.U.L. Darwin MS 114, 37.

[13] There are two other related letters which Francis Darwin could not date precisely: *More Letters C.D.* I, 411, which I can date to January 1846; *Life and Letters C.D.* II, 38, which can be dated to September 1845.

[14] FORBES, E. 1844. Report on the Mollusca and Radiata of the Aegean Sea, and on their distribution as bearing on geology. *Rep. Br. Ass. Advmt Sci.* 1843: 130–194. Darwin was probably referring to Forbes' 'time-notions' of pp.173–174.

[15] The history of Darwin's and Forbes' equally elegant theories to account for arctic and alpine distribution, and Darwin's extension of the idea to encompass worldwide distribution patterns, deserve more space than a footnote and I intend to develop them elsewhere.

[16] Since this paper was read the first Hooker to Darwin letter has been published in ALLAN, Mea. 1977. *Darwin and his flowers.* London, Faber and Faber, p.135, but with no comments about its significance. See BROWNE, E. J. 1977. Review of *Darwin and his flowers. Journal Museums Assoc.* 77(2): 81.

[17] *Life and letters J.D.H.* I, 436.

[18] Hooker to Darwin, C.U.L. Darwin MS 100, I, dated 28 November 1843. This naturally means that the preceding Darwin to Hooker letter was written pre- 28 November.

[19] Hooker to Darwin 4 August 1856, C.U.L. Darwin MS 100, 100.

[20] Hooker to John Tyndall n.d. [? 1860s], Imperial College Archives, Huxley papers H.P.8, 313.

[21] A selection of titles includes the following:

WATSON, H. C. (various editions after 1844). *London catalogue of British plants.*

HENSLOW, J. S. 1835. *A catalogue of British plants arranged according to the natural system.* Cambridge.

LEDEBOUR, C. F. 1842–1853. *Flora Rossica.* Stuttgart.

DE CANDOLLE, Augustin; DE CANDOLLE, Alphonse (and others). 1824–1857. *Prodromus systematis naturalis Regnii vegetabilis.* Paris. Darwin worked through many but not all of the 17 volumes.

GRAY, A. 1856. *Manual of the botany of the Northern United States.* New York, second edition.

BOREAU, A. 1854. *Flore du Centre de la France.* Paris.

GRISEBACH, A. R. H. 1843–1844. *Spicilegium Florae Rumelicae et Bithynicae.* Braunsweig.

WOLLASTON, T. V. 1854. *Insecta Maderensia.* London.

WESTWOOD, J. O. 1839–1840? *An introduction to the modern classification of insects.* London.

GOULD, J. 1848. *Birds of Australia.* London.

The mass of notes and tables relating to these calculations can be found in C.U.L. Darwin MS 15(ii); 16(i) and (ii). Darwin used the results in Chapter II of the *Origin of Species.*

[22] LYELL, C. 1830–1833. *Principles of geology.* London. Lyell used this technique to illustrate his subdivisions of the Tertiary. See also RUDWICK, M. J. 1972. *The meaning of fossils.* London. and RUDWICK, M. J. 1977. *Charles Lyell's dream of a statistical palaeontology.* 20th Annual Address to the Palaeontological Association (to be published).

[23] DE BEER, G. 1958. *Evolution by natural selection.* Cambridge University Press in which he reprints the Darwin *Sketch* of 1842 and the *Essay* of 1844.

[24] *Charles Darwin's Natural Selection* (see note 11).

[25] Reprinted in DE BEER, G. 1958. *Evolution by natural selection* pp.264–267.

[26] BEDDALL, B. G. 1968. Wallace, Darwin and the theory of natural selection. *J. Hist. Biol.* I: 261–323.

[27] Hooker to Francis Darwin, Kew, J.D.H. outgoing letters 3, 12.

[28] Emma Darwin to Hooker, Kew, J.D.H. incoming letters 6, 1.

[29] Imperial College HP 2, 240. Huxley's obituary appeared in *Nature* (1882) **25**: 597.

[30] Francis Darwin to Hooker, Kew, J.D.H. incoming letters 6, 13.

[31] *Nature* (1882) **26**: 49–51, 73–75, 97–100, 145–147, 169–171, with contributions from Romanes on zoology and psychology, Geike on geology, and Thiselton-Dyer on botany.

[32] June 1st 1882. *Nature* (1882) **26**: 104.

[33] Many of the letters which these notices inspired are in C.U.L. Darwin MS 112.

[34] I have not been able to trace this article further than what seems to be an edited translation: HAECKEL, E. H. P. A. 1882. On Darwin, Goethe, and Lamarck. *Nature* **26**: 533–541. This was not only published at the right time (28 Sept) but is overtly concerned with Goethe's, Lamarck's and Darwin's 'religion of nature'. However the Darwin letter is not printed although there are suggestive ellipses on p.540 column two.

[35] Henrietta Huxley to Macmillan. British Museum Macmillan Archives, 55211, vol CDXXVI, 55.

[36] Leonard Huxley to Macmillan. *Ibid* f.143.

[37] HUXLEY, L. 1912. *Thoughts on education chosen from the writings of Matthew Arnold.* London, Macmillan. Leonard Huxley later produced *Jane Welsh Carlyle: letters to her family 1839–1863.* London (1924) and *Elizabeth Barratt Browning: letters to her sister 1846–1859.* London (1929).

[38] FOSTER, M. and LANKESTER, R. 1898–1903. *The scientific memoirs of T. H. Huxley.* London, was to be their 'consolation prize'.

[39] HUXLEY, T. H. 1887. On the reception of the *Origin of Species. Life and letters C.D. II,* 179–204.

[40] Francis Darwin to T. H. Huxley, Imperial College HP 13, 12.

[41] Francis Darwin to Hooker, Kew, J.D.H. incoming letters 6, 20.

[42] Francis Darwin to Hooker, *ibid* 6, 27. This letter is also valuable as evidence that the early Darwin to Hooker letters had, by this date (1885), been bound into the two volumes now at Cambridge. It is possible that Dawson Turner, Hooker's favourite grandfather, bibliophile and noted collector, could have had the first volume of Darwin's letters (1843–1858) bound up, especially as he had already done so for Darwin's letters to Henslow (now at Kew), but this leaves one guessing as to the origin of the second, matching binding on Darwin's letters to Hooker (1859–1866), which must have been bound after Dawson Turner's death in 1858. For notes on Dawson Turner's extraordinary collection of autographs and his distinctive bindings see MUNBY, A. N. L. 1962. *The cult of the autograph letter.* London. pp.46, 58–59. I am grateful to Sydney Smith of St. Catherine's College, Cambridge, for directing my attention to Dawson Turner as the binder and for his knowledge of, and enthusiasm for, such problems.

[43] *Life and letters C.D.* II, 19–27.

[44] Hooker to Francis Darwin 26 Feb 1886, Kew, J.D.H. outgoing letters 3, 137.

[45] *Life and letters C.D.* I, 318 footnote.

[46] Leonard Huxley was later to use these same transcripts for his *Life* of Hooker.

[47] Emma Darwin to Hooker, Kew, J.D.H. incoming letters 6, 1.

[48] I particularly want to thank Mr. Peter Gautrey of the Cambridge University Library for his help and interest in biographical details of H. W. Rutherford, and for his timely production of a signed photograph of the C.U.L. staff of that period.

[49] I am grateful to Miss Brenda Moon for this reference which she cited in her paper on Miss Marianne North, at this same Easter Meeting, 1977, and to Mr. Roger North for his permission to quote from his collection of Miss North's manuscripts.

[50] BLINDERMAN, C. S. 1971. The Great Bone Case. *Perspect. Biol. Med.* **14**: 370–393, and again in FOSKETT, D. J. 1953. Wilberforce and Huxley on evolution. *Nature* **172**: 920.

[51] Hooker to Francis Darwin, Kew, J.D.H. outgoing letters 3, 155, recollecting that the Bishop spoke of Huxley's *mother.* He later retracted and followed the general opinion that it was Huxley's *grandmother.* The larger the generation gap the less indelicate the Bishop's insult.

52 Imperial College HP 15, 117–8, and printed in BLINDERMAN, C. S. 1954. The Oxford Debate and after, *Notes & Queries o.s.* **202:** *n.s.* **4:** 126–128, and again in FOSKETT, D. J. as in note 50.

53 In DARWIN, F. 1892. *Charles Darwin: his life told in an autobiographical chapter and in a selected series of his published letters.* London, Murray. p.240.

54 Balfour Stewart to J. D. Forbes, St. Andrews University Archives, J. D. Forbes papers, incoming letters f.133.

55 GRANT-DUFF, M. E. 1897. *Notes from a Diary 1851–1872.* London, Murray. First two volumes only. I am grateful to David Roos for these two references.

56 *Ibid* I, 139.

57 Hooker to Francis Darwin, Kew, J.D.H. outgoing letters 3, 150.

58 Francis Darwin to T. H. Huxley, Imperial College HP 13, 45.

59 Hooker to Francis Darwin, Kew, J.D.H. outgoing letters 3, 153.

60 Hooker to Francis Darwin, *ibid* 6, 151.

61 Francis Darwin to T. H. Huxley, Imperial College HP 13, 47: "I send you the account of the B. Assoc at Oxford. It begins at p.320 of what I send. The printers have set it up in sheets (against my orders) but it shall be altered to any extent you wish". The account is in *Life and letters C.D.* II, 320.

62 As in note 51.

63 *Ibid* pp. 236–240.

64 Francis Darwin to Leonard Huxley, Imperial College HP 13, 80.

65 *Life and letters T.H.H.* I, 179.

66 *Life and letters J.D.H.* I, 521.

*Archives of natural history* **30** (1): 118–138. 2003.

# Exploring Darwin's correspondence: some important but lesser known correspondents and projects

T. VEAK

St Andrews College, Department of Philosophy, Laurinburg, North Carolina 28352, USA.

ABSTRACT: This paper explores Darwin's 14,000 plus letters and suggests that in spite of the enormous amount of published material on Darwin and his work, there remains much untapped information in his correspondence. A quantitative analysis of his correspondence reveals that many of Darwin's most important sources and projects have not been researched. I provide examples in two of his correspondents, William B. Tegetmeier and John Scott, who were extremely important to Darwin's work in domestic animal breeding and plant hybrid studies, respectively. In addition, Darwin's work on seed viability and distribution are discussed to illustrate both the extent of his correspondence network and the complexity of his many sub-projects. The appendices suggest avenues for the further research of Darwin's correspondence by correlating the amount of correspondence with the amount of published material on the correspondents.

KEY WORDS: Charles Darwin – William B. Tegetmeier – John Scott – seed distribution – seed viability – correspondence.

## INTRODUCTION

Volume **11** of *The correspondence of Charles Darwin* (Burkhardt *et alii*, 2001) was published recently. Correspondence covering the years 1821 through 1864 is now in print. Darwin scholars continue to rave about the research prospects that these volumes hold. In his survey of the Darwin "industry" Ruse (1996: 219) claimed that in comparison to *The correspondence of Charles Darwin* "all other items of Darwin material come across as a bit anti-climactical". In one of the first reviews of the project, Moore (1985: 578) stated that "the *Calendar* [Burkhardt *et alii*, 1985] is, in short, a monumental achievement – one of the most important books to be published in the twentieth century on the culture of science, technology, and medicine ... the opportunities for microdarwinian investigation have redoubled at a stroke". Undoubtedly, there is a phenomenal amount of information contained in these volumes with more to come. Darwin corresponded voluminously until his death in 1882. However, there remains the question of how to approach such an enormous amount of information.

To this end, I offer a quantitative analysis of the more than 14,000 extant letters.[1] Moore (1985) included a graph of the annual quantities of letters to and from Darwin, and discussed some of the research possibilities revealed by his analysis. Montgomery (1987) performed a more in-depth quantitative analysis of Darwin's correspondence, and Garber (1994) gave a thorough analysis of Darwin's network of Pacific correspondents. My approach builds on these works, but also attempts to correlate Darwin's correspondence with his research projects. Such an analysis reveals that there is a large number of lesser-known figures who Darwin relied heavily on for information. Setting the "big names" aside (for example, Joseph Dalton Hooker, Charles Lyell, Asa Gray, Thomas Henry Huxley), there remain more than 50 correspondents with whom Darwin exchanged 30 or more letters. As Moore (1985: 576) suggested, a closer examination of these important, but lesser known, figures seems an appropriate next step for Darwin scholarship.

Another avenue of research made available through the *Correspondence* is the examination of Darwin's research projects. Secord's (1985) and Bartley's (1992) works on Darwin and domestic animal breeding are two such examples. In so doing, it becomes possible, as Garber (1994) demonstrated, to map out Darwin's network of correspondents and the various projects that he was investigating. A host of research questions can be asked. Who did Darwin rely on and for what kind of information? Where were his correspondents located geographically? What was the education and social status of his correspondents, and did this make a difference in the type and legitimacy of responses he received?[2] To what extent, if at all, were Darwin's queries simply rhetorical attempts to enroll others into his projects?[3]

## METHODS

Using the same resources as Garber (1994), a list of correspondents (Appendix 1) was established by the following method. First, *A calendar of the correspondence of Charles Darwin 1821–1882* (Burkhardt and Smith, 1994) was used to determine correspondents with whom Darwin had more than five exchanges. There is no objective basis for making the cut-off at five; however, it seems reasonable to assume that any significant correspondence that Darwin engaged in would necessitate more that a "few" letters. Secondly, correspondents with whom Darwin's communication was primarily personal in nature were eliminated. This was determined by examining the brief biographies in the *Calendar*. For example, scientists, breeders, foreign correspondents and queries to journals were included, whereas family and friends were not unless Darwin clearly drew on them for his projects.[4] Admittedly, it is difficult to make a decision about the nature of Darwin's correspondence from this limited information; nevertheless, the biographies in conjunction with the author's tacit knowledge from working on the Darwin Correspondence Project provide a fairly sound basis for making these decisions.[5] If there was any doubt, the correspondent was included in the list.

After paring down the list by the above methods, *Isis* cumulative bibliographies over a 20 year period (1978–1998) were examined to determine the extent of existing research on the correspondents. Those individuals with more than three entries in the *Isis* bibliographies were eliminated. The intention here was to develop a list of Darwin's correspondents who have received little attention from researchers. Finally, to provide some idea of the significance of the correspondents, the *Dictionary of British and Irish botanists and horticulturists* (Desmond and Ellwood, 1994) was consulted (Appendix 1).

A more difficult task was determining date ranges during which Darwin worked on his various projects. As any Darwin scholar knows, Darwin had a number of ongoing projects at any one time. Some of his investigations spanned the majority of his working life. For example, although Darwin's theory of the transmutation of species was largely formulated by 1842, it can reasonably be argued that the bulk of Darwin's post-1842 projects (possibly with the exception of some of his geological research) were efforts to bolster this theory. Hence, any attempt to narrow a date range on a project as far reaching as the transmutation of species is difficult, to say the least. I have, however, chosen to include the period between Darwin's return from the *Beagle* voyage in 1836 to 1842 when he wrote the first draft of his species theory. During this period Darwin openly discussed his thoughts on transmutation with a number of correspondents: William D. Fox, Charles Lyell, John S. Henslow, Leonard Jenyns and George Waterhouse (Porter, 1993). The years immediately after his voyage were

Table 1. Date ranges for Darwin's projects.

| Project | began | ended |
|---|---|---|
| entomology | 1827 | 1831 |
| zoology of the *Beagle* voyage | 1837 | 1843 |
| transmutation of species | 1837 | 1844 |
| Cirripedia | 1846 | 1854 |
| breeding domestic animals | 1855 | 1861 |
| seed dispersal and viability | 1855 | 1867 |
| insectivorous plants | 1860 | 1875 |
| climbing plants | 1863 | 1865 |
| pangenesis | 1865 | 1881 |
| expressions | 1867 | 1872 |
| man and sexual selection | 1867 | 1871 |
| movement in plants | 1873 | 1880 |
| worms | 1876 | 1881 |

also significant in that during this time Darwin began to establish a network of correspondents for his questions relating to artificial selection (plant and animal breeders), which would ultimately play an important role in his theory of natural selection (Burkhardt and Smith, 1986: xvii–xviii).

Darwin's work on geology and botany (for example, pollination and fertilization, and related research on forms of flowers) is also impossible to demarcate. Darwin's interest in geology began prior to the *Beagle* voyage (Desmond and Moore, 1992), was intensified by his voyage around the world, and culminated in the publication of a number of major works on geology in the 1840s (Darwin, 1842, 1844, 1846). However, Darwin continued to discuss relevant issues in geology with leading figures in the field (for example with Charles Lyell, Thomas F. Jamieson, Andrew C. Ramsay, Joseph B. Jukes, and the noted physicist, John Tyndall) into the 1860s and 1870s (Burkhardt *et alii*, 1994: xviii–xix).

Darwin began investigating variation in cultivated plants, which included crossing and pollination experiments, in the early 1840s. His fascination with plant variation as a mechanism for supporting his theory of transmutation continued off and on throughout his life and resulted in major publications in the 1860s and 1870s (Darwin, 1862, 1876, 1877).

For the purposes of the present analysis, date ranges have been established for those projects that can be fairly clearly demarcated and which are temporally limited to some extent (Table 1). This was accomplished primarily through examination of Darwin's journal (de Beer, 1959), and the introductions to the *Correspondence* volumes (Burkhardt and Smith, 1985, 1986, 1987, 1988, 1989, 1990, 1991; Burkhardt *et alii*, 1993, 1994, 1997).

These date ranges were correlated with the date ranges of Darwin's correspondents in an effort to determine which correspondents Darwin may have drawn on for particular projects (Appendix 1). Obviously, an overlap between the project date ranges and the range of correspondence does not necessarily establish a connection; Appendix 1 is merely a starting point for further research. In order to demonstrate the value of taking a closer look at Darwin's correspondence, a brief discussion of two of his lesser known correspondents (William. B. Tegetmeier and John Scott), and one of his research projects (seed dissemination and viability) are included.

In regard to Tegetmeier and Scott, both played important roles in Darwin's investigations in the transmutation of species; however, little is known about them. There are no entries pertaining to either Tegetmeier or Scott in the past 20 years of *Isis* cumulative bibliographies. There is only one significant work on Tegetmeier, written by his son-in-law (Richardson, 1916), which is dated and largely anecdotal. Of the four most recent biographies on Darwin (Bowler, 1990; Bowlby, 1991; Desmond and Moore, 1992; Browne, 1996), only Browne (1996: 525) mentioned Tegetmeier, but she did not elaborate on his relationship with Darwin. Even less has been written on Scott. Bowlby (1991: 375) briefly mentioned his significance to Darwin's work.

## WILLIAM B. TEGETMEIER (1816–1912)

Tegetmeier began his long career as a naturalist when a youth. Initially, he lived in Colnbrook, Buckinghamshire, where he spent a great deal of time exploring natural history (Richardson, 1916: 2). When he was 12, Tegetmeier's family moved to London, where he maintained his natural history inclinations by raising pigeons – a pursuit that remained with him throughout his life (Richardson, 1916: 6).

Tegetmeier was apprenticed to his father to become a doctor and apothecary, and enrolled at University College London in 1833 at age 17. He was an excellent student and received many honors and medals (Richardson, 1916: 10–11). But after ten years of study and apprenticeship, Tegetmeier forsook medicine for a life as a "Bohemian journalist" (Richardson, 1916: 27). This move was no doubt partly a result of receiving an inheritance from his father; however, he was by no means wealthy and had to work hard most of his life.

In 1859 Tegetmeier began contributing articles to *The field*, a journal devoted to "the farm, the garden, the country gentleman's newspaper". Shortly thereafter Tegetmeier was appointed head of the Poultry and Pigeon Department at *The field*, where he contributed weekly articles for more than 50 years (Richardson, 1916: 140). In addition to his career as a journalist, Tegetmeier also lectured and wrote textbooks on subjects ranging from botany to domestic economy. His *Manual of domestic economy* (1858), which was oriented toward women's education, went through 14 editions (Richardson, 1916: 37). He became widely recognized as one of the leading authorities on domestic fowls, the breeding of domestic animals and bee-keeping in England. However, it was primarily his expertise on fowls that initially drew Darwin to Tegetmeier.

Darwin and Tegetmeier met in 1855 through William Yarrell[6], a mutual friend. Darwin immediately tapped into Tegetmeier's wealth of knowledge on animal breeding and bees. Most importantly, he was a vital link to the pigeon and poultry fancying community (Burkhardt and Smith, 1989: xix). Bartley (1992) points out that Darwin's work on domesticates, while important to his theory of natural selection, was equally important to his interest in inheritance and variability. In this regard, Tegetmeier played a substantial role by performing a variety of sexual selection and inheritance experiments with birds in the 1850s and 1860s (Bartley, 1992).

Darwin (1859: 250, 254) cited Tegetmeier twice in *Origin*, which was admittedly a reference-sparse "abstract" of his theory; and eight times in *The descent of man, and selection in relation to sex* (Darwin, 1871). However, in *Variation of plants and animals under domestication* (Darwin, 1868), Tegetmeier was cited 33 times on such far reaching topics as "a cat with monstrous teeth", "the length of the middle toe in Cochin fowl" and "intercrossing in bees".

Because Darwin relied so heavily on domestic breeding (artificial selection) as a correlate for natural selection in nature, he was anxious to gather data on the types and extent of variation possible in domestic animals. Tegetmeier assisted Darwin in this endeavor by supplying specimens (particularly pigeons), identifying and describing specimens that Darwin had procured from his extensive network of correspondents[7], and by answering queries about the breeding of domestic animals, or directing Darwin to others who could assist him (see Appendix 2 for a complete list of topics discussed).

## JOHN SCOTT (1836–1880)

John Scott was born in Denholm, Scotland, in 1836. His father and mother died when he was quite young and he was brought up by his grandmother. Scott attended parish school, but left at the age of 14 to work as a gardener. In 1859, after serving in several gardening positions, Scott became foreman of the propagating department at the Royal Botanic Garden, Edinburgh (Kennedy, 1908). It was in this position that Scott began corresponding with Darwin in 1862.

Ironically, Scott first wrote to Darwin to point out an error that Darwin had made in the first edition of his orchid book (Darwin, 1862). In a letter to Darwin dated 11 November 1862 (Burkhardt *et alii*, 1997: 516), Scott claimed that Darwin was mistaken about his identification of a particular genus of orchids. Darwin responded appreciatively, and somewhat demurely, stating that "Botany is a new subject to me" (Burkhardt *et alii*, 1997: 522).

Although their most active period of correspondence only lasted a few years (1862–1864), they both benefited immensely. Darwin was instrumental in obtaining a position for Scott as the head of the herbarium department at the Calcutta Botanic Garden, and in encouraging him to publish his work (See Appendix 2 for a list of letters on these subjects). In terms of Scott's assistance to Darwin, he is referenced twice in *Origin* (Darwin, 1859), six times in *On the various contrivances by which British and foreign orchids are fertilised by insects* (Darwin, 1862), nine times in *The variation of animals and plants under domestication* (Darwin, 1868), once in *The descent of man, and selection in relation to sex* (Darwin, 1871), five times in *The effects of cross and self-fertilisation in the vegetable kingdom* (Darwin, 1876), and 19 times in *The different forms of flowers on plants of the same species* (Darwin, 1877). The majority of their correspondence centered around cross-pollination studies of hybrids to determine the effects on sterility/fertility. However, like many of his correspondents, Darwin gleaned a variety of information from Scott. For example, when Darwin was doing research for *The descent of man* in the late 1860s, he requested information on human variation from Scott (See Appendix 2).

Aside from Joseph D. Hooker and Daniel Oliver, Scott is likely the most important correspondent in regard to Darwin's studies on plant sterility and plant varieties. These studies were of great importance to Darwin because he believed that they were the clearest evidence of evolutionary gradation from varieties to species. Huxley had made the issue of cross-sterility of paramount importance in his assertion that species were delineated by mutual sterility (Burkhardt *et alii*, 1997: 700). Darwin disagreed: "Sterility ... has been acquired ... to favour intercrossing. Sterility may ... have been slowly acquired for a distinct object, namely, to prevent two forms ... becoming blended by marriage" (Burkhardt *et alii*, 1997: 702). In an attempt to counter Huxley, Darwin drew on Scott's extensive knowledge and experience in plant propagating. Most significantly, Darwin persuaded Scott to replicate Karl Friedrich von Gärtner's cross-sterility experiments on *Verbascum* (Scott, 1867). The importance of Gärtner's (1849) experiments cannot be overemphasized, as Darwin stated in a letter to Hooker: "I do not think any experiment can be more important on Origin of species; for if [Gärtner] is correct, we certainly have what Huxley calls new physiological species arising" (Burkhardt *et alii*, 1994: 284).

## SEED DISPERSAL AND VIABILITY

Nelson (2000) demonstrated that seed distribution and viability were significant projects of Darwin throughout his working life. Similar plant species were observed to be widely dispersed, but the question was how to explain this phenomenon. Darwin was motivated by the fact that island floras were known to be highly endemic (Murray, 1986: 76–77) and by the observations of his widely traveled botanist friend, Joseph D. Hooker (1847a, 1847b).

Although Darwin's interest in plant dispersal was not unique (Nelson, 2000: 34), his motivations were. Browne (1983: 196) summed up the importance of this project to Darwin: "the crux of Darwin's system was the proposition that species could spread virtually all over the world, given plenty of time and no physical barriers on the way". Since Darwin rejected independent creation, he sought an alternative explanation for the geographical distribution of plant species. Much of the impetus for Darwin's research was fueled by a desire to refute one of the prevailing hypotheses of his time: that distribution occurred via continental land-bridges, a position espoused by noted geologists such as Edward Forbes, and botanists including Joseph D. Hooker (Burkhardt and Smith, 1989: 331, 349). Darwin knew that if the idea of land-bridges could be refuted, creationists would be forced into the unsavory position of espousing "multiple creations" (Browne, 1983: 199–200).[8]

Although Darwin investigated seed viability as early as 1837 (Burkhardt and Smith, 1986: 13), his research began in earnest in 1855 (Darwin, 1855a, 1855b, 1855c, 1855d, 1855e, 1855f), initiated by a comment by Hooker (Burkhardt and Smith, 1989: 299, note 4; 321). Darwin believed that if he could demonstrate that seeds remain viable in salt water long enough to be transported by ocean currents, a plausible mechanism for plant distribution could be established. The manner in which Darwin conducted these experiments is only one instance among many where he was clearly attempting to buttress his projects by using empirical scientific methods.

Darwin had "sea water" artificially mixed by a local chemist. Then, following Hooker's advice on which seeds to test (Burkhardt and Smith, 1989: 304), Darwin placed a number of seeds of each species in small bottles containing 2–4 fluid ounces of sea water. The bottles containing salt water and seeds were then exposed to two different temperature ranges (at 44°–48°F, and at 32°F) to determine if temperature had any effect on germination. The higher temperature range was accomplished simply by keeping the bottles outside in the shade; however, to maintain 32°F, Darwin was forced continually to pack the bottles in snow. He was, in fact, able to find a number of different species that could endure immersion in salt water for a significant period of time (Table 2). Darwin discussed and summarized these experiments in a series of letters to the *Gardeners' chronicle* (Darwin, 1855a, 1855b, 1855c,

Table 2. Summary of Darwin's seed salting experiments; ordered by length of immersion.

| Botanical name | common name | length of immersion | germination success[9] |
|---|---|---|---|
| *Allium cepa* | onion | 42 days | a few |
| *Apium graveolens* var. *dulce* | celery | 42 days | Well |
| *Daucus carota* | carrot | 42 days | Well |
| *Lactuca sativa* | lettuce | 42 days | All |
| *Lepidium sativum* | common cress | 42 days | All |
| *Raphanus sativus* | radishes | 42 days | less well |
| *Phaseolus vulgaris* | kidney beans | 30 days | None |
| *Atriplex hortensis* | orache, or Atriplex | 28 days | Well |
| *Avena sativa* | oats | 28 days | Well |
| *Beta vulgaris* | beet | 28 days | Well |
| *Borago officinalis* | borage | 28 days | Well |
| *Capsicum* | peppers (chili, red, or sweet) | 28 days | Well |
| *Cucurbita ovifera* | gourd | 28 days | Well |
| *Hordeum vulgare* | barley | 28 days | Well |
| *Phalaris canariensis* | canary grass | 28 days | Well |
| *Rheum × hybridum* | rhubarb | 28 days | Well |
| *Satureja hortensis* | savory, or Satureja | 28 days | less well |
| *Brassica oleracea* | cabbage | 14 days | only one |
| *Linum usitatissimum* | flax | 14 days | only one |
| *Phaseolus* species | beans | 14 days | a few |
| *Pisum sativum* | peas | 14 days | None |
| *Ulex europaeus* | furze, or Ulex | 14 days | a few |
| *Trifolium incarnatum* | crimson clover | 7 days | None |

1855d, 1855e, 1855f, 1856).

From the results of this experiment Darwin determined that the majority of the seeds that he tested could survive long enough to travel 1,300 to 1,400 nautical miles[10] in the ocean, a number that he arrived at by multiplying the average ocean current speed (33 nautical miles per day) by the number of days that the majority of seeds survived (42 days) (Darwin, 1855e). Land-bridges were no longer necessary to explain the observed plant distribution patterns.

In addition to his seed-salting experiments, Darwin performed a number of investigations to determine mechanisms of plant dispersal – driftwood, birds, icebergs and fish (see Appendix 3). For example, he obtained clumps of mud from the legs and feet of various types of birds (ducks, pigeons and partridges). The mud frequently contained seeds which Darwin planted and found viable. In a letter to Joseph Hooker on 5 December 1863, Darwin discussed how he had obtained 32 seeds from mud attached to a partridge's foot (Burkhardt *et alii*, 1999: 687).

Another experiment involved birds of prey. Darwin had sparrows with full crops feed to hawks and owls at the Zoological Society gardens (Burkhardt and Smith, 1990: 248, note

2). Darwin extracted the seeds from the boluses that had been expelled 12 to 18 hours later by the hawks and owls, planted them, and determined that the seeds were indeed viable. From this, Darwin (1859: 357) hypothesized yet another means of transporting seeds, this time up to 500 miles (Burkhardt and Smith, 1990: 249).

It is also interesting to examine the diversity of correspondents (Appendix 4) with whom Darwin discussed his seed experiments. These correspondents were spread throughout the world: France, South Africa, United States, Azores, Jamaica and Norway; and they were equally diverse professionally: geologists, botanists, ornithologists, and conchologists. In addition to these "professional" scientists, Darwin also consulted a number of amateur naturalists and gardeners: his cousin William D. Fox; his sister Susan Darwin; his son William E. Darwin; and Miss Holland, the daughter of Henry Holland.[11]

From this analysis one can easily see the lengths undertaken by Darwin for this one, apparently minor, research project which amounted to only five pages of summary in *Origin* (Darwin, 1859: 355–360). However, the results played a crucial role in Darwin's argument for the geographical distribution of plant species (Darwin, 1859: chapter 12). It was in fact no small accomplishment. The results of these experiments were instrumental in convincing Joseph Hooker (at the time the world's foremost authority on the geographic distribution of plants) that mechanisms for distribution existed. From his extensive travels and research, Hooker had long been convinced of the widespread distributions of many plant species; however, aside from the idea of continental extensions, for which there was then little evidence, the means of distribution were virtually unknown. Persuading Hooker encouraged him to speak openly on behalf of Darwin's theory, which he did in his introduction to *Flora Tasmania* (Hooker, 1860). More importantly, Darwin's experiments "proved highly significant for the theory of evolution, because the results established beyond doubt that species were capable of spreading far more widely than had hitherto been supposed" (Browne, 1983: 199).

## DISCUSSION

The primary intent of this paper is to offer fresh research possibilities for Darwin's correspondence. In this regard, several points should be obvious: First, Darwin's over-arching project – a meta-theory of the unity of nature – was composed of numerous sub-projects which were frequently complex and scientific in nature (as defined by Darwin's context). Although Darwin is more frequently thought of as a grand theorizer, a close examination of his sub-projects reveals the extent to which he was a "practicing" scientist in every sense of the word. Darwin's correspondence is, therefore, invaluable because it reveals the intricacies of his work that cannot be found elsewhere. This may appear to some historians of science to be overly "internalistic" but it must also be noted that this science is being articulated via correspondence *between* individuals – a fact that significantly broadens the historical context.

Secondly, the present analysis demonstrates how important relatively unknown figures were to Darwin's work. William B. Tegetmeier and John Scott are just two of numerous examples (see Appendix 1). This does not negate the originality of Darwin's discovery; however, it goes a long way toward supporting the idea that science was, at least in Darwin's context, a highly communal enterprise. A more thorough examination of the networks created by Darwin would be enlightening from a variety of scholarly perspectives.

Lastly, the publication of Darwin's correspondence is opening vast new research opportunities in nineteenth-century natural history. While Darwin remains a significant figure in this research, his correspondence opens doors for viewing numerous other "actors" and "sub-plots" on this particular stage of history.

## ACKNOWLEDGEMENTS

The author wishes to thank Dr Duncan M. Porter, Director of the Darwin Correspondence Project, for his patient guidance, and for the opportunity to work on this project, and an anonymous referee for helpful suggestions.

## NOTES

[1] This immense quantity does not include the majority of his pre-1862 incoming correspondence, which Darwin did not retain (Montgomery, 1987: 15).

[2] Secord (1985: 537) claimed that Darwin distinguished between naturalists and breeders of domestic animal in terms of reliability of data. He did, however, make exceptions (Tegetmeier, for example).

[3] Darwin was keenly aware that the scientific community had to be won over, and that this process was not entirely a matter of "evidence". For example, Darwin's effort to publish Asa Gray's *Atlantic monthly* articles, which argued for a theistic evolution, was clearly a rhetorical device (Dupree, 1968).

[4] William Darwin Fox (1805–1880), Darwin's cousin, is one such example of a relative from whom Darwin solicited information.

[5] In addition, Dr Duncan M. Porter, Director of the Darwin Correspondence Project, was also consulted.

[6] Yarrell (1784–1856) was a zoologist who "engaged in business as newspaper agent and bookseller in London. An original member of the Zoological Society, 1826. Wrote standard works on British birds and fishes" (Burkhardt and Smith, 1989: 657).

[7] For a list of these correspondents, see Darwin's memorandum dated [December] 1855 (Burkhardt and Smith, 1989: 510).

[8] For Darwin's discussion of continental extensions see *Origin* (Darwin, 1859: 353–354).

[9] This was Darwin's terminology.

[10] In his discussion of these experiments in *Origin* Darwin (1859: 355) used the more conservative estimate of 28 days immersion to calculate the number of possible miles traveled (the result being 924 miles).

[11] "Physician. Distant cousin of Darwins and Wedgwoods" (Burkhardt and Smith, 1994: 614).

[12] URL (accessed 15 November 2002): http://www.lib.cam.ac.uk/Departments/Darwin/calintro.html (1999 Online Calendar of the Correspondence of Charles Darwin. The Darwin Correspondence Project. Cambridge University Library).

## REFERENCES

BARTLEY, M., 1992 Darwin and domestication: studies in inheritance. *Journal of the history of biology* 25: 307–333.

BOWLBY, J., 1991 *Charles Darwin: a new life.* New York: W. W. Norton. Pp xiv, 511.

BOWLER, P., 1990 *Charles Darwin: the man and his influence.* Oxford & Cambridge, Massachusetts: Blackwell. Pp xil, 250.

BROWNE, J., 1983 *The secular ark: studies in the history of biogeography.* New Haven & London: Yale University Press. Pp x, 273.

BROWNE, J., 1996 *Charles Darwin: a biography.* Princeton: Princeton University Press. Pp 622.

BURKHARDT, F. *et alii* (editors), 1985 *A calendar of correspondence of Charles Darwin, 1821–1882.* New York & London: Garland Publishing. Pp 690.

BURKHARDT, F. and SMITH, S. (editors), 1985 *The correspondence of Charles Darwin. Volume* **1**: *1821–1836.* Cambridge: Cambridge University Press. Pp 702.

BURKHARDT, F. and SMITH, S. (editors), 1986 *The correspondence of Charles Darwin. Volume* **2**: *1837–1843.* Cambridge: Cambridge University Press. Pp xxxiii, 603.

BURKHARDT, F. and SMITH, S. (editors), 1987 *The correspondence of Charles Darwin. Volume* **3**: *1844–1846.* Cambridge: Cambridge University Press. Pp xxix, 523.

BURKHARDT, F. and SMITH, S. (editors), 1988 *The correspondence of Charles Darwin. Volume* **4**: *1847–1850.* Cambridge: Cambridge University Press. Pp 752.

BURKHARDT, F. and SMITH, S. (editors), 1989 *The correspondence of Charles Darwin. Volume* **5**: *1851–1855.* Cambridge: Cambridge University Press. Pp 705.

BURKHARDT, F. and SMITH, S. (editors), 1990 *The correspondence of Charles Darwin. Volume* **6**: *1856–1857.* Cambridge: Cambridge University Press. Pp xxix, 673.

BURKHARDT, F. and SMITH, S. (editors), 1991 *The correspondence of Charles Darwin. Volume* **7**: *1858–1859.* Cambridge: Cambridge University Press. Pp xxx, 671.

BURKHARDT, F. and SMITH, S. (editors), 1994 *A calendar of the correspondence of Charles Darwin 1821–1882.* Second Edition. Cambridge: Cambridge University Press. Pp viii, 690.

BURKHARDT, F. *et alii* (editors), 1993 *The correspondence of Charles Darwin. Volume* **8**: *1860.* Cambridge: Cambridge University Press. Pp 808.

BURKHARDT, F. *et alii* (editors), 1994 *The correspondence of Charles Darwin. Volume* **9**: *1861.* Cambridge: Cambridge University Press. Pp xxxiii, 609.

BURKHARDT, F. *et alii* (editors), 1997 *The correspondence of Charles Darwin. Volume* **10**: *1862.* Cambridge: Cambridge University Press. Pp xxxx, 936.

BURKHARDT, F. *et alii* (editors), 1999 *The correspondence of Charles Darwin. Volume* **11**: *1863.* Cambridge: Cambridge University Press. Pp xxxix, 1038.

BURKHARDT, F. *et alii* (editors), 2001 *The correspondence of Charles Darwin. Volume* **12**: *1864.* Cambridge: Cambridge University Press. Pp 700.

DARWIN, C., 1842. *The structure and distribution of coral reefs. Being the first part of the geology of the voyage of the Beagle, under the command of Capt. Fitzroy, R.N. during the years 1832 to 1836.* London: Smith Elder & Co. Pp xii, 214.

DARWIN, C., 1844 *Geological observations on the volcanic islands visited during the voyage of H.M.S. Beagle, together with some brief notices of the geology of Australia and the Cape of Good Hope. Being the second part of the geology of the voyage of the Beagle, under the command of Capt. Fitzroy, R.N. during the years 1832 to 1836.* London: Smith Elder & Co. Pp vii, 175.

DARWIN, C., 1846 *Geological observations on South America. Being the third part of the geology of the voyage of the Beagle, under the command of Capt. Fitzroy, R.N. during the years 1832 to 1836.* London: Smith Elder & Co. Pp vii, 279.

DARWIN, C., 1855a Does sea-water kill seeds? *Gardeners' chronicle* no. 15: 242.

DARWIN, C., 1855b Does sea-water kill seeds? *Gardeners' chronicle* no. 21: 356–357.

DARWIN, C., 1855c Vitality of seeds. *Gardeners' chronicle* no. 46: 758.

DARWIN, C., 1855d Effect of salt-water on the germination of seeds. *Gardeners' chronicle* no. 47: 773.

DARWIN, C., 1855e Effect of salt-water on the germination of seeds. *Gardeners' chronicle* no. 48: 789.

DARWIN, C., 1855f Longevity of seeds. *Gardeners' chronicle* no. 52: 854.

DARWIN, C., 1856 On the action of sea-water on the germination of seeds. *The journal of proceedings of the Linnean society of London (botany)* **1**: 130–140.

DARWIN, C., 1859 *On the origin of species by means of natural selection or the preservation of favoured races in the struggle for life.* London: John Murray. Pp ix, 502.

DARWIN, C., 1862 *On the various contrivances by which British and foreign orchids are fertilised by insects, and the good effects of intercrossing.* London: John Murray. Pp vi, 365.

DARWIN, C., 1868 *The variation of animals and plants under domestication.* London: John Murray. 2 volumes.

DARWIN, C., 1871 *The descent of man, and selection in relation to sex.* London: John Murray. 2 volumes.

DARWIN, C., 1876 *The effects of cross and self fertilisation in the vegetable kingdom.* London: John Murray. Pp viii, 482.

DARWIN, C., 1877 *The different forms of flowers on plants of the same species.* London: John Murray. Pp viii, 352.

DARWIN, F., 1909 *The foundations of the origin of species. two essays written in 1842 and 1844.* Cambridge: Cambridge University Press. Pp xxix, 263.

DE BEER, G., 1959 Darwin's journal. *Bulletin of the British Museum (Natural History), historical series,* **2** (1): 1–21.

DESMOND, A. J. and MOORE, J., 1992 *Darwin: a tormented evolutionist.* New York: Warner Books. Pp xxi, 808.

DESMOND, R. and ELLWOOD, C., 1994 *Dictionary of British and Irish botanists and horticulturists: including plant collectors, flower painters, and garden designers.* London: Taylor & Francis. Pp xl, 825.

DUPREE, A. H., 1968 *Asa Gray, 1810–1888.* New York: Athenaeum. Pp x, 505.

GARBER, J., 1994 Darwin's correspondents in the Pacific: through the looking glass to the antipodes, pp 169–211, in MACLEOD, R. and REHBOCK, P. (editors), *Darwin's laboratory: evolutionary theory and natural history in the Pacific.* Honolulu: University of Hawai'i Press. Pp x, 540.

GÄRTNER, K. F. VON, 1849 *Versuche und beobachtungen über die bastarderzeugung im pflanzenreich. mit hinweisung auf die ähnlichen erscheinungen im thierreiche.* Stuttgart: E. Schweizerbart. Pp xvi, 790.

HOOKER, J. D., 1847a. *The botany of the Antarctic voyage of H. M. Discovery Ships Erebus and Terror, in the years 1839–1843, under the command of Captain Sir James Clark Ross.* Part 1. *Flora Antarctica.* London: Reeve Brothers. 2 volumes.

HOOKER, J. D., 1847b.*The botany of the Antarctic voyage of H. M. Discovery Ships Erebus and Terror, in the years 1839–1843, under the command of Captain Sir James Clark Ross.* Part 2. *Flora Novae Zelandiae.* London: Reeve Brothers. 2 volumes.

HOOKER, J. D., 1860. *The botany of the Antarctic voyage of H. M. Discovery Ships Erebus and Terror, in the years 1839–1843, under the command of Captain Sir James Clark Ross.* Part 3. *Flora Tasmania.* London: Reeve Brothers. 2 volumes.

KENNEDY, J. W., 1908 Notable men of Upper Tevoitdale. No. 1. John Scott of Denholm, a distinguished naturalist. *Transactions of the Hawick archaeological society*: 65–72.

MONTGOMERY, W., 1987 Editing the Darwin correspondence: a quantitative perspective. *British journal for the history of science* **20**: 13–27.

MOORE, J., 1985 Darwin's genesis and revelations. Essay review of BURKHARDT, F. *et alii,* (editors) 1985 *A calendar of correspondence of Charles Darwin, 1821–1882.* (New York and London: Garland Publishing. Pp 690), and BURKHARDT, F. *et alii,* (editors) 1985 *The correspondence of Charles Darwin. Volume* **1**: *1821–1836.* (New York: Cambridge University Press. Pp xxix, 702). *Isis* **76** (4): 570–580.

MURRAY, D. R., 1986 Seed dispersal by water, pp 49–85, in MURRAY, D. R. (editor), *Seed dispersal.* North Ryde, N.S.W.: Academic Press Australia. Pp xiv, 322.

NELSON, E. C., 2000 *Sea beans and nickar nuts: a handbook of exotic seeds and fruits stranded on beaches in north-western Europe.* London: Botanical Society of the British Isles. Pp 156.

PORTER, D. M., 1993 On the road to the *Origin* with Darwin, Hooker, and Gray. *Journal of the history of biology* **26** (1): 1–38.

RICHARDSON, E. W., 1916 *A veteran naturalist: being the life and work of W. B. Tegetmeier.* London: Witherby & Co. Pp xxiv, 232.

RUSE, M., 1996 The Darwin industry: a guide. *Victorian studies* **39** (2): 217–235.

SCOTT, J. 1867 On the reproductive functional relations of several species and varieties of *Verbasca. Journal of the asiatic society of Bengal* **36** (2): 145–174.

SECORD, J., 1985 Darwin and the breeders: a social history, pp 519–542, in KOHN, D. (editor), *The Darwinian heritage.* Princeton: Princeton University Press. Pp xii, 1138.

TEGETMEIER, W. B., 1858 *A manual of domestic economy: with hints on domestic medicine and surgery.* London. Home & Colonial School Society. Pp viii, 164.

Received: 29 January 2001. Accepted: 23 July 2002.

**APPENDIX 1.** Correlation between date ranges of Darwin's correspondence and date ranges of Darwin's projects (see Table 1, p. 120): ✕ indicates an overlap.

\* These correspondents are noted in *Dictionary of British and Irish botanists and horticulturists* (Desmond and Ellwood, 1994).

Columns
**A** — entomology
**B** — zoology of the *Beagle* voyage
**C** — transmutation of species
**D** — Cirripedia
**E** — domestic animal breeding
**F** — seed dispersal and transport
**G** — insectivorous plants

**H**— climbing plants
**I** — pangenesis
**J** — expressions
**K** — man and sexual selection
**L** — movement in plants
**M** — worms

| Correspondent | Dates of correspondence | Number of letters | A | B | C | D | E | F | G | H | I | J | K | L | M |
|---|---|---|---|---|---|---|---|---|---|---|---|---|---|---|---|---|
| Abbott, Francis | 1871–1880 | 16 | | | | | | | ✕ | | ✕ | ✕ | ✕ | ✕ | ✕ |
| Agassi, Alexander* | 1869–1881 | 12 | | | | | | | ✕ | | ✕ | ✕ | ✕ | ✕ | ✕ |
| Airy, Hubert | 1871–1876 | 27 | | | | | | | ✕ | | ✕ | ✕ | ✕ | ✕ | ✕ |
| Alglave, Emile | 1869–1877 | 8 | | | | | | | ✕ | | ✕ | ✕ | ✕ | ✕ | ✕ |
| Allen, Charles G.* | 1878–1882 | 13 | | | | | | | | | ✕ | | | ✕ | ✕ |
| Anderson-Henry, Issaac* | 1863–1867 | 17 | | | | | ✕ | ✕ | ✕ | ✕ | ✕ | ✕ | | | |
| Arruda Furtado, Francisco d' | 1881 | 10 | | | | | | | | | ✕ | | | | ✕ |
| Asher, George | 1877–1879 | 7 | | | | | | | | | ✕ | | | ✕ | ✕ |
| Aubertin, Joshua | 1863–1872 | 7 | | | | | | ✕ | ✕ | ✕ | ✕ | ✕ | ✕ | | |
| Aveling, Edward* | 1878–1881 | 7 | | | | | | | | | ✕ | | | ✕ | ✕ |
| Babington, Charles* | 1837–1877 | 20 | | ✕ | ✕ | ✕ | ✕ | ✕ | ✕ | ✕ | ✕ | ✕ | ✕ | ✕ | ✕ |
| Balfour, Francis | 1873–1881 | 17 | | | | | | | ✕ | | ✕ | | | ✕ | ✕ |
| Bartlett, Abraham | 1861–1872 | 14 | | | | | ✕ | ✕ | ✕ | ✕ | ✕ | ✕ | ✕ | | |
| Bartlett, Edward* | 1871 | 6 | | | | | | | ✕ | | ✕ | ✕ | ✕ | | |
| Bate, Charles S. | 1850–1871 | 20 | | | | ✕ | ✕ | ✕ | ✕ | ✕ | ✕ | ✕ | ✕ | | |
| Baxter, William* | 1855–1882 | 31 | | | | | ✕ | ✕ | ✕ | ✕ | ✕ | ✕ | ✕ | ✕ | ✕ |
| Becker, Lydia* | 1863–1877 | 15 | | | | | ✕ | ✕ | ✕ | ✕ | ✕ | ✕ | ✕ | ✕ | ✕ |
| Belt, Thomas* | 1867–1877 | 15 | | | | | ✕ | ✕ | | ✕ | ✕ | ✕ | ✕ | ✕ | ✕ |
| Bennett, Alfred* | 1869–1876 | 18 | | | | | | | ✕ | | ✕ | ✕ | ✕ | ✕ | ✕ |
| Bentham, George* | 1856–1880 | 66 | | | | | ✕ | ✕ | ✕ | ✕ | ✕ | ✕ | ✕ | ✕ | ✕ |
| Berkeley, Miles* | 1840–1875 | 11 | | ✕ | ✕ | ✕ | ✕ | ✕ | ✕ | ✕ | ✕ | ✕ | ✕ | ✕ | |
| Blackley, Charles | 1873–1879 | 7 | | | | | | | ✕ | | ✕ | | | ✕ | ✕ |
| Blair, Rueben | 1877–1881 | 7 | | | | | | | | | ✕ | | | ✕ | ✕ |
| Bonn, Heinrich | 1860–1862 | 18 | | | | | ✕ | ✕ | ✕ | | | | | | |

| Correspondent | Dates of correspondence | Number of letters | A | B | C | D | E | F | G | H | I | J | K | L | M |
|---|---|---|---|---|---|---|---|---|---|---|---|---|---|---|---|
| Bosquet, Joseph | 1852–1856 | 11 | | | | × | × | × | | | | | | | |
| Bowerbank, James* | 1847–1864 | 14 | | | | × | × | × | × | × | | | | | |
| Bowman, William* | 1865–1878 | 30 | | | | | | × | × | × | × | × | × | × | × |
| Breitenbach, Wilhelm | 1876–1882 | 12 | | | | | | | | | × | | | × | × |
| Brent, Bernard | 1857–1864 | 9 | | | | | × | × | × | × | | | | | |
| Browne, Walter | 1880–1881 | 6 | | | | | | | | | × | | | × | × |
| Brunton, Thomas | 1873–1882 | 38 | | | | | | | × | | × | | | × | × |
| Buckland, Francis | 1863–1870 | 14 | | | | | | × | × | × | × | × | × | | |
| Buckley, Arabella | 1868–1881 | 21 | | | | | | | × | | × | × | × | × | × |
| Bunbury, Charles* | 1855–1860 | 10 | | | | | × | × | × | | | | | | |
| Busk, George | 1858–1873 | 18 | | | | | × | × | × | × | × | × | × | × | |
| Butler, Arthur | 1870–1879 | 15 | | | | | | | × | | × | × | × | × | × |
| Caird, James | 1878–1881 | 9 | | | | | | | | | × | | | × | × |
| Canestrini, Giovanni | 1868–1880 | 9 | | | | | | | × | | × | × | × | × | × |
| Carus, Julius | 1866–1881 | 179 | | | | | | × | × | | × | × | × | × | × |
| Caspary, Johann* | 1866–1876 | 11 | | | | | | × | × | | × | × | × | × | × |
| Caton, John | 1868–1877 | 9 | | | | | | | × | | × | × | × | × | × |
| Child, Gilbert | 1868–1869 | 6 | | | | | | | × | | × | × | × | | |
| Clarke, Richard T.* | 1862–1877 | 7 | | | | | | × | × | × | × | × | × | × | × |
| Claus, Carl | 1869–1877 | 8 | | | | | | | × | | × | × | × | × | × |
| Cobbe, Frances | 1870–1872 | 7 | | | | | | | × | | × | × | | | |
| Coe, Henry | 1857–1858 | 6 | | | | | × | × | | | | | | | |
| Cohn, Ferdinand | 1874–1882 | 22 | | | | | | | × | | × | | | × | × |
| Conway, Moncure | 1863–1878 | 8 | | | | | | × | × | × | × | × | × | × | × |
| Covington, Sims | 1843–1859 | 10 | | × | × | × | × | × | | | | | | | |
| Cresy, Edward | 1845–1860 | 45 | | | | × | × | × | × | | | | | | |
| Crichton-Browne, James | 1869–1875 | 41 | | | | | | | × | | × | × | × | × | |
| Crick, Walter D. | 1879–1882 | 11 | | | | | | | | | × | | | × | × |
| Crocker, Charles* | 1862–1863 | 9 | | | | | | × | × | × | | | | | |
| Croll, James | 1968–1881 | 14 | | | | | | | × | | × | × | × | × | × |
| Crüger, Hermann* | 1863–1864 | 6 | | | | | × | × | × | | | | | | |
| Cupples, George | 1868–1878 | 61 | | | | | | | × | | × | × | × | × | × |
| Dallas, William | 1867–1880 | 55 | | | | | | × | × | | × | × | × | × | × |
| Dareste, Gabriel | 1863–1870 | 9 | | | | | | × | × | × | × | × | × | | |
| Darwin, Francis* | 1857–1882 | 194 | | | | | × | × | × | × | × | × | × | × | × |
| Darwin, George H. | 1856–1882 | 263 | | | | | × | × | × | × | × | × | × | × | × |
| Darwin, Horace | 1859–1881 | 17 | | | | | × | × | × | × | × | × | × | × | × |
| Darwin, Leonard | 1859–1881 | 27 | | | | | × | × | × | × | × | × | × | × | × |
| Darwin, Reginald | 1879 | 19 | | | | | | | | | × | | | × | × |

| Correspondent | Dates of correspondence | Number of letters | A | B | C | D | E | F | G | H | I | J | K | L | M |
|---|---|---|---|---|---|---|---|---|---|---|---|---|---|---|---|
| Darwin, William E.* | 1851–1882 | 219 | | | | × | × | × | × | × | × | × | × | × | × |
| Davidson, Thomas | 1856–1873 | 7 | | | | | × | × | × | × | × | × | × | × | |
| Dawkins, William | 1867–1875 | 20 | | | | | | × | × | | × | × | × | × | |
| De Chaumont, Francis | 1871–1875 | 10 | | | | | | | × | | × | × | × | × | |
| Delpino, G.G. Federico | 1867–1880 | 26 | | | | | | × | × | | × | × | × | × | × |
| Denny, Henry* | 1844–1865 | 8 | | | × | × | × | × | × | × | × | | | | |
| Denny, John* | 1872–1879 | 6 | | | | | | | × | | × | × | | × | × |
| Dieffenbach, Ernest* | 1843–1847 | 13 | | × | × | × | | | | | | | | | |
| Dobell, Horace | 1863–1871 | 11 | | | | | | × | × | × | × | × | × | | |
| Dodel-Port, Arnold | 1874–1880 | 13 | | | | | | | × | | × | | | × | × |
| Donders, Francis | 1869–1874 | 29 | | | | | | | × | | × | × | × | × | |
| Doubleday, Henry* | 1857–1868 | 11 | | | | | × | × | × | × | × | × | × | | |
| Elliot, Walter | 1856–1869 | 6 | | | | | × | × | × | × | × | × | × | | |
| Ernst, Adolf | 1878–1882 | 9 | | | | | | | | | × | | | × | × |
| Errera, Leo | 1877–1879 | 10 | | | | | | | | | × | | | × | × |
| Eyton, Thomas* | 1833–1876 | 39 | | × | × | × | × | × | × | × | × | × | × | × | × |
| Falconer, Hugh* | 1855–1865 | 52 | | | | | × | × | × | × | × | | | | |
| Farrer, Thomas* | 1868–1882 | 134 | | | | | | | × | | × | × | × | × | × |
| Fayner, Joseph | 1874–1882 | 12 | | | | | | | × | | × | | | × | × |
| Fiske, John | 1871–1880 | 14 | | | | | | | × | | × | × | × | × | × |
| Fitch, Robert | 1849–1851 | 17 | | | | × | | | | | | | | | |
| Fitzgerald, Robert D.* | 1875–1881 | 7 | | | | | | | × | | × | | | × | × |
| Flower, William | 1863–1880 | 29 | | | | | | × | × | × | × | × | × | × | × |
| Forbes, David | 1860–1872 | 12 | | | | | × | × | × | × | × | × | × | | |
| Forel, August | 1874–1876 | 7 | | | | | | | × | | × | | | × | × |
| Foster, Michael* | 1871–1882 | 14 | | | | | | | × | | × | × | × | × | × |
| Fox, William D. | 1828–1879 | 206 | × | × | × | × | × | × | × | × | × | × | × | × | × |
| Geikie, James | 1876–1881 | 13 | | | | | | | | | × | | | × | × |
| Gilbert, Joseph* | 1869–1882 | 13 | | | | | | | × | | × | × | × | × | × |
| Gladstone, William | 1872–1881 | 11 | | | | | | | × | | × | × | | × | × |
| Goodacre, Francis | 1873–1880 | 18 | | | | | | | × | | × | | | × | × |
| Gould, August | 1848–1859 | 9 | | | | × | × | × | | | | | | | |
| Gray, George R.* | 1838–1869 | 11 | | × | × | × | × | × | × | × | × | × | × | | |
| Gray, John E.* | 1847–1873 | 37 | | | | × | × | × | × | × | × | × | × | | |
| Günther, Albert C. | 1860–1881 | 66 | | | | | × | × | × | × | × | × | × | × | |
| Haast, Sir Johann* | 1862–1879 | 22 | | | | | × | × | × | × | × | × | × | × | |
| Hancock, Albany* | 1849–1869 | 24 | | | | × | × | × | × | × | × | × | × | | |
| Hartogh, Heijs | 1870–1874 | 14 | | | | | | | × | | × | × | × | × | |
| Heckel, Edouard M. | 1876–1881 | 7 | | | | | | | | | × | | | × | × |

| Correspondent | Dates of correspondence | Number of letters | A | B | C | D | E | F | G | H | I | J | K | L | M |
|---|---|---|---|---|---|---|---|---|---|---|---|---|---|---|---|
| Heer, Oswald | 1860–1877 | 7 | | | | | × | × | × | × | × | × | × | × | × |
| Henslow, George* | 1865–1879 | 35 | | | | | | × | × | × | × | × | × | × | × |
| Henslow, John S.* | 1831–1860 | 144 | × | × | × | × | × | × | × | | | | | | |
| Hildebrand, Friedrich | 1862–1880 | 29 | | | | | | × | × | × | × | × | × | × | × |
| Hoffmann, Hermann | 1870–1877 | 6 | | | | | | | × | | × | × | × | × | × |
| Hooker, William J.* | 1843–1858 | 9 | | × | × | × | × | × | | | | | | | |
| Horner, Leonard | 1838–1862 | 13 | | × | × | × | × | × | × | | | | | | |
| Hunt, Robert | 1855–1880 | 6 | | | | | × | × | × | × | × | × | × | × | × |
| Hyatt, Alpheus | 1872–1881 | 11 | | | | | | | × | | × | × | | × | × |
| Jamieson, Thomas* | 1861–1866 | 7 | | | | | × | × | × | × | × | | | | |
| Jenyns, Leonard* | 1837–1877 | 42 | | × | × | × | × | × | × | × | × | × | × | × | × |
| Jesse, George | 1871–1881 | 6 | | | | | | | × | | × | × | × | × | × |
| Judd, John | 1876–1882 | 12 | | | | | | | | | × | | | × | × |
| Jukes, Joseph B. | 1838–1864 | 9 | | × | × | × | × | × | × | × | | | | | |
| Kindt, Hermann | 1864–1865 | 10 | | | | | | × | × | × | × | | | | |
| King, George* | 1871–1881 | 14 | | | | | | | × | | × | × | × | × | × |
| Kingsley, Charles* | 1859–1869 | 17 | | | | | × | × | × | × | × | × | × | | |
| Kippist, Richard* | 1857–1877 | 12 | | | | | × | × | × | × | × | × | × | × | × |
| Krause, Ernst | 1877–1881 | 108 | | | | | | | | | × | | | × | × |
| Krefft, Johann | 1872–1876 | 16 | | | | | | | × | | × | × | | × | × |
| Lankester, Edwin* | 1850–1853 | 13 | | | | × | | | | | | | | | |
| Lankester, Edwin R. | 1869–1881 | 8 | | | | | | | × | | × | × | × | × | × |
| Leighton, William* | 1840–1865 | 6 | | × | × | × | × | × | × | × | × | | | | |
| Lindley, John* | 1843–1862 | 13 | | × | × | × | × | × | × | | | | | | |
| Lonsdale, William | 1837–1868 | 9 | | × | × | × | × | × | × | × | × | × | × | | |
| Lynch, Richard* | 1877–1878 | 11 | | | | | | | | | × | | | × | × |
| MacKintosh, David | 1867–1882 | 12 | | | | | | × | × | | × | × | × | × | × |
| Marsh, O.C. | 1875–1880 | 6 | | | | | | | × | | × | | | × | × |
| Master, Max | 1860–1880 | 44 | | | | | × | × | × | × | × | × | × | × | × |
| Matthew, Patrick | 1862–1871 | 6 | | | | | | | × | × | × | × | × | | |
| Maw, George* | 1861–1880 | 29 | | | | | × | × | × | × | × | × | × | × | × |
| McLennan, John* | 1871–1880 | 12 | | | | | | | × | | × | × | × | × | × |
| Meehan, Thomas* | 1871–1880 | 12 | | | | | | | × | | × | × | × | × | × |
| Meldola, Raphael | 1871–1882 | 88 | | | | | | | × | | × | × | × | × | × |
| Mengozzi, Giovanni | 1880–1881 | 6 | | | | | | | | | × | | | × | × |
| Miller, William H.* | 1839–1859 | 10 | | × | × | × | × | × | | | | | | | |
| Moggridge, John T.* | 1864–1874 | 43 | | | | | | | × | × | × | × | × | × | |
| More, Alexander* | 1860–1881 | 24 | | | | | × | × | × | × | × | × | × | × | × |
| Moschkan, Alfred | 1873–1878 | 6 | | | | | | | × | | × | | | × | × |

| Correspondent | Dates of correspondence | Number of letters | A | B | C | D | E | F | G | H | I | J | K | L | M |
|---|---|---|---|---|---|---|---|---|---|---|---|---|---|---|---|---|
| Moseley, Henry* | 1876–1882 | 21 | | | | | | | | | × | | | × | × |
| Moulinie, Jean | 1867–1872 | 31 | | | | | | × | × | | × | × | × | | |
| Müller, Heinrich L. | 1867–1880 | 52 | | | | | | × | × | | × | × | × | × | × |
| Müller, Johann F. | 1865–1882 | 107 | | | | | | × | × | × | × | × | × | × | × |
| Murie, James* | 1868–1880 | 6 | | | | | | | × | | × | × | × | × | × |
| Murray, Andrew* | 1860–1872 | 16 | | | | | × | × | × | × | × | × | × | | |
| Naudin, Charles V. | 1862–1882 | 9 | | | | | | × | × | × | × | × | × | × | × |
| Neumayr, Melchior | 1877–1882 | 6 | | | | | | | | | × | | | × | × |
| Nevill, Lady Dorothy F.* | 1861–1881 | 30 | | | | | × | × | × | × | × | × | × | × | × |
| Newington, Samuel* | 1875–1880 | 6 | | | | | | | × | | × | | | × | × |
| Newton, Alfred | 1863–1881 | 34 | | | | | × | × | × | × | × | × | × | × | × |
| Nicols, Arthur | 1871–1880 | 14 | | | | | | | × | | × | × | × | × | × |
| Norton, Charles | 1871–1881 | 9 | | | | | | | × | | × | × | × | × | × |
| Ogle, William* | 1867–1882 | 38 | | | | | | × | × | | × | × | × | × | × |
| Oliver, Daniel* | 1860–1877 | 120 | | | | | × | × | × | × | × | × | × | × | × |
| Orton, James | 1869–1870 | 6 | | | | | | | × | | × | × | × | | |
| Oxenden, George | 1862–1872 | 22 | | | | | | × | × | × | × | × | × | | |
| Paget, James* | 1859–1881 | 30 | | | | | × | × | × | × | × | × | × | × | × |
| Palaeontographical Society | 1850–1859 | 6 | | | | × | × | × | | | | | | | |
| Patterson, Robert | 1847–1860 | 8 | | | | × | × | × | × | | | | | | |
| Playfair, Lyon | 1875 | 8 | | | | | | | × | | × | | | × | |
| Preston, Samuel | 1880 | 7 | | | | | | | | | × | | | × | × |
| Preyer, Thierry | 1868–1881 | 18 | | | | | | | | | × | | | | × |
| Price, John* | 1826–1881 | 22 | | | | | | | | | × | | | | × |
| Quatrefages, Jean | 1857–1877 | 25 | | | | | × | × | × | × | × | × | × | × | × |
| Ralston, William | 1875–1881 | 6 | | | | | | | × | | × | | | × | × |
| Ramsay, Andew C. | 1846–1881 | 37 | | | | × | × | × | × | × | × | × | × | × | × |
| Ransome, Frederick | 1864–1866 | 7 | | | | | | × | × | × | × | | | | |
| Reade, Thomas | 1874–1881 | 16 | | | | | | | × | | × | | | × | × |
| Reade, William* | 1868–1875 | 50 | | | | | | | × | | × | × | × | × | |
| Reeks, Henry* | 1871–1879 | 9 | | | | | | | × | | × | × | × | × | × |
| Reuter, Adolf | 1869–1873 | 10 | | | | | | | × | | × | × | × | × | |
| Rich, Anthony | 1878–1882 | 27 | | | | | | | × | | × | | | × | × |
| Rivers, Thomas* | 1862–1872 | 31 | | | | | | × | × | × | × | × | × | | |
| Riviere, Briton | 1871–1872 | 14 | | | | | | | × | | × | × | × | | |
| Robertson, George | 1875–1882 | 8 | | | | | | | × | | × | | | × | × |
| Rolle, Friedrich | 1862–1868 | 13 | | | | | | × | × | × | × | × | × | | |
| Rolleston, George | 1861–1878 | 18 | | | | | × | × | × | × | × | × | × | × | × |

| Correspondent | Dates of correspondence | Number of letters | A | B | C | D | E | F | G | H | I | J | K | L | M |
|---|---|---|---|---|---|---|---|---|---|---|---|---|---|---|---|
| Romanes, George* | 1874–1882 | 136 |  |  |  |  |  |  | × |  | × |  |  | × | × |
| Royle, John* | 1838–1847 | 7 |  | × | × | × |  |  |  |  |  |  |  |  |  |
| Rütimeyer, Karl | 1861–1874 | 7 |  |  |  |  | × | × | × | × | × | × | × | × |  |
| Sabine, Edward* | 1854–1864 | 7 |  |  |  | × | × | × | × | × |  |  |  |  |  |
| Salvin, Osbert* | 1863–1875 | 20 |  |  |  |  |  | × | × | × | × | × | × | × |  |
| Sanderson, John S.* | 1873–1880 | 63 |  |  |  |  |  |  | × |  | × |  |  | × | × |
| Saporta, Louis | 1868–1881 | 21 |  |  |  |  |  |  | × |  | × | × | × | × | × |
| Scherzer, Karl von | 1867–1878 | 9 |  |  |  |  |  | × | × |  | × | × | × | × | × |
| Sclater, Philip | 1860–1882 | 29 |  |  |  |  | × | × | × | × | × | × | × | × | × |
| Scott, John* | 1862–1877 | 91 |  |  |  |  |  | × | × | × | × | × | × | × | × |
| Semper, Carl G. | 1874–1881 | 17 |  |  |  |  |  |  | × |  | × |  |  | × | × |
| Sharpe, Daniel* | 1846–1854 | 7 |  |  |  | × |  |  |  |  |  |  |  |  |  |
| Shaw, James* | 1865–1868 | 10 |  |  |  |  | × | × | × | × | × | × |  |  |  |
| Simpson, J.F. | 1881–1882 | 6 |  |  |  |  |  |  |  |  | × |  |  |  | × |
| Skertchly, Sydney* | 1878–1881 | 6 |  |  |  |  |  |  |  |  | × |  |  | × | × |
| Smith, Andrew | 1839–1871 | 7 |  | × | × | × | × | × | × | × | × | × | × |  |  |
| Smith, Frederick | 1857–1878 | 20 |  |  |  |  | × | × | × | × | × | × | × | × | × |
| Sowerby, George* | 1844–1846 | 12 |  |  | × | × |  |  |  |  |  |  |  |  |  |
| Stainton, Henry | 1855–1881 | 14 |  |  |  |  | × | × | × | × | × | × | × | × | × |
| Steenstrup, Japetus | 1849–1881 | 16 |  |  |  | × | × | × | × | × | × | × | × | × | × |
| Stephen, Leslie | 1879–1882 | 8 |  |  |  |  |  |  |  |  | × |  |  | × | × |
| Strickland, Hugh* | 1842–1849 | 11 |  | × | × | × |  |  |  |  |  |  |  |  |  |
| Sulivan, Bartholomew* | 1832–1881 | 72 |  | × | × | × | × | × | × | × | × | × | × | × | × |
| Swinhoe, Robert* | 1862–1874 | 18 |  |  |  |  |  | × | × | × | × | × | × | × |  |
| Tait, Robert | 1871–1882 | 76 |  |  |  |  |  |  | × |  | × | × | × | × | × |
| Tait, William* | 1869 | 18 |  |  |  |  |  |  | × |  | × | × | × |  |  |
| Tegetmeier, William | 1855–1881 | 188 |  |  |  |  | × | × | × | × | × | × | × | × | × |
| Thiselton-Dyer, William* | 1868–1881 | 148 |  |  |  |  |  |  | × |  | × | × | × | × | × |
| Thomas, Charles | 1870–1878 | 6 |  |  |  |  |  |  | × |  | × | × | × | × | × |
| Thwaites, George* | 1855–1877 | 30 |  |  |  |  | × | × | × | × | × | × | × | × | × |
| Torbitt, James | 1876–1882 | 93 |  |  |  |  |  |  |  |  | × |  |  | × | × |
| Treat, Mary | 1871–1876 | 15 |  |  |  |  |  |  | × |  | × | × | × | × | × |
| Trimen, Roland* | 1863–1877 | 39 |  |  |  |  |  | × | × | × | × | × | × | × | × |
| Voysey, Charles* | 1869–1876 | 6 |  |  |  |  |  |  | × |  | × | × | × | × | × |
| Walker, Francis | 1838–1876 | 6 |  | × | × | × | × | × | × | × | × | × | × | × | × |
| Wallich, George* | 1860–1882 | 7 |  |  |  |  | × | × | × | × | × | × | × | × | × |
| Walsh, Benjamin D. | 1864–1868 | 34 |  |  |  |  |  | × | × | × | × | × | × |  |  |
| Waterhouse, George R. | 1838–1868 | 45 |  | × | × | × | × | × | × | × | × | × | × |  |  |
| Watson, Hewett* | 1847–1864 | 45 |  |  |  | × | × | × | × | × |  |  |  |  |  |

| Correspondent | Dates of correspondence | Number of letters | A | B | C | D | E | F | G | H | I | J | K | L | M |
|---|---|---|---|---|---|---|---|---|---|---|---|---|---|---|---|
| Weale, John | 1865–1874 | 14 | | | | | | × | × | × | × | × | × | × | |
| Weir, John J.* | 1868–1881 | 72 | | | | | | | × | | × | × | × | × | × |
| Westwood, John | 1856–1864 | 9 | | | | | × | × | × | × | | | | | |
| Whitelegge, Thomas* | 1878 | 7 | | | | | | | | | × | | | × | × |
| Williamson, William C.* | 1846–1880 | 16 | | | | × | × | × | × | × | × | × | × | × | × |
| Wilson, Alexander S.* | 1878–1881 | 22 | | | | | | | | | × | | | × | × |
| Wollaston, Thomas* | 1855–1860 | 10 | | | | | × | × | × | | | | | | |
| Woodward, Samuel P.* | 1842–1863 | 25 | | × | × | × | × | × | × | × | | | | | |
| Wright, Chaucey | 1871–1875 | 20 | | | | | | | × | | × | × | × | × | |
| Würtenberger, Leopold | 1879–1881 | 7 | | | | | | | | | × | | | × | × |
| Yarrell, William | 1838–1856 | 6 | | × | × | × | × | × | | | | | | | |
| Zacharias, Otto | 1875–1877 | 10 | | | | | | | × | | × | | | × | × |
| Zincke, Foster | 1876–1881 | 7 | | | | | | | | | × | | | × | × |

**APPENDIX 2.** William B. Tegetmeier and John Scott: correspondence topics with lists of letters (numbers correspond to letter in Burkhardt and Smith, 1994[12]).

**William B. Tegetmeier**

| Topic | Letters |
|---|---|
| Bees | 2270 2271 2280 2281 2289 2325 2332 2384 2762 2883 |
| Crossing/breeding experiments | 2051 2090 2093 2362 2375 2491 2656 2712 3103 4238 4798 4849 4977 5431 8322 |
| Egg incubation time | 2790 2872 2883 6808 6811 |
| General on birds/animals | 2068 3055 3108 4233 4314 4785 4796 5211 7274 7822 10111 |
| Help in identification | 1981 2004 2011 2260 |
| Queries on domestic animals | 2762 3118 3139 3869 3944 3949 5176 |
| Request for particular breeds | 2407 2417 2790 3095 |
| Request to read manuscript | 3095 4533 4787 4806 5507 10118 |
| Sex ratios | 5859 5878 5879 5882 5906 5934 6000 6356 6441 6600 6698 6700 9266 9298 |
| Sexual selection | 3164 3166 5473 5475 6017 7822 |
| Sterility/fertility | 1947 3869 3877 3998 4233 4687 4720 4761 4785 |
| Variation | 1751 1754 1788 1791 1947 2048 3070 3075 1820 2090 2108 2479 2790 5171 6188 6199 6208 6210 6347 6356 6600 6698 6702 6837 6839 6870 6920 8322 |

**John Scott**

| Topic | Letters |
|---|---|
| Assistance with manuscript | 3814 3847 3853 4197 4206 4213 4301 4386 4402 4485 4498 4526 4751 4876 |
| Crossing/pollination experiments | 3400 3805 3808 3844 3865 3868 3904 3921 3932 3934 4021 4031 4055 4060 4073 4084 4086 4114 4229 4252 4332 4343 4432 4498 4526 10555 10864 10928 |
| Expression/emotion in humans | 6160 6815 7030 8045 |
| Ferns | 3853 3847 3808 3991 3997 |
| Help in finding position | 4177 4183 4187 4206 4212 4123 4463 4513 4524 4527 4528 4541 4578 4582 4751 4810 4876 |
| *Linum* and *Lythrum* | 3844 4087 4385 4810 |
| Miscellaneous on plants | 4190 4232 4252 4332 4343 4382 4382 4386 4432 4438 4526 4541 5344 5633a 10864 10928 |
| Natural selection | 3815 3847 3853 4021 4031 4137 4174 4190 4260 |
| Orchids | 3800 3844 3934 4060 4073 4084 4087 4137 4175 4190 4382 4751 |
| Plant sterility and fertility | 3814 3815 3904 3934 4021 4055 4073 4087 4137 4174 4197 4229 4252 4438 4498 4541 |
| *Primula* | 3814 3844 3932 3991 4003 4021 4086 4185 4175 4229 4252 4260 4253 4252 4260 4498 4031 4541 4084 |
| Worms | 8249 8534 |

269

**APPENDIX 3.** Darwin's work on seeds and dispersal with list of letters by topics discussed (numbers correspond to letter in Burkhardt and Smith, 1994[12]).

| Topic | | | | | | | | | | | | | | | |
|---|---|---|---|---|---|---|---|---|---|---|---|---|---|---|---|
| Dispersal by birds | 1608 | 1816 | 1866 | 1948 | 1948 | 1978 | 2328 | 2338 | 2343 | 2395 | 2722 | 3587 | 4054 | 4351 | 4353 |
| | 4435 | 4440 | 4446 | 4453 | 4456 | 5287 | 5300 | 6780 | 10340 | 10341 | 12435 | | | | |
| Dispersal by fish | 1948 | 2042 | 2064 | 2069 | 2081 | | | | | | | | | | |
| Dispersal by ice | 1636 | | | | | | | | | | | | | | |
| Dispersal by insects | 5649 | 5659 | 5714 | | | | | | | | | | | | |
| Floating plants | 1950 | | | | | | | | | | | | | | |
| General on dispersal | 849 | 1948 | 1956 | 2075 | 2315 | 3667 | 4269 | 5007 | 5305 | 5364 | 5373 | 5659 | 7848 | 10370 | |
| General on vitality | 353 | 390 | 579 | 668 | 669 | 690 | 691 | 696 | 699 | 701 | 705 | 706 | 707 | 708 | 711 |
| | 712 | 713 | 716 | 720 | 799 | 10802 | 13331 | | | | | | | | |
| | 1708 | 1777 | | | | | | | | | | | | | |
| Request for seeds | 1660 | 1661 | 1667 | 1669 | 1671 | 1680 | 1681 | 1684 | 1704 | 1708 | 1710 | 1711 | 1733 | 1742 | 1763 |
| Salting experiment | 1783 | 1787 | 1834 | 1911 | 1985 | 1994 | 2074 | 2075 | 2117 | | | | | | |

**APPENDIX 4.** Darwin's work on seeds and dispersal; list of correspondence, their profession, location and list of letters (numbers correspond to letter in Burkhardt and Smith, 1994[12]). ** indicates individuals who are known to have traveled extensively.

| Correspondent | Profession | Location | | | | | | | | | | | | |
|---|---|---|---|---|---|---|---|---|---|---|---|---|---|---|
| Arruda Furtado, F. | naturalist | Azores | 13331 | | | | | | | | | | | |
| Babington, Charles.C. | botanist | England | 708 | | | | | | | | | | | |
| Belt, Thomas | geologist | Central America | 5364 | 1834 | | | | | | | | | | |
| Berkeley, Miles J. | botanist | England | 1710 | | | | | | | | | | | |
| Crowe, John R. | diplomat | Norway | 1777 | | | | | | | | | | | |
| Darwin, Susan | | England | 390 | | | | | | | | | | | |
| Darwin, William E. | banker | England | 1660 | | | | | | | | | | | |
| Dixon, Charles | ornithologist | England | 12435 | | | | | | | | | | | |
| Eyton, Thomas C. | ornithologist | England | 1948 | 2338 | 2343 | | | | | | | | | |
| Fox, William D. | clergyman | England | 1704 | 1733 | 1978 | 2049 | 2057 | 2328 | | | | | | |
| *Gardeners' chronicle* | journal | England | 1666 | 1684 | 1783 | 1787 | 4269 | | | | | | | |
| Garrett, James R. | botanist | Ireland | 1608 | | | | | | | | | | | |
| Gray, Asa | botanist | United States | 3667 | 5649 | 10370 | | | | | | | | | |
| Henslow, John S. | botanist | England | 353 | 579 | 701 | 705 | 707 | 712 | 1708 | 1816 | | | | |
| Hill, Richard | naturalist | Jamaica | 2064 | | | | | | | | | | | |
| Holland, Miss | | England | 2395 | | | | | | | | | | | |
| Hooker, Joseph D. | botanist | England** | 799 | 1661 | 1667 | 1669 | 1671 | 1680 | 1681 | 1711 | 1742 | 1763 | 1911 | 1950 |
| | | | 2042 | 2075 | 2117 | 2315 | 4351 | 4353 | 5300 | 5305 | 5373 | 7848 | 10802 | |
| Kemp, William | unknown | Scotland | 699 | 706 | 711 | 713 | 716 | 720 | | | | | | |
| Lamont, James | geologist | England** | 2722 | | | | | | | | | | | |
| Lindley, John | botanist | England | 668 | 669 | 690 | 691 | | | | | | | | |
| Lyell, Charles | geologist | England | 696 | 899 | 1866 | 2050 | 5007 | 5659 | | | | | | |
| Newton, Alfred | ornithologist | England | 4054 | 4435 | 4440 | 4446 | 4453 | 4456 | | | | | | |
| Norman, Herbert G. | unknown | England | 5287 | | | | | | | | | | | |
| Oliver, Daniel | botanist | England | 3587 | | | | | | | | | | | |
| Rae, John | explorer | England** | 1636 | | | | | | | | | | | |
| Robillard, V. de | zoologist | Mauritius | 1956 | | | | | | | | | | | |
| Swaysland, George | unknown | England | 6780 | | | | | | | | | | | |
| Tenant, James | unknown | England | 2069 | 2081 | | | | | | | | | | |
| Thistleton-Dyer, William | botanist | England | 10340 | 10341 | | | | | | | | | | |
| Watson, Hewett C. | botanist | England | 1985 | 1994 | | | | | | | | | | |
| Weale, John P. | Naturalist | South Africa | 5714 | | | | | | | | | | | |

*Archives of Natural History* (1986) **13** (3): 313–324

# Unveiling Darwin's Roots

MARIO A DI GREGORIO

Darwin College
Cambridge

## INTRODUCTION

In this article I attempt through the annotations made by Charles Darwin in the margins of his own library books to give glimpses of his consideration of particular scientific topics during the course of his career (section 2). Later in the paper I consider his reactions to English-, French- and German-language natural science and related publications, commenting especially on his readings of De Candolle and Humboldt.

## THE MARGINALIA

The invaluable importance of manuscript sources for research on Charles Darwin's views is nowadays recognised by all scholars. A good example of scholarly work based on manuscript sources is David Kohn's 'Theories to work by: rejected theories, reproduction and Darwin's path to natural selection' (Kohn, 1980), which follows step-by-step Darwin's path towards his theory of evolution by natural selection.[2] Through careful analysis of Darwin's notebooks Kohn was able to show that Darwin outlined two different theories before arriving at the one later expounded in *The Origin of Species*. Thanks mainly to Sydney Smith, the bulk of Darwin's manuscripts is now preserved in the Cambridge University Library where it is lovingly looked after by Peter Gautrey; material of lesser importance is still kept at Down House, guarded with equal care by Philip Titheradge. A few things are scattered elsewhere (for example at the Linnean Society of London), but are readily accessible. Darwin's manuscripts fall into three categories: (a) notebooks; (b) letters; and (c) library (books and pamphlets). Such distinguished scholars as Sir Gavin de Beer (1959–67),[3] Peter Vorzimmer (1977),[4] Howard Gruber and Paul Barrett (1974)[5] and Sandra Herbert (1980)[6] have edited the notebooks, which are now used as major references in scholarship on Darwin.

The herculean task of producing a definitive edition of Darwin's letters is occupying a team of scholars led by Frederick Burckhardt and Sydney Smith. This extraordinarily important enterprise will throw new light on Darwin's views and his relationship with his contemporaries. (Several years will elapse before we see all the results; the first three volumes of many have recently been published.[7])

The third, and so far most neglected, kind of material available in the Darwin Archive is Charles Darwin's own library. This consists of a few thousand items, divided into books (including bound journals) and a collection of separate pamphlets (often but mistakenly called reprints). The importance of these two collections lies in the number of annotations which Darwin wrote on the margin of the book he was reading, or on small pieces of paper enclosed in the book, usually stuck inside the back cover. Considerably less scholarship has been based on these marginalia than on either the notebooks or the letters, due to problems of access for scholars not resident in Cambridge.

Three recent works have made use of the marginalia, all produced by scholars from the New World: R A Richardson (1981)[8]; S Sheets Pyenson (1981)[9]; and J A Secord (1981).[10] As Richardson (1981, p 3) points out:

> The (Cambridge) collection thus offers unique and detailed access to most of Darwin's sources of information, and to his comments and reflections on them. It can be used as a key to establish with some precision the general pattern of growth of his ideas on geographical distributions and race formations, and on the origin and meaning of adaptations.

Richardson refers to the marginalia for his understanding of Darwin's approach to Lyell and Blyth (1981, p 35), and quite rightly advocates their wider use in research.

Sheets Pyenson's article is a very detailed study of Darwin's reading of the *Magazine of Natural History*, the *Annals of Botany and Zoology*, the *Annals of Natural History* and the *Annals and Magazine of Natural History*. Her article helps one to understand how influential upon Darwin the British tradition of field natural science was, with its emphasis on animal instincts and the relationship between organisms. For example, one sees to what extent Darwin considered Blyth's work (Sheets Pyenson, 1981, pp 240–1).

Secord's article on Darwin's pigeons emphasises the importance of William Yarrell's work for Darwin's approach to the problems of breeding—again, a vital connection with British natural science is established. Secord would have gathered even more evidence had he had easier access to the marginalia. In fact he quotes only Darwin's remarks on Dixon's *Dovecot*. In Darwin's library there are a great number of books on pigeons and most of them bear a considerable number of annotations, such as J Eaton's *A treatise on the art of breeding and managing tame, domesticated and fancy pigeons* (London, 1852), with annotations concerning variation, domesticated and wild animals, and breeding; Boitard and Corbié's *Les pigeons de volière et de colombier* (London, 1824), with annotations on domesticated and wild animals, and speciation; and E Dixon's *Ornamental and domestic poultry* (London, 1849), with annotations concerning animal behaviour, breeding, domesticated and wild animals, heredity, hybridity, sex, and variation.

I have been working on the production of a complete codified index to the marginalia, with considerable actual transcription; this article represents a collection of my general observations in the course of that project as to what the annotations reveal directly about Darwin's sources and his relations to them, and about his own 'thinking out loud' at various stages in his theoretical development. Knowledge of these aspects of Darwin's work will improve our understanding of his thought and its connections with the natural science of his time—all these aspects, especially when taken in conjunction with the other manuscript material, must in the end converge into a complete mosaic.

Darwin himself was well aware of the potential future importance of the annotations he had made in his personal library. For example, he makes certain, in a letter written to his wife Emma, to prescribe that in the event of his death 'some competent person' should receive 'all my Books on Natural History, which are either scored or have references at end to the pages, begging him carefully to look over & consider such passages, as actually bearing or by possibility bearing' on the subject of the sketch of his species theory, which he had just finished (5 July 1844), when the question of its publication in book form should arise.


## GENERAL TOPICS

Some topics seem to have attracted Darwin's attention to a very great extent. Variation is the commonest topic, and appears in almost every work which he annotated. A remarkably large set of annotations concerns variation in colour, size and reproductive power of animals and plants. The varieties of plants and especially of maize greatly attracted Darwin's interest: see for example Metzger, *Die Getreidearten und Wiesengräser* (Heidelberg, 1841). Needless to say, annotations on variation must be considered in their relation to other kinds of annotations. For example the following remark on J Gould's *Handbook to the Birds of Australia* (2 vols, London, 1865) shows the connection between variation and sex: 'Hence it does not seem as variation had occurred early in life, but it had crawled backwards and invaded the young. & it seems whenever it happens it invades both sexes—& ceases to be limited to one sex' (vol 2, p 56).

The importance of sex is stressed throughout the marginalia, in connection not only with reproduction but also with survival, the struggle for existence, and more broadly all that concerns the relationship of organism to organism; see for example F Hildebrandt, *Die Geschlechter-Verheilung bei den Pflanzen* (Leipzig, 1867), p 13: 'It may however be plants have survived owing to having this advantage of separate sexes'.

Annotations on variation undoubtedly form the bulk of the marginalia and confirm that variation was indeed the starting point of Darwin's research; not surprisingly he opened the *Origin* with two chapters on variation in nature and under domestication. The set of annotations which most immediately relate to variation are those concerning domesticated and wild animals. These topics contribute to the understanding of selection, the principle which makes sense of them; see for example D Low, *On the domesticated animals of the British Islands* (London, 1845), p 394: 'such selection cd. never apply to wild animals, as every parent must be adapted to same conditions'. In fact Darwin had a low opinion of this work: inside the back cover he noted 'a Poor Book—not to be trusted'. Darwin's most interesting annotations are not necessarily in the best books: we learn at least as much from 'bad' books which motivated him to form clear objections.

The relationship of organism to organism constitutes the central focus of Darwin's research—without a clear understanding of the place of that notion in Darwin's thought, including what came to be called the struggle for existence, one cannot understand either the notion of selection of the Darwinian conception of evolution. For example, in J D Hooker's *Flora indica* (2 vols, London, 1855), he notes inside the back cover: 'p 41 Good Remarks on Strife of Plants'; and in E Haeckel, *Generelle Morphologie* (Berlin, 1866): 'good criticism of my theory of struggle for existence— says ought to be confined to struggle between organisms for same end—all other causes dependence—misseltoe depends on apple'. Darwin painstakingly wrote notes concerning the relationship between insects and pollen. In the long run, virtually everything Darwin wrote refers to the relationship between organism and organism, since his is eminently an ecological theory; especially extinction and its precursor, rarity (see for example H G Bronn, *Handbuch einer Geschichte der Natur*, 2 vols, Stuttgart, 1843; and *Flora indica*). Extinction leads one to consider geographical distribution: after all Darwin started as an all-round naturalist according to the expectations of his time, but with a special interest in geology. It was probably geology which during the *Beagle* voyage attracted his attention to questions of distribution. Darwin was later to reflect on what he had seen during the voyage and made many interesting annotations—for example in L von Buch, *Description physique des Iles Canaries* (Paris, 1835); H Lecoq, *Etudes sur la géographie botanique de l'Europe. . .* (9 vols, Paris, 1854); and above all J D Hooker, *The botany of the Antarctic voyage. . .* (London, 1844-7). In Hooker's work we find variation related to distribution: 'A species varies more in one country than in another'. The notion of distribution leads to that of isolation and variation in isolation: for example, 'Plant constant in leaves in Falkland, very variable at R Plata' is his back-cover comment about p 288.

Selection is often related to distribution and isolation—in Frémont, *Report of the exploring expedition to the Rocky Mountains* (Washington, 1845), Darwin notes on the back cover: 'It might well happen, as in Horses of Falkland, that the old animals might live at ease and not be driven to search new countries <u>open</u> to them (as is evidently the case with the Buffalo) and the pressures are chiefly falling on the young—It is important to observe that no selection cd. aid Horse in Falkland'. Selection is also related to variation and the struggle for existence—for example Darwin notes inside the back cover of G Jaeger, *In Sachen Darwins insbesondere contra Wigand* (Stuttgart, 1874): 'Sacral Pigeons killed by Hawks are white or yellow vars'; and on p 9: 'if all species varied equally all wd be in confusion'.

Heredity and hybridity are closely related and again they both relate to variation. P Lucas, *Traité philosophique et physiologique de l'hérédité naturelle* (2 vols, Paris, 1847) bears a large number of important annotations on variation and heredity; for example vol 1, p 181: 'inheritance cannot be cause of variation has nothing to do with it'. Also connections with sex are clear—see the back cover of vol 2: 'one might fancy that in ass crossed with Horse there is a greater potency of race and that this potency is transmitted more by male in this case than in others. Niater cow transmits with more force than Bull—Peuter cock often equally'; and the connection between hybridity and domestication: vol 2, p 184: 'This variability of hybrids is independent of domesticity'. Hybridity is also related to variation, for example in Gallesio, *Traité de citrus* (Paris, 1811); to sex, the relation of organism to organism, fertility and sterility; for example in J G Kölreuter, *Vorläufige Nachricht von einigen das Geschlecht der Pflanzen betreffenden Versuchen* (Leipzig, 1761-8).

A few words must be said about those annotations which concern man. Although many marginalia concern that topic, it seems to me that ethnology and anthropology as such were not amongst Darwin's primary interests. He read a lot of books on these subjects, but referred to them primarily because of their relevance to other matters, principally variation and sex (one should bear in mind that the greater part of the *Descent of Man* is in fact devoted to the study of sexual selection). Inside the back cover of S G Morton, *Types of Mankind* (London, 1854), Darwin notes: 'I am beginning to conclude that it is more difficult to account for small variations of man where there is *no* adaptation than great differences, where adaptation. Consider cases of Rabbits, mere law of growth'; 'Nothing is more odd than similarity of Fuegians and Brazilians. Why Puma shd range continent invaried and Monkeys differ in every province—It is [word faded] in Partridge. I may contrast Man with Monkeys, for on my theory, the Monkeys have varied'. In these annotations Darwin's interest is in variation and distribution, not man—he sees man just as a good test-case for his theory—like partridges and monkeys.

It used to be said that Darwin did not know enough about physiology and morphology and was therefore left out of the mainstream of nineteenth-century biology (see for example E S Russell in his otherwise fundamental *Form and function*[11]). On the other hand modern scholars realise that Darwin wanted to unify both field natural science and laboratory biology through the application of his theory of evolution by natural selection. As far as physiology is concerned, Darwin seems to be especially interested in the effects of light on plants, since this bears on problems related to adaptation: see for example A B Frank, *Die natürliche wagerechte Richtung der Pflanzentheilen* (Leipzig, 1870). A considerable number of annotations on physiology concern Helmholtz's consideration of the imperfection of the eye, since this was directly relevant to Darwin's view of adaptations as non-perfect; Darwin's back-cover notes in Helmholtz, *Popular Lectures on Scientific Subjects* (London, 1873) are all on that question; as they are in L Büchner, *Conférences sur la théorie Darwinienne de la transmutation des espèces* (Paris, 1869). Whilst J Liebig's *Organic Chemistry* (London, 1840) is only slightly annotated, and the annotations in W B Carpenter's *Principles of comparative physiology* (London, 1854) deal more with morphology and embryology than with physiology, we find a great number of annotations on different topics in Johannes Müller's *Elements of physiology* (2 vols, London, 1838–42), showing how Darwin thought of physiology as a fundamental part of natural science and how it related to his theory.[12]

Darwin's approach to morphology, and therefore his relationship with Richard Owen, are also revealed in his annotations. Although in the *Origin* Darwin avoided arguing openly against what Russell called 'transcendental morphology' (1916, pp 103–12), of which Owen was a representative, he took his colleague's works very seriously and annotated them very carefully. Owen was an acknowledged authority on affinity and palaeontology, and the upholder of the concept of the 'archetype' as the central reference concept of natural science. It is interesting to discover Darwin's opinion of the archetype, which he reported on the back of Owen's most influential book *On the Nature of Limbs* (London, 1849): 'I look at Owen's Archetypus as more than ideal, as a real representation as far as the most consummate skill and loftiest generalisations can represent the present forms of vertebrata—I follow him that there is a created archetype, the parent of its class'. (It is worth pointing out here that Darwin uses 'created' as a synonym for 'naturally formed', and does not intend to implicate the Almighty.) This annotation helps to make sense of the theoretical difference between Darwin and the natural scientists who either preceded him or were his contemporaries. Ernst Mayr (1964, p xx)[13] argues that a necessary step to the founding of a new view of organic phenomena was the elimination of Platonism from scientific explanation—the banishment of the *eidos*:

> According to (the concept of *eidos*) there are a limited number of fixed, unchangeable 'ideas' underlying the observed variability, with the *eidos* (idea) being the only thing that is fixed and real, while the observed variability has no more reality than the shadows of an object on a cave wall, as it is stated

in Plato's allegory. . . . Most of the great philosophers of the 17th, 18th and 19th centuries were
influenced by the idealistic philosophy of Plato, and the school dominated the thinking of the period.

Any attachment to metaphysical idealism, any commitment to an unchanging *eidos*,
precludes belief in descent with modifications. The concept of evolution rejects the
*eidos*. From a Darwinian viewpoint, the 'type' is simply the ancestor of evolving,
living forms, while the emphasis is on variety, that is, the diversity of life rather
than its unity as with Owen.

Darwin was interested not only in practical natural science but also in its theoretical
aspects. J Abercrombie's *Inquiries Concerning the Intellectual Powers and the Investigation of Truth* (London, 1838) is a good example. It must be noted that, as in the
case of man, Darwin was not here interested in philosophy per se, but only insofar
as it related to his interests in natural science, especially in respect of instincts. A
revealing example is J Mackintosh's *Dissertations on the Progress of Ethical Philosophy* (2nd edn, 1837), with a preface by W Whewell; here Darwin's marginalia relate
conscience to habit, both in man and animals. The moral sense is seen from the
viewpoint of what we would call 'animal behaviour'—for example the love of parents
for their children is related to adaptation and selection.[14] Such an attitude might be
of considerable interest to sociobiologists.

## ENGLISH-LANGUAGE SCIENCE AND PUBLICATIONS

Darwin's theory was an ecological one—and as such it sprang from natural science
rather than laboratory biology. We are accustomed to considering laboratory biology
(physiology, biochemistry, microbiology, etc) as the mainstream of research into
living organisms; but that was not the case in Britain just before Darwin. A vital
science of natural history which had its roots in the works of Ray and Willoughby
reached its height immediately before and during Darwin's youth—such authors as
W Kirby, W Spence, J Fleming, R Strickland, J Henslow, E Blyth, J Bicheno,
W O Westwood, L Jenyns and W Roscoe were familiar to the young Darwin and
deeply influenced his approach to the questions of living beings. The views of both
Wallace and Darwin lie firmly within the British natural science tradition. Darwin's
approach focusses on instincts (like Fleming and Blyth) and the relationship of
organism to organism (like Fleming, Westwood and Strickland), and therefore tends
to be an ecological theory in the manner of Strickland.[15] In Ray's *The Wisdom of
God manifested in the Works of the Creation* (London, 1692) Darwin discerned the
ecological approach he made his own in the *Origin*; in Ray we find annotations
concerning animal behaviour, adaption, sex, morphology and of course the relationship of organism to organism.

The relationship between instinct and acquisition by habit is the main topic of the
marginalia to be found in W Kirkby and W Spence, *An Introduction to Entomology*
(4 vols, different edns, London, 1818–26); here Darwin focused on the problems of
neuter insects which was to intrigue him in the *Origin*: for example, on vol 2, p 513,
he notes: 'one may suppose that originally many queens were ordinarily thus reared
and a few workers and the instinct is thus retained'; much is to be found on 'reason'
in animals as related to instinct—the names of Lamarck, Huber, Gould and Latreille

on such topics are carefully marked. Annotations on the struggle for existence, selection, speciation, and distribution, all interrelated, are found in this book.

Another author who contributed to attracting Darwin's attention to the role of instincts was John Fleming with *The Philosophy of Zoology* (2 vols, Edinburgh, 1822). For example, on vol 1, p 241, he notes: 'it is strange according to my theory that habit which results of intellectual process, is related to instinct, which analogy of plants leads one to believe to exist independently of intellect'; and 'The individual who by long intellectual study acquires a habit, and can perform action almost instinctively, does that in his lifetime, which successive generations do in acquiring true instinct: instinct is a habit of generations,—each step in each generation, being intellectual.'

The emphasis on the influence of British natural science in the first half of the nineteenth century requires a mention of Darwin's relationship with natural theology and especially its central tenet of perfect adaptation. Darwin read and annotated H Brougham, *Dissertation on Subjects of Science Connected with Natural Theology* (2 vols, London, 1839); but again the markings mainly concern animal behaviour and especially the rearing of pigeons—another piece of evidence for the influence of this kind of science on him. But in J Henslow's *Descriptive and Physiological Botany* (new edn, London, 1837) he made clear his disagreement with perfect adaptation on the back cover: 'People constantly speak about every organism being *perfectly* adapted to circumstances, if so how can there be a rare . . . species breeding power being efficient and food not sufficiently (in some cases probably) abundant is answer'.

It is clear from the intensity of annotation that Lyell was of paramount importance to Darwin's development; in fact Lyell is the most annotated author (although De Candolle's *Géographie botanique* (2 vols, Paris, 1855) is the most heavily annotated single book—see below). However, Darwin seems to have been much less sympathetic to Edward Forbes. After reading 'On the Asteridae. . .' (*Edin Mem Wern Soc*, 8, 1829: 114–28), which he had as a separate bound article, he commented 'absolutely unintelligible' (p 532).

The major repositories of early nineteenth-century British natural science were such specialised journals as the *Annals of Natural Science*, the *Magazine of Natural Science*, and so forth (see Sheets Pyenson's article[9]). If, for example, we consider the *Magazine of Natural History*, we find a large number of important marginalia concerning the relation of organism to organism, extinction, hybridity, animal behaviour, sex, variation, and breeding habits—all central themes in both the natural science of the time and Darwin's theory. Articles by Blyth, Blackwall, Newman, Jenyns and Henslow are marked and annotated. C Waterton's articles on the habits of birds attracted Darwin's interest, as well as those by W Yarrell, again on birds, and by Blyth on the habits of the cuckoo, and hybridity. Blyth's 'On the physiological distinctions between man and all other animals' (*Mag Nat Hist*, 1, ns, 1837: 1–9) is important not only for Darwin's study of animal behaviour but also for his view of the relationship between man and the other animals. See for example, p 2: 'Child fears the dark—before reason has told it'. The list of other authors whose works were annotated includes W Ogilby on affinities and classification, J O Westwood on the habits of the ant-lion, G R Waterhouse on the importance of characters for classification, W Weissenhorn on the influence of man modifying the natural world, and Lund on the fauna of Brazil (important for the connection between extinction and distribution), and Strickland on analogy and affinity, also containing references to Blyth.

Also important was the *Magazine of Zoology and Botany*, especially W Thompson, 'Contributions to the natural history of Iceland' (*Mag Zool Bot*, 2, 1838); see for example these annotations inside the back cover, which link extinction, variation and speciation: 'on changes of rarity in Birds', 'increase in numbers first step to variation and formation of new species—decrease of numbers first step towards extinction'.

## FRENCH-LANGUAGE SCIENCE AND PUBLICATIONS

By observing which books are annotated, we may deduce that Darwin was confident with the French language, less confident but still reasonably conversant with German, and seems to have occasionally read some Italian and Spanish, especially works published in Italian by Delpino on dichogamy and Mantegazza on man.

Only a few marginalia are found in Cuvier, *Leçons d'anatomie comparée* (5 vols, Paris, 1799-1805), and all of them concern morphology. There are a few more in *La Règne animal* (5 vols, Paris, 1829), concerning animal behaviour, sex, speciation, morphology and variation. Darwin also possessed the English translation of *The theory of the Earth* (5th edn, Edinburgh and London, 1827). Cuvier is often mentioned and his name marked in other people's books; but Geoffroy St Hilaire's work, to judge by the degree of annotation, was much more important to him than Cuvier's.

Of H Milne-Edwards, Darwin annotated the *Histoire des crustacées* (3 vols, Paris, 1834-40)—for example in vol 3 we find a very important annotation which explains that Darwin was opposed to the idea of individual acts of creation rather than to creation as such—he often used that word as a synonym of 'naturally formed', as we saw before—p 555: 'How explains this, except by single creations'. On the same page there is another important annotation concerning isolation: 'without regard to anything else—Make a Barrier and you will have different species on opposite sides'. Other Milne-Edwards marginalia are found in *Introduction à la zoologie générale* (Paris, 1851)—for example inside the front cover: 'p 7—Diversity of organisms first condition of nature'; 'p 9—Law of "economy of nature" "sober in innovations" has not recourse to any new creation of organ'; 'p 13—nature varies degree of perfection'; 'p 172—on value of characters in classification'; and in the text itself on p 21: 'Best way of putting superiority.—though each perfectly (?) (can young be said to be perfectly?) adapted to conditions'.

As far as Lamarck is concerned, his *Histoire naturelle des animaux sans vertèbres* (Paris, 1817) bears very few annotations. More are found in the *Zoologie philosophique* (Paris, 1809). Darwin's relationship with Lamarck is very complex, and one should not take the disparaging remarks found here as Darwin's only view of the French evolutionist—'very poor and useless book'. Basically Darwin charged Lamarck with failure to understand the roles of extinction and distribution: p 80: 'Therefore every fossil species direct father of existing analogies and no extinction except through man—(hence cause of innumerable errors in Lamarck)'; p 266: 'Does not pursue this into Geographical Distribution'. Moreover Darwin tends to equate Lamarck with

Swainson and MacLeay: p 151: 'as bad as Swainson'; but takes him seriously on instincts: p 68: 'The case of acquired hereditary instincts shows that instincts *can* be acquired'; and in his connection with Lyell: p 71: 'Like Lyell in Geology'. (Not surprisingly, Darwin's opinion of the *Vestiges of Creation* (6th edn, London, 1847), the other hotly-debated pre-Darwinian espousal of evolution, is not high: for example, inside the back cover he notes: 'The idea of a fish passing into a Reptile—(his idea)— monstrous'; and on p 286: 'It is strange error that generally he looks at every form as having started from some known form'. The occasional exasperated—or merry— 'oh!!' escapes from his pencil.)

Other important French-language authors are C L Bonaparte, especially on the connection between distribution and the struggle for existence; and F Huber on insects' instincts in *Nouvelles observations sur les abeilles* (Geneva, 1814).


## GERMAN-LANGUAGE SCIENCE AND PUBLICATIONS

The myth that Darwin did not read German is definitely exploded by the number of annotations in books written in that language—Gaertner, Kölreuter, Ehrenberg, Haeckel, to mention only some of the best known, are well represented in his library and their works carefully annotated wherever Darwin expected to find something relevant to his own concerns. See for example Gaertner, *Beiträge zur Kenntnis der Befruchtung der vollkommenen Gewächse* (1. Theil, Stuttgart, 1844), which is very heavily annotated with a great number of marginalia on variation, fertility, hybrids, and the relation of organism to organism, very often interrelated: for example, on p 137 he comments that castrated flowers are seldom visited by bees. Many annotations concern contabescence and refer to Kölreuter: for example p 137 again: 'most important to compare Kolreuters experiments and Gaertner's'; some markings concern dichogamy as seen by Sprengel and Delpino.

That Darwin read German when necessary is also proved by the fact that he read and annotated the original edition of E Haeckel, *Natürliche Schöpfungsgeschichte* (Berlin, 1868) rather than waiting for the English translation. Incidentally, Haeckel sent Darwin most of his publications, but not many of them greatly interested Darwin. Very often in their inscriptions to Darwin in their books German scientists including Haeckel himself wrote 'Sir' or 'Professor', not being able to believe that someone as distinguished as Darwin would not be either one or the other—or both.

It is interesting that there is no annotational evidence that Darwin read K E von Baer's *Entwickelungsgeschichte der Thiere* (Königsberg, 1828–37), which is not even in his list of 'Books to Read' (see Vorzimmer[4]). But he certainly read Huxley's translation of the fundamental fifth *Scholium* (T H Huxley, 'Fragments relating to philosophical zoology, selected from the works of K E von Baer', (*Taylor's*) *scientific memoirs, natural history*, 3: 176–238).


## DE CANDOLLE AND HUMBOLDT

Judging by the quality and quantity of annotation, it emerges that two authors who greatly inspired and provoked Darwin at different times in his development were Alphonse De Candolle and Alexander von Humboldt.

De Candolle's *Géographie botanique* is probably the most heavily annotated book in the whole library. This book seems to have been the catalyst for many of Darwin's views on distribution, the struggle for existence, isolation, and consequently selection. It is impossible that this book suggested to Darwin the solution to problems related to those topics—Darwin had reached his major conclusions long before 1855—but it refined his thinking and presentation considerably. See for example on the struggle for existence, vol 1, p 419: 'All used in the Chapter on Sociability—Struggle for existence' and on rarity of species as related to distribution: 'A species might abound in some spot and yet be rare over all England, but is this so?' (p 460). Here the marginalia in De Candolle must be related to those in Hooker's *Antarctic Voyage* (London, 1844-7), which deals with similar topics. While reading De Candolle's book Darwin must have thought of the inseparable connection between life and death: 'much life causes much decay. . .' (p 462); as well as that between confined range and rarity: '"species with restricted range are not common" i.e. confined range and rarity go together' (p 470); and between adaptation and the struggle for existence: 'If sociability depends on other species and not on external conditions, then very slight change might determine their existence' (p 472). A large number of annotations concern isolation and distribution: for example, p 545: 'Here isolation clearly comes into play; but this does not account for smaller range of plants within large district'; p 550: 'As far as I can see (which is very little) isolation of area seems to have little to do with confinement of species!! In this family'; p 561: 'I never shd look at it under this light; yet perhaps agree with Herbert's views. When there only few species, we must suppose either others extinct, or then few only are yet introduced'; p 588: 'This bears on . . . few species inhabiting 2 areas, when there are many species—Does it not come to this, that widely extended species break into varieties and then become species with confined range—Anyhow this shows how complicated a question it is'. In vol 2 Darwin annotated inside the back cover: 'struggle between Fish and Water Plants'; p 201: 'England formerly connected hence most plants which could live in England would have immigrated. If any species had been introduced by birds within the last century, and was not mentioned by old Books, it wd have been thought to have been overlooked'. This annotation, apart from throwing light on Darwin's understanding of the connections between geology, distribution and the struggle for existence, also hints at his interest in old books as sources of information. He did not disdain to quote the Bible about the antiquity of mules, as in mentioned in D A Godson's *De l'espèce et des races* (2 vols, Paris, 1859), vol 1. Very often he quotes sources on ancient Egypt.

Returning to considerations of De Candolle's *Géographie botanique*, we find a reference in vol 2, p 886 to the necessity of reading Gaertner (which indeed he did); another provides further evidence that Darwin was coming to see man just as a crucial test case: p 935: 'The more I reflect, the more I come to the conclusion that *antiquity* of man one of the most important elements in history of variation'. Around p 1000 Darwin picks up the crucial problem of extinction and starts being quite critical of the Swiss naturalist: p 1023: 'He always leaves out struggle with other species'; p 1062: 'He looks at extinction as due all to Deluges &c!!'. Darwin is by now disappointed with De Candolle and realises that his colleague has approached the right problems but lacks one of the central insights into the process of speciation: without the idea of selection it is impossible to make sense of variation, extinction, isolation, distribution and the struggle for existence forming a single complex nexus. Thus p 1085: '(Always this) he has not the Key'.

On the other hand Humboldt's works focused Darwin's attention upon distribution and the relation of organism to organism in relation to isolation, extinction and the breeding of wild and domesticated animals. See for example *Personal Narrative of Travels to the Equinoctial Regions of the New Continent during the years 1799–1804* (7 vols, difft edns). For example in vol 1 he notes inside the front cover: 'p 274–Camels abundant in Fortaventura and vegetation different from . . . other Islands—NB Numerous <u>wild</u> asses formerly in Fortaventura vide early part of Chapter'. One annotation suggests Darwin's refusal of the rather easy-going optimism 'à la Pangloss' which underlies much of Humboldt's writing and in turn leads Humboldt to adopt a falsely a priori harmonious world where adaptations are basically perfect. As such I rate the following annotation as one of the most indicative of the development of Darwin's view of the relationship between organisms in nature and possibly of his entire world-view: vol 5 (1st edn), p 590: 'to show how animals prey on each other— what a "positive check"—Smaller Carnivora—Hawks—what hourly carnage in the magnificent calm picture of tropical forests—Let him from some pinnacle view one of these Tropical how peaceful and full of life' and 'think of death only in terrestrial vertebrates'.

## CONCLUSION

If the Notebooks represent Darwin talking potentially to a public, and the Letters represent Darwin talking to his friends and colleagues, then the Marginalia represent Darwin talking to himself, and thus probably constitute the least guarded and filtered commentary available on the development of his thought—often no doubt giving us the benefit of his immediate reactions to individual statements made by his contemporaries and predecessors. But furthermore, especially in the remarks reserved for the bookcovers, the Marginalia pick out in a potentially revealing fashion those aspects which at the time of reading Darwin thought particularly important, and thus also represent a more considered inventory of the succession of topics to which he turned his attention during the course of his development. Collation of the completed index will help to identify these matters in greater detail and their changes over time, their relation to periods of ideation and periods of writing-up, and so forth. Meanwhile I hope that this article has given an indication of the flavour and scope of the 'raw material' of Darwin's thought.

## NOTES AND REFERENCES

[1] I should like to record my debt of gratitude to all those who have helped me in various different ways in my work on the marginalia; but especially to David Kohn, Sydney Smith, Stephen Pocock, Nick Gill, Peter Gautrey, and the Staff of the Manuscripts Room of the Cambridge University Library; and to the Syndics of Cambridge University for permission to quote from the books and margins cited in this article

[2] Kohn, D, Theories to work by: rejected theories, reproduction and Darwin's path to natural selection, *Studies in the history of biology*, 4, 1980: 67–170

[3] de Beer, G, Darwin's notebooks on the transmutation of species, *Bulletin of the British Museum (Natural History), historical series*, 2, 3, 5, (1959–67)

[4] Vorzimmer, P, The Darwin reading notebooks 1838-1860, *J Hist Biol*, 10 (1977): 107-53

[5] Gruber, H and Barrett, P, *Darwin on man* (London, 1974)

[6] Herbert, S, The red notebook of Charles Darwin, *Bulletin of the British Museum* (*Natural Science*), *historical series*, 7 (1980)

[7] Burckhardt, F and Smith, S (ed), *The correspondence of Charles Darwin* vol 1 1821-1836 (Cambridge, 1985); vol 2 1837-43 (Cambridge, 1986); vol 3 (Cambridge, 1987, in press)

[8] Richardson, R A, Biogeography and the genesis of Darwin's ideas on transmutation, *J Hist Biol*, 14 (1981): 1-42

[9] Sheets Pyenson, S, Darwin's data: his reading of natural history journals, 1837-42, *J Hist Biol*, 14 (1981): 231-48

[10] Secord, J A, Nature's fancy: Charles Darwin and the breeding of pigeons, *Isis,* 72, (1981): 163-86

[11] Russell, E S, *Form and function* (London, 1916)

[12] See also Ospovat, D, *The development of Darwin's theory* (Cambridge, 1981). Other authors who have made use of the marginalia include: Smith, S, The Darwin collection at Cambridge, with one example of its use: Charles Darwin and Cirripedes, *Actes du xie Congrès International d'Histoire des Sciences* (Warsaw, 1965): 96-100
Ghiselin, M, Darwin and evolutionary psychology, *Science*, 179 (1973): 964-968
Vorzimmer, P J, Mr Darwin's critics old and new, *J Hist Biol*, 6 (1973) 155-165
Greene, J C, Darwin as a social evolutionist, *J Hist Biol*, 10 (1977): 1-27
Stauffer, R C, *Charles Darwin's natural selection* (Cambridge, 1975)

[13] Mayr, E, Introduction to Charles Darwin, *On the Origin of Species*, facsimile of the first edition (Cambridge, Mass, 1964)

[14] Manier, E, *The young Darwin and his cultural circle* (Dordrecht, 1978)

[15] For Strickland see di Gregorio, M.A., Hugh Edwin Strickland (1811-1853) on affinities and analogies: or, the case of the missing key, *Ideas and Production,* 7 (1987): 35-50